PLANT COLD HARDINESS

PLANT COLD HARDINESS

Gene Regulation and Genetic Engineering

Edited by

Paul H. Li

University of Minnesota
St. Paul, Minnesota

and

E. Tapio Palva

University of Helsinki
Helsinki, Finland

Kluwer Academic / Plenum Publishers
New York, Boston, Dordrecht, London, Moscow

Library of Congress Cataloging-in-Publication Data

Plant cold hardiness: gene regulation and genetic engineering/edited by Paul H. Li and Tapio Palva.
p. cm.
Includes bibliographical references (p.).
ISBN 0-306-47286-4
1. Plants—Frost resistance—Congresses. 2. Plant genetic regulation—Congresses. 3. Plant genetic engineering—Congresses. I. Li, P. H. (Paul H.), 1933- II. Palva, Tapio. III. International Plant Cold Hardiness Seminar (6th: 2001: Helsinki, Finland)

QK756 .P55 2002
631'.11—dc21

2002066910

Proceedings of the 6th International Plant Cold Hardiness Seminar, held July 1–5, 2001, in Helsinki, Finland

ISBN 0-306-47286-4

233 Spring Street, New York, N.Y. 10013

http://www.wkap.nl/

10 9 8 7 6 5 4 3 2 1

A C.I.P. record for this book is available from the Library of Congress

Printed in the United States of America

PREFACE

We compiled this volume mostly from presentations at the 6th International Plant Cold Hardiness Seminar (PCHS) after consulting with Professor Tony H. H. Chen, Oregon State University, USA, Professor Pekka Heino, University of Helsinki, Finland, and Dr. Gareth J. Warren, University of London, Surrey, UK. The 6th International PCHS was held at the Unitas Congress Center, Helsinki, Finland from July 1–5, 2001. There were 110 registered scientists at the seminar representing 20 countries: Australia, Belgium, Canada, Chile, the Czech Republic, Denmark, Estonia, Finland, Germany, Hungary, Iceland, Italy, Japan, Norway, Poland, Spain, Sweden, Taiwan, United Kingdom, and United States of America. The information compiled represents the state of the art of research in plant cold hardiness in terms of gene regulation, gene expression, signal transduction, the physiology of cold hardiness and, ultimately, the genetic engineering for cold tolerant plants.

The International PCHS was initiated in 1977 at the University of Minnesota, St. Paul, Minnesota. It has been traditionally held at 5-year intervals at various locations. Because of the rapid advances of research in plant cold hardiness, attendees at the 6th meeting unanimously adopted a resolution to hold the seminar in 3-year intervals instead of 5 in the future. Consequently, the next seminar will be held in 2004 in Sapporo, Japan, and Professor Seizo Fujikawa from Hokkaido University will serve as the host.

As always, the overall goal of the seminar is to update the research findings and to provide a forum for the exchange of ideas among scientists with diverse disciplines who are actively engaged in studying plant freezing and chilling stress. The chapter authors are researchers who have made significant contributions to the knowledge of plant cold hardiness. It is our sincere hope that this volume will provide readers with the most recent advances in plant cold hardiness research since the 1997 proceedings, Plant Cold Hardiness: Molecular Biology, Biochemistry and Physiology, which was compiled from the presentations at the 1996 seminar.

We wish to thank the following for their financial support of the 6th seminar: Biocenter Viikki, University of Helsinki, Helsinki Graduate School in Biotechnology and Molecular Biology, Research Programme in Plant Biology, Viikki Graduate School in Biosciences, Plant Molecular Biology and Forest Biotechnology Center of Excellence 2001-2005 and Merck Eurolab. Our special thanks also go to Dr. Karen Sims-Huopaniemi, Department of Biosciences, Division of Genetics, University of Helsinki, for her help in managing the daily activities during the meeting. We would also like to acknowledge Dr. Wen-Ping Chen's assistance in preparing the final version of this volume for publication.

Lastly, a few words from PHL: For everything there is a beginning and always an end. I consider myself to have been very fortunate to organize the PCHS series since 1977. I am very grateful to colleagues and friends for their help, understanding, and patience through my entire professional career. I am also pleased to see that the seminar event will be continued after I pass the baton.

Paul H. Li
E. Tapio Palva

CONTENTS

Part I: Gene Regulation and Signal Transduction

Part II: Physiology Aspects of Plant Cold Hardiness

Part III: Genetic Engineering

Part I

Gene Regulation and Signal Transduction

1

MOLECULAR GENETICS OF PLANT RESPONSES TO LOW TEMPERATURES

Byeong-ha Lee, YongSig Kim, and Jian-Kang Zhu*

1. INTRODUCTION

Living organisms are always interacting with their environments. Those that achieve successful adaptations to unfavorable environmental conditions will survive and prevail. During evolution, organisms have developed mechanisms to cope with unfavorable environments. Cold stress is one of the adverse environmental factors that determines ecological habitat of organisms. Being sessile, plants are greatly affected by low temperatures in their geographical distribution and productivity. Therefore, plants are not exempt from developing unique mechanisms to protect themselves from cold stress damage. Some moderately hardy woody plants use supercooling of water to avoid freezing damage. Without ice nucleation, pure water can be supercooled or remain unfrozen until a certain point below 0°C. However, this supercooling is effective only within a narrow temperature window (George et al., 1982). The most common cold stress adaptation is cold acclimation. That is, the process by which plants acquire freezing tolerance after being exposed to cold, but non-freezing temperatures (Guy, 1990). Due to cold acclimation, plants from temperate regions are able to survive winter after experiencing cold, but non-freezing temperatures of the fall. Cold acclimation is correlated with several changes in plants such as changes in leaf ultrastructure (Ristic and Ashworth, 1993), membrane lipid composition (Lynch and Steponkus, 1987; Miquel et al., 1993), enzyme activities, ion channel activities (Knight et al., 1996), levels of sugars and polyamines (Levitt, 1980; Strand et al., 1997) and gene expression (Thomashow, 1994). Attempts to understand cold tolerance mechanisms have long been made because of scientific interest as well as its economic importance in agriculture. Thanks to current advances in plant molecular genetics, our understanding is expanding fast. This chapter will include an overview of our current understanding of cold tolerance and focus mostly on studies with cold-stress mutants.

*Byeong-ha Lee, YongSig Kim, and Jian-Kang Zhu, Department of Plant Sciences, University of Arizona, Tucson, Arizona 85721.

Plant Cold Hardiness, edited by Li and Palva
Kluwer Academic/Plenum Publishers, 2002

2. COLD INJURIES TO CELLULAR MEMBRANES

Cellular membrane is the primary site of cold stress damage. As the temperature drops to below freezing, ice formation occurs first in intercellular spaces. The intercellular ice formation brings about a water potential gradient across the plasma membrane, with lower water potential outside and higher water potential inside. Due to this freezing-induced osmotic gradient, unfrozen water inside the cell moves out into the intercellular space, resulting in concentrated intracellular fluid and cell contraction. In non-acclimated rye protoplasts, low temperature treatment over the range of 0°C to -5°C caused freezing-induced protoplast contraction with irreversible vesiculation of the plasma membrane, and the subsequent thawing resulted in lysis of protoplasts (Dowgert and Steponkus, 1984; Gordon-Kamm and Steponkus, 1984). In addition, freezing temperatures below -5°C cause dehydration-induced destabilization of plasma membrane in the rye protoplast system. Injured protoplasts completely lost their ability of osmotic responsiveness during thawing (Dowgert and Steponkus, 1984; Gordon-Kamm and Steponkus, 1984). Loss of osmotic responsiveness might be due to several changes in the ultrastructure of the plasma membrane (Gordon-Kamm and Steponkus, 1984). However these injuries were not observed in cold-acclimated rye protoplasts (Gordonkamm and Steponkus, 1984). These different responses of plasma membrane with or without cold acclimation are thought to be due to differences in lipid composition of the plasma membrane (Lynch and Steponkus, 1987). In studies with highly-enriched plasma membrane fractions from rye seedlings, the ratio of unsaturated phosphatidylcholine and phosphatidylethanolamine increased after cold acclimation (Lynch and Steponkus, 1987). Increased unsaturation of the membrane lipids in response to cold was also observed in cyanobacteria (Sato and Murata, 1980; Wada and Murata, 1990). The increase in the degree of unsaturation is likely because of up-regulation of desaturase genes upon cold stress (Los et al., 1993; Sakamoto and Bryant, 1997). As membrane fluidity affects the functions of integral membrane proteins (Squier et al., 1988; Gasser et al., 1990), maintaining normal membrane fluidity is important to avoid low-temperature induced phase separation/transition of the plasma membrane that causes leakage of the cytosolic solutes, disrupted gradient and perturbed metabolism (Levitt, 1980; Wang, 1982; Murata, 1989). Thus, elevated lipid unsaturation would compensate for the membrane fluidity decrease caused by low temperatures (Nishida and Murata, 1996). Overexpression of a desaturase gene in cyanobacteria or tobacco resulted in a higher ratio of unsaturated lipids and thus improved cold tolerance (Wada et al., 1990; Ishizaki-Nishizawa et al., 1996).

3. Ca^{2+} AND COLD SENSING

In many plants including tobacco, alfalfa and Arabidopsis, it was shown that cold stress triggers a transient cytosolic Ca^{2+} increase that is required to induce cold-responsive gene expression (Knight et al., 1991; Monroy et al., 1993; Monroy and Dhindsa, 1995; Knight et al., 1996). Örvar et al (2000) demonstrated that pharmacological manipulation of membrane fluidity status, independently of temperature, was sufficient to induce a cold-regulated gene, *cas30*, in alfalfa protoplasts. Treatment with a membrane fluidizer, benzyl alcohol, or an actin microfilament stabilizer, jasplakinnolide inhibited Ca^{2+} increase, and blocked *cas30* gene induction and development of cold tolerance at 4°C. In contrast, Ca^{2+} increase, *cas30* gene induction and cold tolerance were triggered by a membrane rigidifier, DMSO, or an actin microfilament destabilizer, cytochalasin D at 25°C. These studies suggested that membrane and cytoskeleton are involved in cold signaling. Furthermore, the studies suggested a model where changes in membrane fluidity induced by cold bring about the modification of actin microfilament organization, which then induces Ca^{2+} influx into the cytosol. Thion et al (1996) and Mazars et al (1997) also showed that Ca^{2+} increase is dependent on rearrangement of cytoskeleton structure in carrot protoplasts and aequorin-expressing tobacco. Correlations between membrane fluidity and cold-responsive gene expression were also reported in *Synechocystis* PCC6803 (Vigh et al., 1993) and a transgenic *Brassica napus* expressing the transgene of *BN115* promoter fused to the GUS reporter (Sangwan et al., 2001).

A question arises: what senses the membrane rigidification induced by cold? Mechanosensitive Ca^{2+} channel in the plasma membrane was proposed as a candidate by Monroy and Dhindsa (1995). Very recently they provided evidence in *Brassica napus* with a cold-inducible *BN115* promoter::*GUS* transgene. Gd^{3+}, a specific blocker of mechanosensitive Ca^{2+} channels, inhibited the endogenous *BN115* expression as well as the transgene expression more effectively than La^{3+}, a blocker of voltage-gated Ca^{2+} channels. Since destablization of microfilaments and microtubules (removal of tension on membrane by cytoskeleton) induced increase in the *BN115* transcripts at 25°C, the hypothesis of mechanosensitive Ca^{2+} channels as cold sensors appears plausible. However, other studies suggested that Ca^{2+} increase requires both extracellular and intracellular sources for full-strength expression of cold-responsive genes (Knight et al., 1996; Sangwan et al., 2001). Thus, mechanosensitive Ca^{2+} channels alone may not be sufficient to sense cold signals. Murata and Los (1997) proposed histidine kinases as cold sensors in plants based on analogy with an osmosensor histidine kinase in *S. cerevisiae* (Maeda et al., 1994) and cold sensor histidine kinase in *Synechocystis* (Suzuki et al., 2000) and *Bacillus subtilis* (Aguilar et al., 2001). In the *Synechocystis* system, deletion of *hik33*, a gene encoding a histidine kinase with membrane spanning domain, and *hik19*, a gene encoding a soluble histidine kinase resulted in lower expression of cold-inducible genes, *desB*, *desD*, and *crh*, but not *desA*. Mutation in *rer1* (Response regulator1) also reduced the expression of *desB*. Rer1 is a typical response element of a two-component regulator. Thus, a cold signaling model was proposed, whereHik33 is activated by reduced membrane fluidity, and in the form of phosphoryl group the signal passes through Hik19 to reach Rer1. This pathway only induces *desB* gene expression. For *desD* and *crh* expression, other unknown signal receivers may participate in transferring the signal from Hik33 and Hik19. *desA* gene expression is thought to be regulated through another different thermosensor system. Similarly, a two-component system including DesK (YocF), an integral membrane histidine kinase and DesR (YocG), a DNA binding response regulator, was characterized in *Bacillus subtilis* (Aguilar et al., 2001).

4. SIGNALING COMPONENTS

Despite intensive studies, signaling components that perceive Ca^{2+} signatures and translate them into the expression of cold-regulated genes are largely unknown. However, pharmacological studies with inhibitors of protein kinases and phosphatases suggested that phosophorylation/dephosphorylation is involved in cold signal transduction (Monroy et al., 1993; Monroy and Dhindsa, 1995; Tahtiharju et al., 1997; Sangwan et al., 2001). Most protein phosphorylations following cytosolic Ca^{2+} increase are mediated by Ca^{2+} dependent protein kinases (CDPK) in plants (Sanders et al., 1999). CDPKs are serine/threonine protein kinases with a calmodulin-like domain at the C-terminus, and are found only in plants and protozoa (Harper et al., 1991; Zhang and Choi, 2001). Using maize protoplast transient expression system, it was demonstrated that activated AtCDPK1 induced the stress-responsive *HVA1* promoter-driven luciferase expression (Sheen, 1996). Also, in plants such as alfalfa and rice, CDPK transcript expression was induced by cold (Monroy and Dhindsa, 1995; Saijo et al., 2000). Although *OsCDPK7* expression in rice was increased under cold, OsCDPK7 protein level remained at the basal constitutive level independent of cold-stress duration (Saijo et al., 2000). This result indicates that posttranscriptional modification, not only gene induction, is important in cold signal transduction, with the possibility of a positive feedback regulation of *OsCDPK7* by itself.

As in animals and yeast, mitogen-activated protein kinase (MAPK) pathways play pivotal roles in stress signaling in plants. A MAPK cascade involves the sequential activation of MAPK kinase kinase (MAPKKK), MAPK kinase (MAPKK) and MAPK by phosphorylation (Jouannic et al., 2000). In the Arabidopsis genome, 21 different genes may encode potential MAPK module components (Morris, 2001). Studies thus far showed that gene expression or protein activities of most MAPK/MAPKK/MAPKKK in plants are regulated by several stress conditions including cold (Morris, 2001) indicating that MAPK pathways may provide a node for cross-talk or a common module that connects responses to abiotic stresses. In Arabidopsis, *AtMPK3* (*MAPK*), *AtPK19* (*S6 ribosomal protein kinase*), and *AtMEKK1* (*MAPKKK*) gene expression is induced in response to cold, touch, or salt stress (Mizoguchi et al., 1996). In alfalfa, MMK4 (or SAM kinase, stress-activated MAP kinase) transcript and the protein activity are induced by cold, drought or wounding stress (Jonak et al., 1996; Bogre et al., 1997). Similar to OsCDPK7, alfalfa MMK4 protein level did not change after stress applications, although its transcript level and protein activity increased. Studies of MMK4 under wounding stress showed that deactivation of activated MMK4 (or phosphorylated MMK4) is sensitive to transcription inhibitors or translation inhibitors suggesting that the synthesis of a new protein, probably a protein phosphatase is required for deactivation (or dephosphorylation) (Bogre et al., 1997). Indeed, Meskiene et al. (1998) showed that the alfalfa *MP2C* gene product, a protein phosphatase 2C, negatively regulates stress-activated MAPK pathway. Although these studies were done with wounding stress, they provide some clues about MAPK-mediated cold signaling.

Studies with *aba1* mutant (ABA biosynthesis mutant) showed that cold-acclimation in the mutant was hampered (Heino et al., 1990; Gilmour and Thomashow, 1991) suggesting ABA involvement in cold acclimation. Tahtiharju and Palva (2001) reported that antisense suppression of *AtPP2CA*, a protein phosphatase 2C gene in Arabidopsis plants resulted in the up-regulation of cold and ABA-induced genes and better freezing tolerance. Based on these results, the authors proposed that AtPP2CA is a negative regulator of ABA responses during cold acclimation. In the context of the *MP2C* studies, it is tempting to speculate that *AtPP2CA* may also modulate the stress-induced MAPK pathway that regulates cold-induced gene expression, which brings about freezing tolerance. However, in ABA-treated alfalfa plants, no activation of MMK4 was observed, indicating the cold- and drought-induced activations of MMK4 are independent of ABA (Jonak et al., 1996). Thus, *AtPP2CA* might be involved in different pathways that include ABA, rather than MMK4-like pathway. Sheen (1996) reported that a constitutively active AtPP2C decreased *HVA1::LUC* expression activated by AtCDPK1, suggesting that AtPP2C may play a role in the CDPK pathway.

5. COLD-INDUCIBLE GENE EXPRESSION

Many genes are induced by cold stress (Guy et al., 1985; Thomashow, 1999). Among these, some genes encode proteins with known functions including desaturases (Gibson et al., 1994), molecular chaperones (Anderson et al., 1994; Krishna et al., 1995), and signaling molecules such as a MAPK module components and CDPK (Monroy and Dhindsa, 1995; Jonak et al., 1996; Mizoguchi et al., 1996; Saijo et al., 2000). Given the changes in membranes under cold stress, it is not surprising that desaturases and molecular chaperones are up-regulated upon cold conditions where they might play roles in stabilizing membranes and proteins.

In contrast to the small number of cold-induced genes with known functions, most of the cold-induced genes have unknown functions. Most of these genes were isolated by differential hybridization screens and were named either as *COR* (cold-regulated), *LTI* (low-temperature-induced), or *KIN* (kykna-indusoitu; Finish for cold-induced) by different research groups (Kurkela and Franck, 1990; Lin and Thomashow, 1992; Nordin et al., 1993). These gene products are very hydrophilic, and have relatively simple amino acid compositions (Thomashow, 1999). Among them, only the function of COR15A in cold-tolerance has been tested biochemically (Artus et al., 1996). COR15A is also very hydrophilic and is targeted to chloroplasts (Lin and Thomashow, 1992). When constitutively expressed in Arabidopsis, COR15A increased freezing tolerance at the chloroplast and protoplast levels (Artus et al., 1996). COR15A appears to function by decreasing the tendency of membranes to form the lamella-to-hexagonal II phase that leads to membrane damage under freezing (Steponkus et al., 1998).

Correlations between the up-regulation of *COR* genes and cold tolerance were also observed in studies in which transcription activators of cold-responsive genes were overexpressed in Arabidopsis. In the promoter region of many *COR* genes, regulatory elements that confer gene specific expression have been identified and named dehydration-responsive element (DRE, 5'-TACCGACAT-3') or C-repeat (CRT, 5'-TGGCCGAC-3') which share the common core motif, 5'-CCGAC-3'(Baker et al., 1994; Yamaguchi-Shinozaki and Shinozaki, 1994; Seki et al., 2001). Using a yeast one-hybrid system, DRE/CRT binding proteins, *CBF1/DREB1B*, *CBF2/DREB1C*, and *CBF3/DREB1A* have been cloned and shown to function as transcriptional activators (Stockinger et al., 1997; Gilmour et al., 1998; Liu et al., 1998). *CBFs/DREB1s* are induced early upon cold stress and are involved in the regulation of the DRE/CRT class of *COR* genes. Overexpression of *CBF1* resulted in constitutive expression of the DRE/CRT class of genes and enhanced freezing tolerance (Jaglo-Ottosen et al., 1998). Overexpression of a *CBF1* isolog, *DREB1A/CBF3* also brought about expression of the DRE/CRT class of genes and increased drought tolerance as well as freezing tolerance (Kasuga et al., 1999). Therefore, the DRE/CRT class of *COR* genes are induced through CBFs/DREB1s transcription activators upon cold stress and are correlated with freezing tolerance.

6. ARABIDOPSIS COLD-STRESS MUTANTS

Arabidopsis is a chilling-tolerant plant and capable of cold acclimation within a few days to acquire freezing tolerance. These features along with copious genetic information made many researchers choose Arabidopsis as a model system for cold stress studies. The efforts have yielded several Arabidopsis mutants altered in cold responses.

6.1. *esk1* Mutant

Xin and Browse (1998) isolated a freezing tolerant mutant, *esk1* (*eskimo1*) that is able to survive freezing at -8°C without prior cold acclimation. The constitutive freezing tolerance may be due to enhanced accumulation of proline and sugars in *esk1*, as proline and sugars act as osmolytes to protect cells from dehydration (Delauney and Verma, 1993; Zhu et al., 1997). Higher proline levels in *esk1* mutant plants might be caused by a higher rate of proline biosynthesis in *esk1* as a result of up-regulation of a proline biosynthesis gene, pyrroline-5-carboxylate synthetase (P5CS). This suggests that ESK1 might be a negative regulator of proline biosynthesis under cold stress (Xin and Browse, 1998). As there was no changes in the expression of four major DRE/CRT class of *COR* genes (*COR6.6*, *COR15A*, *COR47*, and *COR78*) in *esk1* under cold stress, ESK1 may not be involved in CBF-mediated pathway, which is represented by mutants defective in 'damage repair' as suggested by Xiong et al (2001c). The result also suggests that signaling for cold tolerance is not simple, but complicated with parallel or branched pathways (Xin and Browse, 2000; Browse and Xin, 2001).

6.2. *sfr* Mutants

Warren and colleagues have isolated several freezing sensitive mutants named *sfr* (*sensitive to freezing*) (Warren et al., 1996). *sfr4* displayed an opposite phenotype of *esk1*, i.e., freezing sensitive phenotype with defect in cold-induced sugar accumulation, while *sfr1*, *sfr2* and *sfr5* did not show defects in sugar accumulation, fatty acid composition, or cold-induced anthocyanin accumulation, but were still sensitive to freezing (McKown et al., 1996). Thus, the results suggest that multiple factors are required to achieve cold tolerance. Another *sfr* mutant, *sfr6*, showed a decreased level of expression of *KIN1*, *COR15A* and *LTI78 (RD29A)* in response to cold. However, there was no difference in the expression of *CBF* genes and of *P5CS*. Cytosolic calcium levels were also not different between wild type and the *sfr6* mutant under low temperatures (Knight et al., 1999). Thus, SFR6 appears to act as a signaling component intermediate between CBF and cold-inducible genes such as *KIN1*, *COR15A* and *LTI78 (RD29A)* or as a cofactor of CBF to activate the downstream genes (Knight et al., 1999).

6.3. *hos1* Mutant

While *esk1* and *sfr* mutants were selected by its freezing tolerant or sensitive phenotypes, we have taken a different approach to screening for mutants with altered response to cold and other stresses (Ishitani et al., 1997; Xiong et al., 1999; Lee et al., 2002). As noted before, one of the changes under cold stress is the induction of certain genes. To monitor the changes in cold-inducible gene expression, we developed a mutant screening method that utilizes Arabidopsis plants transformed with the firefly luciferase reporter gene under control of a stress responsive gene (*RD29A*) promoter. The promoter of *RD29A* has two cis-acting elements: DRE/CRT and ABRE, an ABA-responsive element (Yamaguchi-Shinozaki and Shinozaki, 1994). In response to either cold, ABA or NaCl, the *RD29A::LUC* transgenic plants are capable of emitting bioluminescence that can be detected by a CCD camera (Ishitani et al., 1997). The *RD29A::LUC* transgenic Arabidopsis was mutagenized and seedlings with altered luminescence under different stresses were screened for. Resultant mutants have been categorized into several groups: *hos* (high expression of osmotically responsive genes) or *los* (low expression of osmotically responsive genes) to cold, *hos* or *los* to ABA, *hos* or *los* to NaCl, and *cos* (constitutive expression of osmotically responsive genes). Some mutants fell into more than one category, indicating cross-talks between different stress signal pathways (Ishitani et al., 1997; Xiong et al., 1999).

Among these mutants, *hos1* showed higher luminescence only under cold stress (Ishitani et al., 1998). In other stress conditions such as ABA, NaCl or PEG treatment, *hos1* did not display altered *RD29A::LUC* expression. Cold-induced gene expressions in *hos1* and wild type plants were compared by RNA blot hybridization. In response to cold, the *hos1* mutant showed increased expression of stress-responsive genes such as *RD29A*, *COR47*, *COR15A*, *KIN1* and *ADH* as well as *CBF2* and *CBF3* (Ishitani et al., 1998; Lee et al., 2001). In contrast, expression of these stress-responsive genes in *hos1* was similar or lower in response to ABA, NaCl or PEG treatment. Additionally, during vernalization, *hos1* exhibited decreased expression of the *FLC* gene (flowering locus C gene, a negative regulator of flowering), which is consistent with its early flowering phenotype. Because the altered gene expression pattern in *hos1* includes the non-DRE/CRT class genes (e.g. *ADH*), HOS1 appears to play a role in early cold signaling. Positional cloning revealed that *HOS1* encodes a novel protein with a RING finger motif similar to that of IAP (inhibitors of apoptosis) proteins that may act as an E3 ubiquitin protein ligase (Lee et al., 2001). Thus, Lee et al (2001) proposed that HOS1 may regulate the CBF pathway through selective degradation of positive regulators such as ICE (inducer of CBF expression) (Thomashow, 1999). One intriguing observation was made with cellular localization studies of the HOS1:GFP fusion protein. The HOS1:GFP proteins, which were localized in cytosol under normal conditions, were observed to accumulate in nuclei after cold treatment (0°C, 1-2 day) implicating that HOS1 may participate in cytosol-nucleus communication in cold signal transduction (Lee et al., 2001).

Non-cold acclimated *hos1* mutants appeared to be more sensitive to cold stress than the wild type, while no significant differences were observed in ion leakage test over the freezing temperatures between the *hos1* mutant and wild type after cold acclimation (Ishitani et al., 1998). This may be because *hos1* mutation affects plant vigor, which might result in cold sensitivity. However, superinduction of the DRE/CRT class genes after cold acclimation may compensate for the reduced tolerance.

6.4. *hos2* Mutant

In another recessive mutant named *hos2*, categorized also as hos_{cold}, endogenous *RD29A* as well as the *RD29A::LUC* transgene expression was at higher levels in response to cold stress (Lee et al., 1999). Other treatment such as ABA or NaCl treatment did not affect either the luminescence intensity or the *RD29A* transcript level. Similar to *hos1*, *hos2* showed higher induction of cold-regulated genes such as *RD29A*, *KIN1*, *COR15A*, *P5CS*, *COR47* and *ADH* in response to cold stress. Because *P5CS* and *ADH* are not the DRE/CRT class of stress-responsive genes, HOS2 may also represent an early cold-signaling component. Interestingly, despite the enhanced cold-regulated gene expressions, *hos2* was more sensitive to freezing temperatures than wild type, both before and after cold acclimation as shown by electrolyte leakage tests (Lee et al., 1999). In fact, the freezing sensitivity in *hos2* was more pronounced after cold acclimation. These results appear inconsistent with previous studies with *CBF1*- or *DREB1A*-overexpressing Arabidopsis that was more tolerant to stresses possibly due to increased DRE/CRT class gene expression (Jaglo-Ottosen et al., 1998; Kasuga et al., 1999). Therefore, it was proposed that the *hos2* mutation might increase the cold-regulated gene induction through a mechanism different from the one in the *hos1* mutant, and might negatively affect other cellular responses (e.g. osmolyte metabolism or membrane components), which compromises freezing tolerance (Lee et al., 1999). The *hos2* mutant provides a lesson that multiple factors should be considered to achieve freezing tolerance in plants.

6.5. *fro1* Mutant

The *fro1* (*frostbite1*) mutant displays a luminescence phenotype opposite to that of *hos1* or *hos2*, i.e. it belongs to the los_{cold} category (Lee et al., 2001). Consistent with its lower luminescence intensity under cold conditions, *fro1* mutant plants displayed decreased expression level of cold-inducible genes such as *KIN1* and *COR15A* as well as the endogenous *RD29A* gene. In addition, *fro1* mutant leaves showed a water-soaked phenotype that resembles wild-type leaves that have been subjected to freezing, hence the mutant was named *frostbite1*. *fro1* mutant leaves were constitutively leaky and were less capable of cold acclimation. The *FRO1* gene encodes a protein with high similarity to the 18 kD Fe-S subunit of Complex I (NADH dehydrogenase, EC 1.6.5.3) in the mitochondrial electron transfer chain. As shown in previous studies with mitochondria isolated from skin fibroblast (Pitkanen and Robinson, 1996), the Complex I defect in *fro1* resulted in constitutive accumulation of reactive oxygen species (ROS). Oxidative stress has been suggested to be a secondary stress induced by primary biotic and abiotic stresses (Prasad et al., 1994). ROS effects on biological systems are dependent on its severity and duration (Prasad, 2001). For instance, in maize ROS was induced in cold-acclimated seedlings at 14°C. This in turn induced catalase gene expression that plays a protective role by scavenging the ROS generated during cold stress (Prasad et al., 1994). In contrast, severe and prolonged ROS has cytotoxic effects such as lipid peroxidation in membranes, protein denaturation, and DNA damage (Halliwell and Gutteridge, 1986). Thus, ROS plays a dual role in response to stresses including cold (Prasad, 2001). In the case of *fro1*, ROS seemed to damage the plasma membrane, possibly through lipid peroxidation. This hypothesis is supported by its water-soaked phenotype and higher level of electrolyte leakage. Also, ROS can bring about changes in cytosol Ca^{2+} concentration (Knight, 2000). Exposure of Arabidopsis to H_2O_2 before cold stress changed cold-induced Ca^{2+} signature, although a 3-day successive treatment with H_2O_2 resulted in a Ca^{2+} signature similar to that induced by low temperature in 3-day cold-acclimated Arabidopsis (Knight et al., 1996). It has also been shown that lowering Ca^{2+} amplitude by inhibiting plasma membrane Ca^{2+} channels resulted in lower *KIN1* gene expression (Knight et al., 1996). Therefore, ROS in *fro1* might affect Ca^{2+} homeostasis in cells and desensitize cold responses, thus lowering the expression of cold-regulated genes (Lee et al., 2001). Interestingly, *CBF* gene induction in *fro1* was not significantly different from that in the wild type. The only difference was a slight change in *CBF* gene expression kinetics. It is not known whether this different kinetics might be causally related to the lower levels of *COR* gene expression in *fro1*.

6.6. *fry1* Mutant

Because of their enhanced *RD29A::LUC* expression under either cold, salt, osmotic stress or ABA treatment, *fry1* (*fiery1*) mutants were isolated and characterized (Xiong et al., 2001a). RNA blot hybridization showed that in *fry1* mutant plants the expression levels of several stress-responsive genes such as *KIN1*, *COR15A*, *HSP70*, and *ADH* as well as the endogenous *RD29A* gene were higher than those in the wild type. The expression of *CBF2* gene in *fry1* was sustained longer than in the wild type. Map-based cloning revealed that *FRY1* encodes an inositol polyphosphate 1-phosphatase that plays a role in inositol 1,4,5-trisphosphate (IP_3) catabolism. Many reports have shown that IP_3 triggers Ca^{2+} release from internal stores in many cells including plant cells (Berridge, 1993; Sanders et al., 1999). IP_3 has been proposed to be involved in ABA and stress signaling (Lee et al., 1996; Staxen et al., 1999; DeWald et al., 2001). As predicted, the IP_3 level in *fry1* was higher than in the wild type following ABA treatment (Xiong et al., 2001a). Thus, it is reasonable to speculate that the sustained high level of IP_3 in *fry1* is responsible for the enhanced expression of stress genes. Therefore, FRY1 attenuates IP_3 signaling by regulating IP_3 turnover rate (Xiong et al., 2001a). Surprisingly, despite the higher expression levels of stress-responsive genes, cold acclimation in *fry1* appeared impaired as suggested by electrolyte leakage analysis. In fact, *fry1* mutant plants were also very sensitive to ABA, drought, or salt stress indicating again that FRY1 might be involved in early general abiotic signaling. Before cold acclimation, wild type and *fry1* showed similar LT_{50} (the temperature at which 50 % electrolyte leakage occurs) values (-3.8°C for wild type and -2.5°C for *fry1*). However, after a 7-day cold acclimation at 4°C, the freezing tolerance in *fry1* was not improved as much as in wild type (LT_{50} for wild type, -9.4°C; for *fry1*, -5.0°C). The stress sensitivities were thought to be partly due to disturbed cellular ion homeostasis in *fry1* as a consequence of the accumulation of Ins(1,3,4)P_3 and Ins(1,3,4,5)P_4 (Xiong et al., 2001a). Alternatively, the *fry1* mutation might cause changes in the turnover of the highly phosporylated inositols such as IP_5 and IP_6, which then might affect mRNA export and gene expression (Xiong et al., 2001a). These results again suggested that there are many factors, not only the cold-regulated genes, that are important for stress tolerance in plants.

6.7. *los5* Mutant

Mutant screens using the *RD29A::LUC* Arabidopsis have also generated alleles of known stress mutants. Genetic analyses of *los5* (*los*$_{cold/salt}$) and *los6* (*los*$_{salt}$) mutants revealed that they are new alleles of *aba3* and *aba1*, respectively (Xiong et al., 2001b). In *los5* mutant plants, expression of stress-responsive genes such as *RD29A*, *COR15A*, *COR47*, *KIN1*, *RD22* and *P5CS* in response to cold (0°C) or osmotic stress (NaCl or PEG) was substantially lower (Xiong et al., 2001b). In particular, osmotic stress induction of the genes was almost blocked by *los5*. In agreement with these low levels of the stress-gene induction, *los5* mutant plants are impaired in freezing tolerance after cold acclimation and are more damaged under osmotic stress (Xiong et al., 2001b). Although both *los5/aba3* and *los6/aba1* are ABA-deficient mutants, reduced *RD29A::LUC* expression in *los5/aba3* under cold stress was not recovered to the wild type level when exogenous ABA was applied along with cold stress. In contrast, exogenous ABA application was able to restore cold-induced *RD29A::LUC* expression in *los6/aba1* to the wild-type level, suggesting that the reduced cold induction in *los5/aba3* is not due to ABA deficiency (Xiong et al., 2001b). These findings thus implicate a possible involvement of LOS5/ABA3 in cold signaling independent of ABA biosynthesis. Although the cold-responsive genes are reduced in *los5*, the mutation did not significantly affect *CBF2* expression. Therefore, LOS5 may mediate a non-CBF signaling pathway or may be a cofactor/modifier of CBF proteins. The gene induction pattern in *los5* again adds another example of changing the DRE/CRT class gene induction by cold without affecting CBF gene expression. Through map-based cloning, the *LOS5* gene was cloned and shown to encode a putative molybdopterin cofactor (MoCo) sulfurase (Xiong et al., 2001b) with a NifS-like N-terminal domain and a C-terminal domain with unknown function. It was shown that *in vitro* the recombinant MoCo sulfurase was able to activate ABA-aldehyde oxidase (AAO), the last enzyme in ABA biosynthesis by transferring sulfur to MoCo that functions as an AAO cofactor in a sulfurylated form (Bittner et al., 2001).

7. CLOSING COMMENTS

Freezing tolerance is a quantitative trait. Although it seems that the induction of DRE/CRT class genes are sufficient to achieve some level of cold tolerance as shown in the studies by Jaglo-Ottosen et al (1998) and Kasuga et al (1999), studies with cold-stress mutants suggest that there are multiple pathways/factors involved in cold signaling and freezing tolerance. Since gene expression changes are key to freezing tolerance, mutants defective in cold-responsive gene expression are valuable tools to dissect cold signaling and tolerance. More insights can be expected from continued analysis of the mutants already isolated and from future studies with new mutants from new genetic screens.

8. ACKNOWLEDGEMENTS

We thank Drs. Liming Xiong and Viswanathan Chinnusamy for helpful discussions and critical reading of the manuscript. Work in our laboratory has been supported by grants from the National Science Foundation, US Department of Agriculture, and the Southwest Consortium for Plant Genetics and Water Resources.

REFERENCES

Aguilar, P.S., Hernandez-Arriaga, A.M., Cybulski, L.E., Erazo, A.C., and de Mendoza, D., 2001, Molecular biasis of thermosensing: a two-component signal transduction thermometer in *Bacillus subtilis*, *EMBO J.* **20**: 1681.

Anderson, J.V., Li, Q.B., Haskell, D.W., and Guy, C.L., 1994, Structural organization of the spinach endoplasmic reticulum luminal 70-kilodalton heat-shock cognate gene and expression of 70-kilodalton heat-shock genes during cold acclimation, *Plant Physiol.* **104**: 1359.

Artus, N.N., Uemura, M., Steponkus, P.L., Gilmour, S.J., Lin, C.T., and Thomashow, M.F., 1996, Constitutive expression of the cold-regulated *Arabidopsis thaliana COR15a* gene affects both chloroplast and protoplast freezing tolerance, *Proc. Natl. Acad. Sci. U. S. A.* **93**: 13404.

Baker, S.S., Wilhelm, K.S., and Thomashow, M.F., 1994, The 5'-Region of *Arabidopsis thaliana COR15a* has cis-acting elements that confer cold-regulated, drought-regulated and ABA-regulated gene expression, *Plant Mol.Biol.* **24**: 701.

Berridge, M.J., 1993, Inositol trisphosphate and calcium signaling, *Nature*. **361**: 315.

Bittner, F., Oreb, M., and Mendel, R.R., 2001, ABA3 is a molybdenum cofactor sulfurase required for activation of aldehyde oxidase and xanthine dehydrogenase in *Arabidopsis thaliana*, *J. Biol. Chem.* **276**: 40381.

Bogre, L., Ligterink, W., Meskiene, I., Barker, P.J., HeberleBors, E., Huskisson, N.S., and Hirt, H., 1997, Wounding induces the rapid and transient activation of a specific MAP kinase pathway, *Plant Cell*. **9**: 75.

Browse, J., and Xin, Z.G., 2001, Temperature sensing and cold acclimation, *Curr. Opin. Plant Biol.* **4**: 241.

Delauney, A.J., and Verma, D.P.S., 1993, Proline biosynthesis and osmoregulation in plants, *Plnat J.* **4**: 215.

DeWald, D.B., Torabinejad, J., Jones, C.A., Shope, J.C., Cangelosi, A.R., Thompson, J.E., Prestwich, G.D., and Hama, H., 2001, Rapid accumulation of phosphatidylinositol 4,5-bisphosphate and inositol 1,4,5-trisphosphate correlates with calcium mobilization in salt-stressed Arabidopsis, *Plant Physiol.* **126**: 759.

Dowgert, M.F., and Steponkus, P.L., 1984, Behavior of the plasma membrane of isolated protoplasts during a freeze-thaw cycle, *Plant Physiol.* **75**: 1139.

Gasser, K.W., Goldsmith, A., and Hopfer, U., 1990, Regulation of chloride transport in parotid secretory granules by membrane fluidity, *Biochemistry*. **29**: 7282.

George, M.F., Becwar, M.R., and Burke, M.J., 1982, Freezing avoidance by deep undercooling of tissue water in winter-hardy plants, *Cryobiology*. **19**: 628.

Gibson, S., Arondel, V., Iba, K., and Somerville, C., 1994, Cloning of a temperature regulated gene encoding a chloroplast omega-3 desaturase from Arabidopsis thaliana, *Plant Physiol.* **106**: 1615.

Gilmour, S.J., and Thomashow, M.F., 1991, Cold acclimation and cold-regulated gene expression in ABA mutants of *Arabidopsis thaliana*, *Plant Mol.Biol.* **17**: 1233.

Gilmour, S.J., Zarka, D.G., Stockinger, E.J., Salazar, M.P., Houghton, J.M., and Thomashow, M.F., 1998, Low temperature regulation of the *Arabidopsis* CBF family of AP2 transcriptional activators as an early step in cold-induced *COR* gene expression, *Plnat J.* **16**: 433.

Gordonkamm, W.J., and Steponkus, P.L., 1984, The influence of cold-acclimation on the behavior of the plasma membrane following osmotic contraction of isolated protoplasts, *Protoplasma*. **123**: 161.

Gordon-Kamm, W.J., and Steponkus, P.L., 1984, The behavior of the plasma membrane following osmotic contraction of isolated protoplasts: Implications in freezing injury, *Protoplasma*. **123**: 83.

Gordon-Kamm, W.J., and Steponkus, P.L., 1984, Lamellar-to-hexagonal II phase transitions in the plasma membrane of isolated protoplasts after freeze-induced dehydration, *Proc. Natl. Acad. Sci. U. S. A.* **81**: 6373.

Guy, C.L., 1990, Cold acclimation and freezing stress tolerance: Role of protein metabolism, *Annu. Rev. Plant Physiol. Plant Mol. Biol.* **41**: 187.

Guy, C.L., Niemi, K.J., and Brambl, R., 1985, Altered gene expression during cold acclimation of spinach, *Proc. Natl. Acad. Sci. U. S. A.* **82**: 3673.

Halliwell, B., and Gutteridge, J.M.C., 1986, Oxygen free radicals and iron in relation to biology and medicine: Some problems and concepts, *Arch. Biochem. Biophys.* **246**: 501.

Harper, J.F., Sussman, M.R., Schaller, G.E., Putnamevans, C., Charbonneau, H., and Harmon, A.C., 1991, A calcium-dependent protein kinase with a regulatory domain similar to calmodulin, *Science*. **252**: 951.

Heino, P., Sandman, G., Lang, V., Nordin, K., and Palva, E.T., 1990, Abscisic acid deficiency prevents development of freezing tolerance in *Arabidopsis thaliana* (L.) Heynh, *Theor. Appl. Genet.* **79**: 801.

Ishitani, M., Xiong, L., Stevenson, B., and Zhu, J.K., 1997, Genetic analysis of osmotic and cold stress signal transduction in *Arabidopsis*: interactions and convergence of abscisic acid-dependent and abscisic acid-independent pathways, *Plant Cell*. **9**: 1935.

Ishitani, M., Xiong, L.M., Lee, H.J., Stevenson, B., and Zhu, J.K., 1998, *HOS1*, a genetic locus involved in cold-responsive gene expression in *Arabidopsis*, *Plant Cell*. **10**: 1151.

Ishizaki-Nishizawa, O., Fujii, T., Azuma, M., Sekiguchi, K., Murata, N., Ohtani, T., and Toguri, T., 1996, Low-temperature resistance of higher plants is significantly enhanced by a nonspecific cyanobacterial desaturase, *Nat. Biotechnol.* **14**: 1003.

Jaglo-Ottosen, K.R., Gilmour, S.J., Zarka, D.G., Schabenberger, O., and Thomashow, M.F., 1998, *Arabidopsis CBF1* overexpression induces *COR* genes and enhances freezing tolerance, *Science*. **280**: 104.

Jonak, C., Kiegerl, S., Ligterink, W., Barker, P.J., Huskisson, N.S., and Hirt, H., 1996, Stress signaling in plants: A mitogen-activated protein kinase pathway is activated by cold and drought, *Proc. Natl. Acad. Sci. U. S. A.* **93**: 11274.

Jouannic, S., Leprince, A.S., Hamal, A., Picaud, A., Kreis, M., and Henry, Y., 2000, Plant mitogen-activated protein kinase signalling pathways in the limelight, in: *Advances in Botanical Research Incorporating Advances in Plant Pathology, Vol 32*, M. Kreis and J.C. Walker, ed., Academic Press Inc, San Diego, pp. 299-354.

Kasuga, M., Liu, Q., Miura, S., Yamaguchi-Shinozaki, K., and Shinozaki, K., 1999, Improving plant drought, salt, and freezing tolerance by gene transfer of a single stress-inducible transcription factor, *Nat. Biotechnol.* **17**: 287.

Knight, H., 2000, Calcium signaling during abiotic stress in plants, in: *International Review of Cytology - a Survey of Cell Biology, Vol 195*, K.W. Jeon, ed., pp. 269-324.

Knight, H., Trewavas, A.J., and Knight, M.R., 1996, Cold calcium signaling in *Arabidopsis* involves two cellular pools and a change in calcium signature after acclimation, *Plant Cell.* **8**: 489.

Knight, H., Veale, E.L., Warren, G.J., and Knight, M.R., 1999, The *sfr6* mutation in Arabidopsis suppresses low-temperature induction of genes dependent on the CRT/DRE sequence motif, *Plant Cell.* **11**: 875.

Knight, M.R., Campbell, A.K., Smith, S.M., and Trewavas, A.J., 1991, Transgenic plant aequorin reports the effects of touch and cold-shock and elicitors on cytoplasmic calcium, *Nature.* **352**: 524.

Krishna, P., Sacco, M., Cherutti, J.F., and Hill, S., 1995, Cold-induced accumulation of *hsp90* transcripts in *Brassica napus*, *Plant Physiol.* **107**: 915.

Kurkela, S., and Franck, M., 1990, Cloning and characterization of a cold- and ABA-inducible *Arabidopsis* gene, *Plant Mol.Biol.* **15**: 137.

Lee, B.-h., Lee, H., Xiong, L., and Zhu, J.K., 2001, A mitochondrial complex I defect impairs cold regulated nuclear gene expression, *submitted.*

Lee, B.-h., Stevenson, B., and Zhu, J.K., 2002, High-throughput screening of *Arabidopsis* mutants with deregulated stress-responsive luciferase gene expression using a CCD camera, in: *Luminescence Biotechnology: Instruments and Applications*, K. Van Dyke, C. Van Dyke and K. Woodfork, ed., CRC press LLC, Boca Raton, FL, pp. 557-564.

Lee, H., Xiong, L.M., Ishitani, M., Stevenson, B., and Zhu, J.K., 1999, Cold-regulated gene expression and freezing tolerance in an *Arabidopsis thaliana* mutant, *Plnat J.* **17**: 301.

Lee, H.J., Xiong, L.M., Gong, Z.Z., Ishitani, M., Stevenson, B., and Zhu, J.K., 2001, The Arabidopsis *HOS1* gene negatively regulates cold signal transduction and encodes a RING finger protein that displays cold-regulated nucleo-cytoplasmic partitioning, *Genes Dev.* **15**: 912.

Lee, Y.S., Choi, Y.B., Suh, S., Lee, J., Assmann, S.M., Joe, C.O., Kelleher, J.F., and Crain, R.C., 1996, Abscisic acid-induced phosphoinositide turnover in guard cell protoplasts of Vicia faba, *Plant Physiol.* **110**: 987.

Levitt, J., 1980, *Responses of plants to environmental stress*, Academic press, New York.

Lin, C.T., and Thomashow, M.F., 1992, DNA sequence analysis of a complementary DNA for cold-regulated *Arabidopsis* gene *Cor15* and characterization of the Cor15 polypeptide, *Plant Physiol.* **99**: 519.

Liu, Q., Kasuga, M., Sakuma, Y., Abe, H., Miura, S., Yamaguchi-Shinozaki, K., and Shinozaki, K., 1998, Two transcription factors, DREB1 and DREB2, with an EREBP/AP2 DNA binding domain separate two cellular signal transduction pathways in drought- and low-temperature-responsive gene expression, respectively, in Arabidopsis, *Plant Cell.* **10**: 1391.

Los, D.A., Horvath, I., Vigh, L., and Murata, N., 1993, The temperature-dependent expression of the desaturase gene *desA* in *Synechocystis* PCC6803, *FEBS Lett.* **318**: 57.

Lynch, D.V., and Steponkus, P.L., 1987, Plasma membrane lipid alterations associated with cold acclimation of winter rye seedlings (*Secale cereale* L. cv Puma), *Plant Physiol.* **83**: 761.

Maeda, M., Wurgler-Murphy, S.M., and Saito, H., 1994, A two-component system that regulates an osmosensing MAP kinase cascade in yeat, *Naure.* **309**: 242.

Mazars, C., Thion, L., Thuleau, P., Graziana, A., Knight, M.R., Moreau, M., and Ranjeva, R., 1997, Organization of cytoskeleton controls the changes in cytosolic calcium of cold-shocked Nicotiana plumbaginifolia protoplasts, *Cell Calcium.* **22**: 413.

McKown, R., Kuroki, G., and Warren, G., 1996, Cold responses of Arabidopsis mutants impaired in freezing tolerance, *J. Exp. Bot.* **47**: 1919.

Meskiene, I., Bogre, L., Glaser, W., Balog, J., Brandstotter, M., Zwerger, K., Ammerer, G., and Hirt, H., 1998, MP2C, a plant protein phosphatase 2C, functions as a negative regulator of mitogen-activated protein kinase pathways in yeast and plants, *Proc. Natl. Acad. Sci. U. S. A.* **95**: 1938.

Miquel, M., James, D., Dooner, H., and Browse, J., 1993, *Arabidopsis* requires polyunsaturated lipids for low-temperature survival, *Proc. Natl. Acad. Sci. U. S. A.* **90**: 6208.

Mizoguchi, T., Irie, K., Hirayama, T., Hayashida, N., YamaguchiShinozaki, K., Matsumoto, K., and Shinozaki, K., 1996, A gene encoding a mitogen-activated protein kinase kinase kinase is induced simultaneously with genes for a mitogen-activated protein kinase and an S6 ribosomal protein kinase by touch, cold, and water stress in *Arabidopsis thaliana*, *Proc. Natl. Acad. Sci. U. S. A.* **93**: 765.

Monroy, A.F., and Dhindsa, R.S., 1995, Low-temperature signal transduction: induction of cold acclimation-specific genes of alfalfa by calcium at 25 degrees C, *Plant Cell.* 7: 321.

Monroy, A.F., Sarhan, F., and Dhindsa, R.S., 1993, Cold-iduced changes in freezing tolerance, protein phosphorylation, and gene expression: Evidence for a role of calcium, *Plant Physiol.* **102**: 1227.

Morris, P.C., 2001, MAP kinase signal transduction pathways in plants, *New Phytol.* **151**: 67.

Murata, N., 1989, Low-temperature effects on cyanobacterial membranes, *J. Bioenerg. Biomembr.* **21**: 61.

Murata, N., and Los, D.A., 1997, Membrane fluidity and temperature perception, *Plant Physiol.* **115**: 875.

Nishida, I., and Murata, N., 1996, Chilling sensitivity in plants and cyanobacteria: The crucial contribution of membrane lipids, *Annu. Rev. Plant Physiol. Plant Mol. Biol.* **47**: 541.

Nordin, K., Vahala, T., and Palva, E.T., 1993, Differential expression of two related, low-temperature induced genes in *Arabidopsis thaliana* (L) Heynh, *Plant Mol.Biol.* **21**: 641.

Orvar, B.L., Sangwan, V., Omann, F., and Dhindsa, R.S., 2000, Early steps in cold sensing by plant cells: the role of actin cytoskeleton and membrane fluidity, *Plnat J.* **23**: 785.

Pitkanen, S., and Robinson, B.H., 1996, Mitochondrial complex I deficiency leads to increased production of superoxide radicals and induction of superoxide dismutase, *J. Clin. Invest.* **98**: 345.

Prasad, T.K., 2001, Mechanisms of chilling injury and tolerance, in: *Crop responses and adaptations to temperature stress*, B. AS, ed., Food Products Press, Binghamton, NY, pp. 1-52.

Prasad, T.K., Anderson, M.D., Martin, B.A., and Stewart, C.R., 1994, Evidence for chilling-induced oxidative stress in maize seedlings and a regulatory role for hydrogen peroxide, *Plant Cell.* **6**: 65.

Ristic, Z., and Ashworth, E.N., 1993, Changes in leaf ultrastructure and carbohydrates in *Arabidopsis thaliana* L (Heyn) cv. Columbia during rapid cold acclimation, *Protoplasma.* **172**: 111.

Saijo, Y., Hata, S., Kyozuka, J., Shimamoto, K., and Izui, K., 2000, Over-expression of a single Ca2+-dependent protein kinase confers both cold and salt/drought tolerance on rice plants, *Plnat J.* **23**: 319.

Sakamoto, T., and Bryant, D.A., 1997, Temperature-regulated mRNA accumulation and stabilization for fatty acid desaturase genes in the cyanobacterium *Synechococcus* sp. strain PCC 7002, *Mol. Microbiol.* **23**: 1281.

Sanders, D., Brownlee, C., and Harper, J.F., 1999, Communicating with calcium, *Plant Cell.* **11**: 691.

Sangwan, V., Foulds, I., Singh, J., and Dhindsa, R.S., 2001, Cold-activation of Brassica napus BN115 promoter is mediated by structural changes in membranes and cytoskeleton, and requires Ca2+ influx, *Plnat J.* **27**: 1.

Sato, N., and Murata, N., 1980, Temperature shift induced responses in lipids in the blue-green alga, *Anabaena variabilis*: The central role of diacylmonogalactosylglycerol in thermo-adaptation, *Biochimica Et Biophysica Acta.* **619**: 353.

Seki, M., Narusaka, M., Abe, H., Kasuga, M., Yamaguchi-Shinozaki, K., Carninci, P., Hayashizaki, Y., and Shinozaki, K., 2001, Monitoring the expression pattern of 1300 Arabidopsis genes under drought and cold stresses by using a full-length cDNA microarray, *Plant Cell.* **13**: 61.

Sheen, J., 1996, Ca2+-dependent protein kinases and stress signal transduction in plants, *Science.* **274**: 1900.

Squier, T.C., Bigelow, D.J., and Thomas, D.D., 1988, Lipid fluidity directly modulates the overall protein rotational mobility of the Ca2+-ATPase in sarcoplasmic reticulum, *J. Biol. Chem.* **263**: 9178.

Staxen, I., Pical, C., Montgomery, L.T., Gray, J.E., Hetherington, A.M., and McAinsh, M.R., 1999, Abscisic acid induces oscillations in guard-cell cytosolic free calcium that involve phosphoinositide-specific phospholipase C, *Proc. Natl. Acad. Sci. U. S. A.* **96**: 1779.

Steponkus, P.L., Uemura, M., Joseph, R.A., Gilmour, S.J., and Thomashow, M.F., 1998, Mode of action of the COR15a gene on the freezing tolerance of Arabidopsis thaliana, *Proc. Natl. Acad. Sci. U. S. A.* **95**: 14570.

Stockinger, E.J., Gilmour, S.J., and Thomashow, M.F., 1997, *Arabidopsis thaliana* CBF1 encodes an AP2 domain-containing transcriptional activator that binds to the C-repeat/DRE, a cis-acting DNA regulatory element that stimulates transcription in response to low temperature and water deficit, *Proc. Natl. Acad. Sci. U. S. A.* **94**: 1035.

Strand, A., Hurry, V., Gustafsson, P., and Gardestrom, P., 1997, Development of Arabidopsis thaliana leaves at low temperatures releases the suppression of photosynthesis and photosynthetic gene expression despite the accumulation of soluble carbohydrates, *Plnat J.* **12**: 605.

Suzuki, I., Los, D.A., Kanesaki, Y., Mikami, K., and Murata, N., 2000, The pathway for perception and transduction of low-temperature signals in Synechocystis, *EMBO J.* **19**: 1327.

Tahtiharju, S., and Palva, T., 2001, Antisense inhibition of protein phosphatase 2C accelerates cold acclimation in Arabidopsis thaliana, *Plnat J.* **26**: 461.

Tahtiharju, S., Sangwan, V., Monroy, A.F., Dhindsa, R.S., and Borg, M., 1997, The induction of *kin* genes in cold-acclimating *Arabidopsis thaliana*. Evidence of a role for calcium, *Planta.* **203**: 442.

Thion, L., Mazars, C., Thuleau, P., Graziana, A., Rossignol, M., Moreau, M., and Ranjeva, R., 1996, Activation of plasma membrane voltage-dependent calcium- permeable channels by disruption of microtubules in carrot cells, *FEBS Lett.* **393**: 13.

Thomashow, M.F., 1994, *Arabidopsis thaliana* as a model for studying mechanisms of plant cold tolerance, in: *Arabidopsis*, E.M. Meyerowitz and C.R. Somerville, ed., Cold Spring Harbor Laboratory Press, Cold Spring Harbor, NY, pp. 807-834.

Thomashow, M.F., 1999, Plant cold acclimation: Freezing tolerance genes and regulatory mechanisms, *Annu. Rev. Plant Physiol. Plant Mol. Biol.* **50**: 571.

Vigh, L., Los, D.A., Horvath, I., and Murata, N., 1993, The primary signal in the biological perception of temperature: Pd-catalyzed hydrogenation of membrane lipids stimulated the expression of the *desA* Gene in *Synechocystis* PCC6803, *Proc. Natl. Acad. Sci. U. S. A.* **90**: 9090.

Wada, H., Gombos, Z., and Murata, N., 1990, Enhancement of chilling tolerance of a cyanobacterium by genetic manipulation of fatty acid desaturation, *Nature*. **347**: 200.

Wada, H., and Murata, N., 1990, Temperature-induced changes in the fatty acid composition of the cyanobacterium, *Synechocystis* PCC6803, *Plant Physiol.* **92**: 1062.

Wang, C.Y., 1982, Physiological and biochemical responses of plants to chilling stress, *Hortscience*. **17**: 173.

Warren, G., McKown, R., Marin, A., and Teutonico, R., 1996, Isolation of mutations affecting the development of freezing tolerance in Arabidopsis thaliana (L) Heynh, *Plant Physiol.* **111**: 1011.

Xin, Z., and Browse, J., 2000, Cold comfort farm: the acclimation of plants to freezing temperatures, *Plant Cell Environ.* **23**: 893.

Xin, Z.G., and Browse, J., 1998, *eskimo1* mutants of *Arabidopsis* are constitutively freezing-tolerant, *Proc. Natl. Acad. Sci. U. S. A.* **95**: 7799.

Xiong, L., Lee, B.-h., Ishitani, M., Lee, H., Zhang, C.Q., and Zhu, J.K., 2001a, *FIERY1* encoding an inositol polyphosphate 1-phosphatase is a negative regulator of abscisic acid and stress signaling in *Arabidopsis*, *Genes Dev.* **15**: 1971.

Xiong, L., Schumaker, K., and Zhu, J.K., 2001c, Cell signaling during cold, drought and salt stresses, *Plant Cell.* **13**: *in press*.

Xiong, L.M., David, L., Stevenson, B., and Zhu, J.K., 1999, High throughput screening of signal transduction mutants with luciferase imaging, *Plant Mol. Biol. Rep.* **17**: 159.

Xiong, L.M., Ishitani, M., Lee, H., and Zhu, J.K., 2001b, The Arabidopsis LOS5/ABA3 locus encodes a molybdenum cofactor sulfurase and modulates cold stress- and osmotic stress- responsive gene expression, *Plant Cell.* **13**: 2063.

Xiong, L.M., Ishitani, M., and Zhu, J.K., 1999, Interaction of osmotic stress, temperature, and abscisic acid in the regulation of gene expression in *Arabidopsis*, *Plant Physiol.* **119**: 205.

Yamaguchi-Shinozaki, K., and Shinozaki, K., 1994, A novel *cis*-acting element in an *Arabidopsis* gene is involved in responsiveness to drought, low-temperature, or high-salt stress, *Plant Cell.* **6**: 251.

Zhang, X.R.S., and Choi, J.H., 2001, Molecular evolution of calmodulin-like domain protein kinases (CDPKs) in plants and protists, *J. Mol. Evol.* **53**: 214.

Zhu, J.K., Hasegawa, P.M., and Bressan, R.A., 1997, Molecular aspects of osmotic stress in plants, *Crit. Rev. Plant Sci.* **16**: 253.

MUTANTS DEFICIENT IN COLD HARDINESS

What can they reveal about freezing tolerance?

Gareth J. Warren, Glenn J. Thorlby, and Irene Bramke*

1. OVERVIEW

Mutations that affect cold hardiness, if their effects are specific, identify proteins that help to protect plant cells against chilling or freezing. This offers an approach to cold hardiness complementary to the study of low temperature-induced genes and the cold signal transduction pathway. The approaches have different limitations. The mutational approach will not discover functions for which there is genetic redundancy — for example, it would be unlikely to have identified the *CBF* transcription factor genes in *Arabidopsis*. On the other hand, the inducibility approach will miss any mechanisms of hardiness that are constitutive.

The *sfr* (sensitive to freezing) mutations of *Arabidopsis* display distinguishable behaviors after freezing. Their various phenotypes may be reflecting deficiencies in different protective mechanisms. At the molecular level, the *sfr6* and *cls8* mutations have broad effects on the cold-inducible expression of other genes. Identification of the affected genes will help to elucidate the cold signal transduction pathway. On the other hand, because the *sfr3* mutation appears not to affect cold-inducible expression, knowledge of the *SFR3* gene may elucidate a specific individual mechanism for preventing freezing injury.

The ease of isolating the genes defined by mutation is highly dependent on the clarity of the phenotype in individual plants, and the absence of modifier genes segregating in a mapping population. Thus the strong and unmodified phenotype of the *cls8* mutation, and its pleiotropic effects, have facilitated its mapping to a region of 250 kb. In contrast, the *sfr6* mutation, because its phenotype appears to be affected by segregating modifier genes, and also because it has deleterious effects on plant viability and seed set, requires several-fold more effort to achieve an equivalent degree of progress in genetic mapping.

---

* Gareth J. Warren, Glenn J. Thorlby, Irene Bramke, School of Biological Sciences, Royal Holloway and Bedford New College, University of London, Egham, Surrey TW20 0EX, U.K.

A reverse genetic approach has been sampled by isolating and examining insertional mutations in two cold-inducible genes, *LTI30* and *COR47*. In neither type of mutant were we able to detect any degradation of freezing tolerance.

2. ALTERNATIVE APPROACHES

A powerful approach to the understanding of cold hardiness has been to observe the changes that occur during cold acclimation, try to understand the chain of events that brings about such changes, and discover whether and how the changes contribute to the increase in freezing tolerance. Early on, changes such as alterations in leaf morphology, and increases in soluble carbohydrates, were readily apparent. In the last fifteen years, the acclimation-related changes that have been studied most intensively have been in the realm of molecular genetics: changes in the expression of genes and the levels of specific proteins (reviewed by Thomashow, 1999). In its application to genes and proteins, the approach has been successful in several ways. It has provided an increasingly detailed quantitative picture of cold acclimation in a number of species. It has indicated elements in common between the cold responses of different species, and in common between the cold response and the drought response (Shinozaki and Yamaguchi-Shinozaki, 2000). It has shown us parts of the mechanism of response: the chain of events leading from perception of low temperature to the final ensemble of responses (Thomashow, 2001). The approach led to tests of the role of a particular component of the response mechanism, the CBF factors. The spectacular results of those tests (Jaglo-Ottosen et al., 1998; Kasuga et al., 1999) demonstrated beyond doubt their relevance to cold hardiness.

So successful has been the study of cold responses that a few of its practitioners and, worryingly, even national grant agencies, have appeared to equate this approach with the study of cold hardiness. We will leave them unreferenced so as not to encourage them. In this chapter we discuss the classical genetic approach towards the same goal, which has been successful in the study of other phenomena but has only lately been applied to cold hardiness. We begin by considering what each approach is likely to discover.

3. SCOPE OF THE APPROACHES

In understanding cold hardiness, a fundamentally interesting question is the following: What are the essential differences between the cold-acclimated plant of a hardy species, and the plant that is alive and healthy but freezing-sensitive by virtue of belonging to a tender species? This is pertinent to the goal of engineering freezing tolerance in non-hardy species.

At the genetic level, we know that at least some part of the functional difference results from gene expression that is induced in the hardy species during cold acclimation. This may be the major part of the difference. However, any remaining difference would be due to genes whose expression was not cold-induced: *constitutive* freezing-tolerance genes. It is reasonable to ask why such genes should be constitutive rather than cold-inducible, if their only function is protection against frost. At present any answers would be speculative — but this cannot be a basis for concluding that constitutive freezing-tolerance genes do not exist. If they do exist, the study of cold-induced gene expression will certainly overlook them. On the other hand, the expression profile of a hardiness gene has no bearing whatever on our ability to detect it by mutation. Null mutations in freezing tolerance genes should, whether constitutive or inducible, cause deficiency in freezing tolerance.

How great will be the deficiency in freezing tolerance due to such mutations? It is generally accepted that hardiness is a complex phenomenon dependent on the combined action of many gene products (Thomashow, 1999). But in what way do the gene products combine their actions? A common simplifying assumption is that their effects will be additive, each gene contributing a certain amount of tolerance, quantifiable in degrees Celsius as a difference in LT_{50}. Additive effects are supported by the result that elevating osmolyte levels (see for example the chapter by Tahtiharju) can increase freezing tolerance, in the absence of the other phenomena of cold acclimation. Mutations in genes having additive effects will be detectable only if those genes' contributions to freezing tolerance are sufficiently large. If freezing tolerance in *Arabidopsis thaliana* was provided by the additive effects of 10 equally-contributing genes, for technical reasons we would be very unlikely to detect any of them by mutant screens. They might be mappable as quantitative trait loci (assuming that strong variants existed in natural populations), but huge technical difficulties would stand in the way of molecular identification.

The outlook for the mutant approach appears even worse when functionally redundant genes are considered. The loss of one member of a redundant set will have no impact on the organism's freezing tolerance. Even in *Arabidopsis thaliana*, with its unusually small genome, a high proportion of genes have a closely-related copy elsewhere in the genome, often in tandem repeat (The Arabidopsis Genome Initiative, 2000). Genes responsible for freezing tolerance, but existing in functionally-redundant families, will be refractory to the mutant approach.

At the opposite extreme of mutational detectability would be genes which, like enzymes for successive steps in a metabolic pathway, have cooperative effects. This is reasonable if we surmise that some genes will be specific in the types of freezing damage that they prevent, but that prevention of most or all types of damage will be necessary for survival (Warren, 1998).

In summary, the mutant approach should be very good at detecting genes with cooperative or synergistic effects on freezing tolerance; it should be capable of detecting only the major genes with additive effects; it should not be capable of detecting genes with redundant functions. Although its useful range overlaps with the study of induced genes (Figure 1), the approaches have distinct blind spots and should be considered complementary.

We note in passing how two other variant approaches fit into this scheme. The approach described in Chapter 2, of isolating and analyzing mutations that affect transduction of the cold signal (Ishitani et al., 1997; Xiong et al., 1999; Zhu, 2001), will see subsets of both cold-induced and constitutive freezing-tolerance genes, since both types are potentially involved in signalling. The approach described in Chapter 4, of isolating mutants that mimic cold acclimation or otherwise enhance freezing tolerance (Xin and Browse, 1998), may indicate cold-inducible genes that make individual contributions to freezing tolerance. It may even reveal latent or alternative mechanisms for achieving freezing tolerance, which are not active in the wild-type hardy plant.

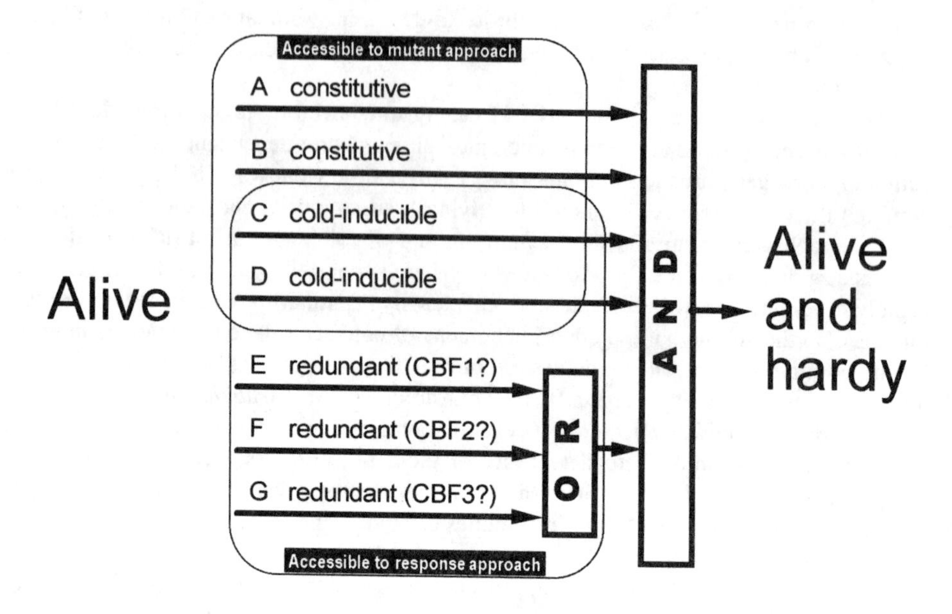

Figure 1. Logical 'AND' indicates that all inputs are required for the output; logical 'OR' indicates that any input is sufficient to cause an output. The *CBF* genes are suggested as exemplars of redundant functions; redundancy of their functions is indicated but not proved by current knowledge.

4. PHENOTYPES OF COLD- AND FREEZING-SENSITIVE MUTANTS

We initially isolated a series of true-breeding mutant lines of *Arabidopsis thaliana*, based on reproducible observations of death or morbidity after freezing of cold-acclimated plants (Warren et al., 1996). Subsequently we observed that most of the lines also exhibited symptoms of injury *during* cold acclimation, even in the absence of freezing: the majority were chilling-sensitive as well as freezing-sensitive. Because freezing stress is complex, survival can be expected to integrate many aspects of metabolism, physiology and perhaps morphology. The prior existence of sublethal injury, or of reduced vigor, would probably reduce the resources available for multiple survival mechanisms. It would be extremely difficult to disentangle such effects, and so to draw novel inferences from the freezing-sensitivity of chilling-sensitive mutants. For this reason, we did not immediately investigate the chilling-sensitive mutants.

The minority of mutants, which although sensitive to freezing were tolerant to chilling, were designated *sfr* (*s*ensitive to *f*reezing). Their sensitivity was quantified by the electrolyte leakage test. Within the limits of accuracy of the test, three classes of sensitivity could be distinguished. *sfr4* showed the greatest sensitivity, its EL_{50} being closer to that of nonacclimated wild type than to cold-acclimated wild type. *sfr2* showed the least sensitivity, the mutation affecting its EL_{50} by only approximately 1 degree. The remainder of the mutants were quantitatively intermediate in their sensitivity by this measure (Warren et al., 1996).

The mutants were also distinct in their freezing-sensitivity in at least three other objectively describable ways. The timing with which visible injury was manifested, following a marginal freezing condition, varied reproducibly between mutants. The color and appearance of damaged leaf tissue varied: the majority of mutants showed a dark, water-soaked appearance in leaf tissue that was destined to die, but *sfr2* leaves showed a lightening of color distinct from senescence, and did not exhibit water-soaking except when frozen to near the viability limit temperature of wild type. Finally, the tissues that were most sensitive to marginal freezes varied. The rosette center was most sensitive in *sfr3* and *sfr5*; in contrast, the expanded leaves of *sfr2* and *sfr4* were killed before other tissues. Refinement of these observations may enable us to understand what roles the corresponding wild-type genes play, may indicate the tissues in which their activities are most vital, and may point to the timescale of the injurious events that they are responsible for preventing.

A few simple biochemical analyses were carried out on the mutants (McKown et al., 1996). One line, *sfr4*, was observed to fail to elevate its levels of soluble sugars in response to cold. Since many other lines of evidence have indicated that soluble sugars make a major contribution to freezing tolerance, their depression in *sfr4* provides sufficient explanation for the freezing-sensitivity of the mutant. Therefore, identification of the *SFR4* gene will probably not elucidate a *novel* mechanism of damage repair or avoidance. However, the *sfr4* mutant's alteration in cold-acclimated sugar levels has been used as a tool in the laboratory of the late Dr Peter Steponkus to investigate the biochemistry of damage prevention by osmolytes (M. Uemura, G. Warren and P. Steponkus, manuscript in preparation). None of the mutants except *sfr4* showed a detectable alteration in soluble carbohydrates, indicating that other mechanisms of tolerance operate and are essential for maximal freezing tolerance in the wild type.

The gross fatty acyl content of leaf lipids showed minor variations from wild type in two of the *sfr* mutants. Fatty acyl unsaturation has been correlated with changes in freezing tolerance that occur during cold acclimation (Palta et al., 1993), and specific molecular species of lipids have been experimentally shown to affect freeze-induced lesions in some experimental systems (Steponkus et al., 1988). Nevertheless, freezing sensitivity can not necessarily be attributed to changes in fatty acyl content in the two *sfr* mutants that show such changes. The fatty acid effects could simply be pleiotropic effects of the mutations. When the cognate *SFR* genes are identified, their identities may indicate, for example, whether those genes have more direct effects on other cellular components: such effects could then be tested for their direct contributions to freezing tolerance.

A number of other gross phenotypes have been observed in the *sfr6* mutant: (i) erratic elongation of racemes between pedicels, (ii) low seed set, (iii) death in response to dipping in *Agrobacterium* suspensions, (iv) slightly paler leaf color, and (v) mottled accumulation of anthocyanin after exposure to high light at low temperature. The diversity of pleiotropic effects of *sfr6* initially concerned us, since it would seem to make it very difficult to trace the mutation's effect on freezing tolerance to a particular aspect of its phenotype.

The inducibility of three cold-regulated (*COR*) genes has been compared among the *sfr* mutants (Knight et al., 1999). Mutants *sfr2*, *3*, *4*, *5* and *7* showed normal cold-inducibility. This indicates that mechanisms of tolerance, other than those due to the previously-described *COR* genes (and other than sugar level elevation), operate and are essential for maximal freezing tolerance in the wild type. The *sfr6* mutant, on the other hand, showed a drastic reduction in the levels of *COR* gene expression in cold-acclimated plants. It was known independently that concerted upregulation of the genes in the *COR* regulon can elicit freezing tolerance (Jaglo-Ottosen et al., 1998). The freezing-sensitivity of *sfr6* is therefore explicable by the assumption that upregulation of the *COR* regulon is not only sufficient but also necessary for freezing tolerance, so that its failure to be expressed in *sfr6* mutant plants provides an adequate explanation for the freezing-sensitivity phenotype of *sfr6*.

The effect of *sfr6* on cold-induced genes was found to be limited to those whose promoters contain the CRT/DRE enhancer element (Knight et al., 1999). Cold-induced genes controlled by other (currently unknown) promoter elements continued to show normal levels of induction. These included the *CBF1* gene, which encodes a member of the family of transcription factors that are responsible for upregulating *COR* genes via the CRT/DRE promoter element (Stockinger et al., 1997). Since *CBF1* undergoes normal cold-induction in *sfr6*, the likelihood is that wild-type SFR6 protein interacts with the CBF proteins, either by modifying them or by mediating the interaction (Stockinger et al., 2000) between them and the RNA polymerase. Although the transcriptional effect could theoretically be a pleiotropic effect of the *sfr6* mutation, transcriptional disruption seems likely to be the primary effect, since it provides a simple and plausible explanation for the extent of pleiotropic effects.

Given what we now know about *sfr6*, it is rather surprising that it did not exhibit chilling sensitivity in the original mutant screen. (It would not then have been classified as an *sfr* mutation, and so would not have been included in the screen by Knight et al. (1999) for expression effects). Other mutants affecting cold-induced gene expression might well manifest deficiencies in chilling, as well as freezing, tolerance by disrupting the induction of genes adaptive for growth at low positive temperatures. Therefore, the full set of 40 chilling-sensitive mutant lines, isolated during the screen for freezing tolerance, has been tested for effects on cold-induced gene expression. One line, designated *cls8*, was found to affect cold induction of *COR78* (H. Knight and M. Knight, personal communication). This is therefore of interest to freezing as well as chilling tolerance.

5. ISOLATION OF COLD-TOLERANCE GENES DEFINED BY MUTATION

The general method for isolating genes defined only by mutation in *Arabidopsis* has been to map them with high accuracy, and then detect the wild type gene by its ability to restore function when cloned DNA from the mapped region is transformed into a mutant line. Several variants of the second part of this method are becoming viable: (i) examination of the phenotypes of insertional mutations in local genes; (ii) targeted transformation with individual genes whose sequence annotation suggests involvement in the phenomenon of interest; (iii) direct detection of the mutation. We present our experience with three genes: *CLS8*, *SFR3*, and *SFR6*.

5.1. Towards Isolation of *CLS8*

By the identification of the *CLS8* gene, we hope to elucidate a part of the signal transduction pathway leading from low temperature to the expression of the *COR78* gene (and possibly other *COR* genes). Given that the *cls8* mutation affects gene expression, there may be multiple reasons for its chilling- and freezing-sensitivity, so that elucidation of individual tolerance mechanisms is not likely to result, and is not part of the rationale for its identification.

To map *CLS8*, a line homozygous for the *cls8* mutation in the genetic background of the Columbia ecotype was crossed to a wild type of the Landsberg ecotype. The heterozygous F1 was selfed, and a small number of F2 individuals were genotyped by chill-testing their F3 progeny. (Heterozygous F2's were revealed by segregation of chilling-sensitivity in the F3 progeny). DNA from each F2 plant was analysed for physical markers mapping at 22 distributed points throughout the genome. Such analysis indicated the presence of *CLS8* on chromosome II, with linkage to markers PHYB and nga1126 at high probability. At the same time, three-point analysis eliminated all potential locations on other chromosomes. Our initial mapping of every *SFR* gene has followed a directly analogous path, but substituting freeze-testing for chill-testing (Thorlby et al., 1999; Knight et al., 1999).

Fine-mapping of a gene for subsequent cloning requires examination of large numbers of F2 plants, very approximately of the order of 500 such plants. Here, the properties of the mutation being mapped have a significant impact on the strategy that can be applied, and the efficiency with which it is executed. The properties of *cls8* come close to representing a best case.

The recessive *cls8* mutation causes not only chilling-sensitivity but also a visible abnormality in leaf morphology at normal growth temperatures. Leaves are slightly distorted, apparently as a result of the spontaneous bleaching of small, irregular patches of the lamina. The mutation has little or no impact on vigor or seed set at normal growth temperatures. Thus mutant homozygotes can be putatively identified, and then reliably grown on to seed, without treatments that jeopardize their survival. By selecting only putative mutant homozygotes for further analysis, we reduced the labor required subsequently to determine each F2 plant's *CLS8* genotype.

Before the F2 plants set seed, a small amount of DNA was isolated from each, and tested to reveal the genotype at each of the flanking markers, PHYB and nga1126. F2 plants in which the flanking markers had equivalent genotypes (eg both markers homozygous for the Columbia alleles) were very unlikely to be informative in fine mapping, and could therefore be discarded at this stage. Plants in which the flanking markers had non-equivalent genotypes (eg one heterozygous and the other homozygous) derived from gametes that had undergone crossing-over in the region of *CLS8*. These F2 individuals ("crossover lines") were grown to seed, and their *CLS8* genotypes examined by chill-testing a small number of F3 progeny from each. In most cases such testing confirmed that the line was a true-breeding mutant (ie that the F2 had been homozygous for the *cls8* mutation). In the minority of cases where this was not true, 15-40 more progeny were chill-tested to distinguish between the possible genotypes (homozygous wild type versus heterozygous) in the F2 individual.

A database of single nucleotide polymorphisms between Columbia and Landsberg has been made available for public use by Cereon™ Genomics (The Arabidopsis Information Resource, 2000). The database was interrogated for polymorphisms between the flanking markers, and those which were *not* readily detectable by restriction enzyme digestion were rejected. From the remaining polymorphisms, we chose a series that were spaced at approximately regular intervals, designed primer pairs (Rosen and Skaletsky, 1997) capable of amplifying each polymorphic region in a PCR reaction, and confirmed the detectability of the polymorphisms between Columbia and Landsberg DNAs. The F2 "crossover lines" were then examined by the same methodology, to determine their genotype at each polymorphic locus. This analysis narrows down the position of the crossover in the recombinant gamete that gave rise to each crossover line. For example, line 135 in Table 1 has a C (homozygous Columbia-type) genotype at all the markers from PHYB to 442838, and then an H (heterozygous) genotype at all the markers from 448063 to nga1126. We infer that one of the progenitor gametes of line 135 inherited markers PHYB to 442838 from the original parent of Columbia background (the *cls8* mutant) but inherited the remaining markers from the original parent of Landsberg background (the wild type): the crossover occurred between 442838 and 448063. (The other gamete forming line 135 can be inferred to have inherited markers throughout the region from the parent of Columbia background.)

Table 1. Mapping data for CLS8 from illustrative lines

Line no.	**Segr in F3**		Genotypes							
	wt	**mut**	**CLS8**	PHYB	430828	432502	442838	448063	PLS7	nga1126
115	**22**	**8**	**H**	H	H	H	H	H	C	C
135	**0**	**5**	**C**	C	C	C	C	H	H	H
151	**0**	**4**	**C**	C					C	H
208	**0**	**7**	**C**	H	C				C	C
224	**26**	**11**	**H**	C	H				H	H
263	**0**	**4**	**C**	C	C	C	H	H	H	H
274	**0**	**6**	**C**	H	H	C	C		C	C
279	**36**	**0**	**L**	L					L	H
293	**0**	**3**	**C**	H	C					C
318	**0**	**4**	**C**	H	C					C
319	**0**	**3**	**C**	C	C	C		H	H	H
324	**0**	**4**	**C**	C					C	H
339	**0**	**6**	**C**	C	C	C	C	H	H	H
352	**0**	**4**	**C**	H	C					C
403	**0**	**4**	**C**	H	H	C	C		C	C

The illustrated lines are selected from mapping data on 335 F2 lines from the mapping cross of *cls8*/*cls8* (Col) x *CLS8*$^+$/*CLS8*$^+$ (Ler). In a total of 670 gametes, 47 (11+36) were recombinant for the flanking markers PHYB and nga1126, approximately half as many as predicted from their separation on the recombinant inbred map (Nottingham Arabidopsis Stock Centre, 2001).

When the implications of the above analysis were reconciled with the physical map of chromosome II, they placed *CLS8* within a region of approximately 250 kb, which is contained in five overlapping BAC clones (Figure 2). We are screening further F2 lines, because further refinement of the map position is still cost-effective in reducing the labor that will be required for the next stage of gene identification. When more crossovers in the critical region become available for analysis, we will obtain further markers with which to interrogate them from the Cereon database: by these means we hope to delimit *CLS8* within one or at most two BAC clones. These will then be subcloned and transformed into the *cls8* mutant line, to screen for restoration of the wild type phenotype.

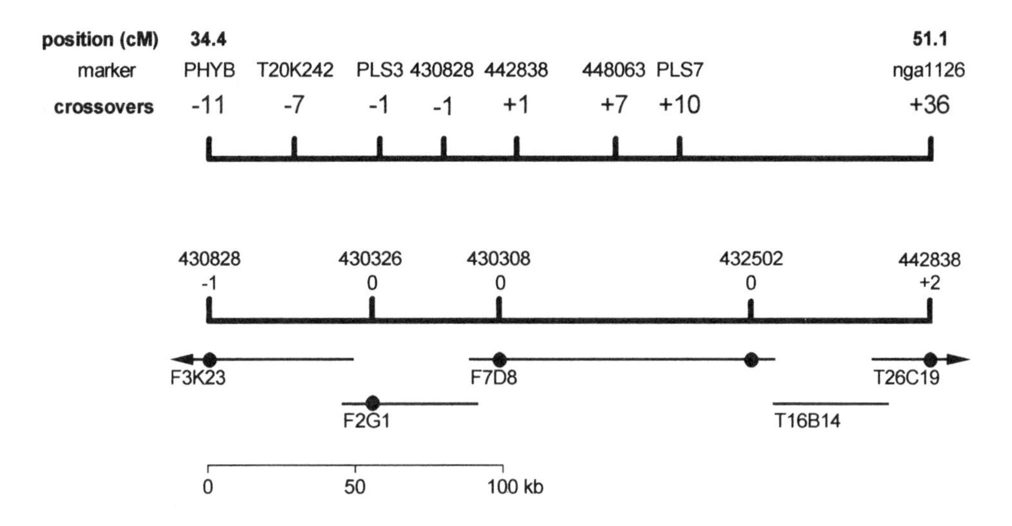

Figure 2. Genetic mapping of the *CLS8* gene to a region of chromosome II that is present in five overlapping BAC clones.

The following properties of the *cls8* mutant have allowed us to rely heavily on mapping for its identification: (i) the strong mutant phenotype, and absence of significant modifiers of that phenotype in the Columbia x Landsberg cross, (ii) the pleiotropic effect of *cls8*, enabling us to identify mutant homozygotes without tests that jeopardize their survival, and (iii) the health and fecundity of the *cls8* mutant under normal growth conditions, enabling us to recover F3 seeds from the majority of the F2 mutant plants identified as crossover lines. These properties are not universal to cold-sensitive mutants.

5.2. Towards Isolation of *SFR3*

The *sfr3* mutant, besides affecting freezing tolerance, has relatively minor pleiotropic effects. As far as we know, they do not include any effect on cold-induced gene expression. Therefore, by identifying the *SFR3* gene, we hope to elucidate a single mechanism by which wild-type cold-acclimated plants protect themselves against freezing injury.

As described by Thorlby et al. (1999), initial mapping placed *SFR3* on chromosome I, using an approach analogous to that described above for *CLS8*. In the subsequent fine-mapping of *SFR3*, the properties of the *sfr3* mutant have made a difference that has moderately hindered progress towards gene isolation.

From the *sfr3*/*sfr3* (Col) x *SFR3*$^{+}$/*SFR3*$^{+}$ (Ler) cross, we attempted to preselect F2 plants that were homozygous for the *sfr3* mutation, based on its documented pleiotropic effects (McKown et al., 1996). These putative homozygotes were then tested for their genotypes at the flanking loci UFO and GAPB. As before, F2 plants with no indication of a crossover in the region (ie, those in which the flanking markers showed matching genotypes) were discarded.

The remaining F2 plants (crossover lines) were grown to seed and their F3 progeny were grown and tested in the freezing assay. Freezing provides the defining test of whether each plant has a mutant or wild type phenotype. The test is destructive, precluding its use in the previous generation to identify F2 plants with a mutant phenotype. It turned out that the effectiveness with which we had pre-selected *sfr3* homozygotes in the F2 had varied between batches. In some batches, better than 90% of the crossover lines turned out to be *sfr3* homozygotes (revealed by apparent freezing-sensitivity of all or nearly all of the F3 progeny). In other batches, the success of pre-selection was not significantly better than would be expected by random choice. This seemed to indicate that the pleiotropic effect, on which we had attempted to rely, had been subject to variation due to factors that we were unaware of. The consequence was that for a majority of lines, F3 plants had to be replanted in sufficient numbers to distinguish heterozygotes from wild type homozygotes by the segregation of freezing-sensitives.

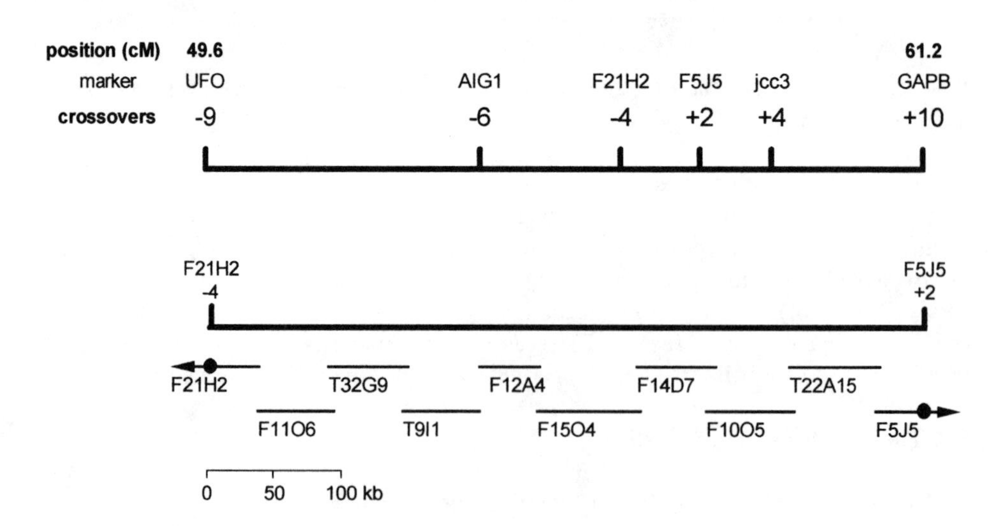

Figure 3. Mapping of the *SFR3* gene to a region of chromosome I that is present in ten overlapping BAC clones.

As in Section 5.1, additional polymorphic markers in the region of interest have been obtained and qualified. The current status, after screening approximately 250 F2 plants, is shown in Figure 3. This places *SFR3* in a region more than twice the size of that in which *CLS8* has been localized. The resolving power of the mapping can still be increased by screening more F2 plants, and mapping them with additional markers in the region of interest. However, this gene isolation project has proceeded more slowly than that for *CLS8* because of the greater difficulty of discovering the *SFR3* genotypes of F2 plants.

5.3. Towards Isolation of *SFR6*

By the identification of the *SFR6* gene, we hope to elucidate the signal transduction pathway leading from low temperature to the induction of genes with CRT/DRE elements in their promoters. Such genes are numerous, and their coordinated induction has been shown to mimic several of the biochemical changes normally associated with cold acclimation (Gilmour et al., 2000). Thus, by a similar argument to that used for *cls8*, there are likely to be multiple reasons for the *sfr6* mutant's freezing-sensitivity. The rationale for identifying *SFR6* does *not* include the elucidation of individual tolerance mechanisms, because knowledge of individual mechanisms is unlikely to result from the study of a mutation that affects a number of them in parallel.

Initial mapping of *SFR6* followed a similar path to that described for *CLS8*, and has been reported by Knight et al. (1999). *SFR6* was mapped to the vicinity of the *MEKK1* gene, a cold-induced MAP kinase kinase kinase (Mizoguchi et al., 1996), on chromosome IV. Because *MEKK1* had been implicated in low temperature signal transduction, the very phenomenon affected by the *sfr6* mutation, we amplified *MEKK1* from the *sfr6* mutant line (in three overlapping amplicons) and sequenced it. The sequence we obtained was identical to that in the public databases, eliminating the possibility that *sfr6* represented a mutation in *MEKK1*. Eliminating *MEKK1* in this way has contributed only minimally to the ongoing project to identify *SFR6*; in retrospect the investigation of this gene may have been premature.

In fine mapping of *SFR6*, the properties of the *sfr6* mutation, including its interactions with other genes segregating in the mapping cross, have had a detrimental effect on progress towards isolation of the gene.

The pleiotropic effects of the *sfr6* mutation, particularly its effect on leaf pigmentation, led us to attempt a similar approach to that described in the previous sections. We pre-selected putative mutant homozygotes among the F2 of the mapping cross, and grew them on to extract and test DNA (using markers GA1 and nga8) to identify crossover lines. The putative *sfr6* homozygotes were significantly more rare than the expected 25% proportion of F2 plants. We inferred that either *sfr6* had a deleterious effect on seed viability, or its visible pleitropic effects were becoming masked in some of the F2 individuals.

Very few of the putative *sfr6* homozygotes went on to produce seed. Poor survival and very low seed set both contributed to this unfortunate outcome. For this reason, we grew further F2 individuals and identified crossover lines without regard to their apparent

SFR6 genotypes. We subsequently attempted to determine the *SFR6* genotype of each F2 individual by observing the segregation of freezing-tolerance in its F3 progeny.

Whereas our controls in the freezing assays (the *sfr6* mutant on one hand, and the Columbia and Landsberg wild types on the other) generally gave clear-cut results, a number of F3 individuals showed phenotypes that were difficult to classify. This had not been the case when the *sfr6* mutant homozygote (Columbia background) was crossed to the Columbia wild type for dominance testing (Warren et al., 1996), so we inferred that the mutant phenotype was being affected by modifier genes whose alleles were segregating in the Columbia x Landsberg mapping cross. While we still felt able to distinguish between heterozygotes and mutant homozygotes, the distinction between heterozygotes and wild type homozygotes was particularly problematic. Therefore, we identified "improved" individual F3 plants in which a flanking marker, heterozygous in the F2, had become homozygous. This is exemplified in Table 2: whereas F2 plant 151 was heterozygous at nga8, MEKK1 and F9M13, its progeny 151-6 was homozygous for the Columbia (C) allele of those markers. The *SFR6* genotype of this "improved" F3, although still indeterminate (H or L), now indicated that the region between F9M13 and nga8 could not contain *SFR6*.

It will be apparent from the data in Table 2 that the results from lines 151-6 and 224-1 conflict with the result from line 296. The most exact determination of a gene's position by genetic fine mapping always depends on results from a comparatively small number of lines, so conflicting results (such as that described here) seriously degrade the resolution of the mapping. It will be necessary to repeat all the results obtained from the conflicting lines, and to eliminate possibilities such as phenotype suppression by modifiers. This will inhibit progress.

Table 2. Mapping data for *SFR6* from illustrative lines

Line No.	**Segr in F3**			Genotypes						
	mut	**wt**	**u***	**SFR6**	GA1	nga12	F4G16	F9M13	MEKK1	nga8
22	**0**	**8**	**4**	**H or L**	H	L			L	L
22-2	**0**	**8**	**2**	**H or L**	C	L			L	L
151	**2**	**16**	**3**	**H or L**	L	L	L	H	H	H
151-6	**2**	**13**	**5**	**H or L**	L	L	L	C	C	C
224	**0**	**6**	**0**	**H or L**	L	L	L	H	H	H
224-1	**0**	**1**	**0**	**H or L**	L	L	L	C	C	C
284	**0**	**8**	**1**	**H or L**	L	L	L	L	H	H
284-3	**0**	**3**	**0**	**H or L**	L	L	L	L	C	C
286	**2**	**4**	**0**	**H or L**	H	H	H	H	C	C
296	**0**	**6**	**0**	**H or L**	C	C	C	C	H	H
365	**6**	**19**	**0**	**H**	H	H	H	H	H	C

The illustrated lines are selected from mapping data on approx 270 F2 lines from the mapping cross of *sfr6*/*sfr6* (Col) x $SFR6^+/SFR6^+$ (Ler). Shading is applied as in Table 1. The names of "improved" lines (eg *22-2*) are shown in italics. *u denotes segregants where the classification of freezing-sensitivity is uncertain.

Since fine-mapping of *sfr6* is more difficult due to the properties of this mutation (including its possible interactions with modifier genes), it will be advantageous to begin the next phase of gene isolation at an earlier stage than was desirable for *CLS8* or *SFR3*. It is impractical by current technology to make stable transformants containing clones representing the region between GA1 and *MEKK1* (the current map definition of *SFR6*) — the region is too large. However, we are contemplating three alternatives: (i) transient transformation, detecting restoration of *SFR6* function by expression of a reporter gene, (ii) enzymatic detection of the *sfr6* mutation in heteroduplexes, following the approach described by Colbert et al. (2001), and (iii) examination of insertional mutants of genes in this region. Ongoing technical advances are likely to increase the relative advantage of such approaches, and reduce the reliance of positional cloning on genetic mapping.

6. REVERSE GENETICS

As a result of functional genomics initiatives, large numbers of *Arabidopsis* lines have become available that carry insertional mutations at essentially random locations in the genome. The identities of the affected genes in many lines are being determined. This will ultimately make it possible to screen the phenotypes of mutations in every gene.

We have taken advantage of an early publication of mutation identities (Jones, 1999), to obtain and qualify homozygous lines with mutations in two cold-induced genes, *LTI30* (also known as *XERO2*) and *COR47* (also known as *RD17*) (Gilmour et al., 1992; Rouse et al., 1996; Takahashi et al., 2001; Welin et al., 1994).

The insertional mutants were available in pooled seed batches deriving from 20 lines each. PCR primers were designed to amplify the respective insertions: in each case one primer had homology to the inserted sequence and the other homology to the target gene. Among plants growing from the pooled seed, these primers detected individuals containing the mutations in *LTI30* (*lti30::dSpm*) and *COR47* (*cor47::dSpm*) respectively. A third insertional mutation from the database, in gene *KIN1*, could not be detected in the respective batch of seedlings.

A second primer pair was designed for each gene, to amplify the wild type gene but but not the mutant allele; this discrimination was permitted by the additional length created (and hence the additional extension time required) by the insertion. Plants that were positive for an insertion were tested by PCR with the cognate second pair of primers. Presence of the wild type amplicon, following PCR, indicated that the insertional mutation was heterozygous; absence of the wild type amplicon indicated that the line was homozygous for the mutation.

After propagation, plants from homozygous mutant lines were tested for freezing tolerance by two methods: (i) freezing of whole plants, similar to the original screen for *sfr* mutants (Warren et al., 1996), and (ii) electrolyte leakage, similar to the protocol used to quantify freezing tolerance in the *sfr* mutants (Warren et al., 1996). In parallel to the insertional mutants, and wild type controls, *sfr2* and *sfr6* mutant plants provided positive controls for the detectability of freezing-sensitive mutations.

Figure 4. Appearance of the insertional mutants and control plants, 5 days after freezing to a nadir temperature of -6.8° C for 16 hours.

Freezing of whole cold-acclimated plants, to a range of nadir temperatures, elicited no visible difference between the wild type and the *lti30::dSpm* and *cor47::dSpm* mutants, whereas the *sfr2* and *sfr6* mutants showed strong differences from wild type: a sample result is shown in Figure 4. Electrolyte leakage assays, on leaves from cold-acclimated plants, indicated EL_{50} values for the mutants that did not differ significantly from those of the wild type (data not shown). Thus the *lti30::dSpm* and *cor47::dSpm* mutants were indistinguishable from wild type by both methods. While neither of the assay methods is highly sensitive, and small differences in freezing tolerance can not be eliminated, we can certainly conclude that the individual contributions (to freezing tolerance) of the *LTI30* and *COR47* genes are less than those of the *SFR2* and *SFR6* genes. It is possible that *LTI30* and *COR47* fall into the category of genes that contribute collectively to freezing tolerance, but whose individual functions are redundant.

7. DISCUSSION

The *sfr* mutations have identified freezing-tolerance genes not previously identified by the cold-inducibility approach (Thorlby et al., 1999). Independently of our arguments about the scope of the alternative approaches, this justifies the classical genetic approach to freezing-tolerance. It is also true that the classical genetic approach immediately demonstrates the involvement of a gene in freezing tolerance, whereas with the cold-inducibility approach this remains to be proved. However, with the advent of functional genomics, this advantage grows smaller as, for example, pre-made knockouts in every gene become available.

Mutations that are deleterious to cold stress tolerance bring the unusual problem that the authentic test for them will, by definition, jeopardize the survival of the mutant individual. This leads us to rely on pleiotropic effects, where feasible, and the reliability of such tests has an influence on the efficiency of fine mapping in practise.

We have described how two genes, *CLS8* and *SFR3*, have been mapped at or near the resolution required for the classical second stage of positional cloning: the detection of gene function in cloned DNA by its stable transformation into the mutant host. In the case of a third gene, *SFR6*, the properties of the defining mutation have inhibited the progress of fine mapping. With genes (such as *SFR6*) that are relatively refractory to genetic fine mapping, it becomes desirable to bring forward the second stage of positional cloning. Alternatives to stable transformation are becoming available for the second stage, and technical advances in these are likely to facilitate the task of positional cloning for many or most genes defined by mutation.

REFERENCES

Colbert, T., Till, B. J., Tompa, R., Reynolds, S., Steine, M. N., Yeung, A. T., McCallum, C. M., Comai, L., and Henikoff, S., 2001, High-throughput screening for induced point mutations, *Plant Physiol.* **126:**480.

Gilmour, S. J., Artus, N. N., and Thomashow, M. F., 1992, cDNA sequence analysis and expression of two cold-regulated genes of *Arabidopsis thaliana*, *Plant Mol. Biol.* **18:**13.

Gilmour, S. J., Sebolt, A. M., Salazar, M. P., Everard, J. D., and Thomashow, M. F., 2000, Overexpression of the *Arabidopsis CBF3* transcriptional activator mimics multiple biochemical changes associated with cold acclimation, *Plant Physiol.* **124:**1854.

Ishitani, M., Xiong, L., Stevenson, B., and Zhu, J. K., 1997, Genetic analysis of osmotic and cold stress signal transduction in *Arabidopsis*: interactions and convergence of abscisic acid-dependent and abscisic acid-independent pathways, *Plant Cell* **9:**1935.

Jaglo-Ottosen, K. R., Gilmour, S. J., Zarka, D. G., Schabenberger, O., and Thomashow, M. F., 1998, *Arabidopsis CBF1* overexpression induces *COR* genes and enhances freezing tolerance, *Science* **280:**104.

Jones, J. D. G., 1999, John Innes Centre, Norwich (November 15, 1999); http://www.jic.bbsrc.ac.uk/ sainsbury-lab/jonathan-jones/SINS-database/sins.htm.

Kasuga, M., Liu, Q., Miura, S., Yamaguchi, S. K., and Shinozaki, K., 1999, Improving plant drought, salt, and freezing tolerance by gene transfer of a single stress-inducible transcription factor, *Nature Biotechnol.* **17:**287.

Knight, H., Veale, E. L., Warren, G. J., and Knight, M. R., 1999, The *sfr6* mutation in *Arabidopsis* suppresses low-temperature induction of genes dependent on the CRT/DRE sequence motif, *Plant Cell* **11:**875.

McKown, R., Kuroki, G., and Warren, G., 1996, Cold responses of *Arabidopsis* mutants impaired in freezing tolerance, *J. Exp. Bot.* **47:**1919.

Mizoguchi, T., Irie, K., Hirayama, T., Hayashida, N., Yamaguchi-Shinozaki, K., Matsumoto, K., and Shinozaki, K., 1996, A gene encoding a mitogen-activated protein kinase kinase kinase is induced simultaneously with genes for a mitogen-activated protein kinase and an S6 ribosomal protein kinase by touch, cold, and water stress in *Arabidopsis thaliana*., *Proc. Natl. Acad. Sci. USA* **93:**765.

Nottingham Arabidopsis Stock Centre, 2001, Nottingham (December 20, 2001); http://nasc.nott.ac.uk.

Palta, J. P., Whitaker, B. D., and Weiss, L. S., 1993, Plasma membrane lipids associated with genetic variability in freezing tolerance and cold acclimation of Solanum species, *Plant Physiol.* **103:**793.

Rouse, D. T., Marotta, R., and Parish, R. W., 1996, Promoter and expression studies on an *Arabidopsis thaliana* dehydrin gene, *FEBS Lett.* **381:**252.

Rozen, S., and Skaletsky, H. J., 1997, Massachusetts Institute of Technology, Boston (December 20, 2001); http://www-genome.wi.mit.edu/genome_software/other/primer3.html.

Shinozaki, K., and Yamaguchi-Shinozaki, K., 2000, Molecular responses to dehydration and low temperature: differences and cross-talk between two stress signaling pathways, *Curr. Opin. Plant Biol.* **3:**217.

Steponkus, P. L., Uemura, M., Balsamo, R. A., Arvinte, T., and Lynch, D. V., 1988, Transformation of the cryobehavior of rye protoplasts by modification of the plasma membrane lipid composition, *Proc. Natl. Acad. Sci. USA* **85:**9026.

Stockinger, E. J., Gilmour, S. J., and Thomashow, M. F., 1997, *Arabidopsis thaliana CBF1* encodes an AP2 domain-containing transcriptional activator that binds to the C-repeat/DRE, a cis-acting DNA regulatory element that stimulates transcription in response to low temperature and water deficit, *Proc. Natl. Acad. Sci. USA* **94:**1035.

Stockinger, E. J., Mao, Y. P., Regier, M. K., Triezenberg, S. J., and Thomashow, M. F., 2001, Transcriptional adaptor and histone acetyltransferase proteins in *Arabidopsis* and their interactions with CBF1, a transcriptional activator involved in cold-regulated gene expression, *Nucleic Acids Research* **29:**1524.

Takahashi, S., Katagiri, T., Hirayama, T., Yamaguchi-Shinozaki, K., and Shinozaki, K., 2001, Hyperosmotic stress induces a rapid and transient increase in inositol 1,4,5-trisphosphate independent of abscisic acid in *Arabidopsis* cell culture, *Plant and Cell Physiology* **42:**214.

The Arabidopsis Genome Initiative, 2000, Analysis of the genome sequence of the flowering plant *Arabidopsis thaliana*, *Nature* **408:**796.

The Arabidopsis Information Resource, 2000, Carnegie Institute, Stanford (December 20, 2001); http://www.arabidopsis.org/cereon/.

Thomashow, M., 1999, Plant cold acclimation: Freezing tolerance genes and regulatory mechanisms, *Ann. Rev. Plant Physiol. Plant Mol. Biol.* **50:**571.

Thomashow, M. F., 2001, So what's new in the field of plant cold acclimation? Lots! *Plant Physiol.* **125:**89.

Thorlby, G., Veale, E., Butcher, K., and Warren, G., 1999, Map positions of *SFR* genes in relation to other freezing-related genes of *Arabidopsis thaliana*, *Plant J.* **17:**445.

Warren, G., McKown, R., Marin, A., and Teutonico, R., 1996, Isolation of mutations affecting the development of freezing tolerance in *Arabidopsis thaliana* (L.) Heynh, *Plant Physiol.* **111:**1011.

Warren, G. J., 1998, Cold stress: Manipulating freezing tolerance in plants, *Current Biol.* **8:**R514.

Welin, B. V., Olson, A., Nylander, M., and Palva, E. T., 1994, Characterization and differential expression of *dhn/lea/rab*-like genes during cold acclimation and drought stress in *Arabidopsis thaliana*, *Plant Mol. Biol.* **26:**131.

Xin, Z., and Browse, J., 1998, *eskimo1* mutants of *Arabidopsis* are constitutively freezing-tolerant, *Proc. Natl. Acad. Sci. USA* **95:**7799.

Xiong, L., Ishitani, M., and Zhu, J., 1999, Interaction of osmotic stress, temperature, and abscisic acid in the regulation of gene expression in *Arabidopsis*, *Plant Physiol.* **119:**205.

Zhu, J. K., 2001, Cell signaling under salt, water and cold stresses, *Curr. Opin. Plant Biol.* **4:**401.

3

MOLECULAR CLONING OF *ESKIMO1* GENE OF ARABIDOPSIS REVEALS NOVEL MECHANISM OF FREEZING TOLERANCE

Zhanguo Xin*

1. INTRODUCTION

Many temperate plant species, including *Arabidopsis*, acquire a greater ability to withstand freezing in response to a period of low nonfreezing temperatures. This process is known as cold acclimation. Cold acclimation is very complex and involves numerous physiological and biochemical changes (Xin and Browse, 2000). The most notable changes include reduction or cessation of growth, reduction of tissue water content, a transient increase in abscisic acid (ABA) (Chen et al., 1983; Lang et al., 1994), changes in membrane lipid composition (Webb et al., 1994; Uemura et al., 1995), and accumulation of compatible osmolytes such as proline, betaine, and soluble sugars (Koster and Lynch, 1992; Ristic and Ashworth, 1993). In short, almost every cellular process is altered during cold acclimation. However, it has been a great challenge to separate the processes that are critical to enhanced freezing tolerance from those merely responsive to low temperatures.

Over the last decade, it has been well established that cold acclimation brings about complex changes in gene expression (Thomashow, 1999; Seki et al., 2001). Many cold inducible genes have been cloned. Some genes, such as *COR15*, have been demonstrated to confer freezing tolerance (Artus et al., 1996). The mature polypeptide of COR15am is localized in chloroplast stroma and is thought to enhance freezing tolerance by deferring the formation of hexagonal II phase to a lower temperature (Steponkus et al., 1998). The functions of most other cold inducible genes in freezing tolerance are unknown. On the other hand, some cold inducible genes have been shown not essential for freezing tolerance but are merely responsive to low nonfreezing temperatures (Jarillo et al., 1993; Leyva et al., 1995). Transcripts for alcohol dehydrogenase, phenylalanine ammonia-lyase

*Zhanguo Xin, Plant Stress and Water Conservation Unit, Cropping Systems Research Laboratory, USDA, Agricultural Research Service, 3810 4th Street, Lubbock, TX 79415.

Plant Cold Hardiness, edited by Li and Palva
Kluwer Academic/Plenum Publishers, 2002

and chalcone synthase are all strongly induced by low temperature in *Arabidopsis*, but mutants deficient in these enzyme activities are able to cold acclimate as fully as wild-type plants (Jarillo et al., 1993; Leyva et al., 1995). Knowledge of the relative importance of cold inducible genes to freezing tolerance is essential to understanding of cold acclimation and to rational design of transgenic approaches for improving freezing tolerance in crops.

Furthermore, some genes that are important to freezing tolerance may be constitutively expressed. Developing freezing tolerance during cold acclimation may also involve repression of certain genes. Such components are difficult to uncover with molecular strategies based on differential screening. To circumvent these limitations and to identify genes that are important to freezing tolerance, we have taken a mutational genetic approach and isolated a series of *Arabidopsis* mutants that are constitutively freezing tolerant in the absence of any low-temperature treatment (Xin and Browse, 1997; 1998). Characterization and molecular cloning of one such mutant, *eskimo1*(*esk1*), revealed novel mechanisms of freezing tolerance.

2. A RECESSIVE MUTATION AT *ESK1* LOCUS CONFERRED A 5°C IMPROVEMENT IN FREEZING TOLERANCE

The freezing survival of *esk1* mutant plants was compared with wild-type plants in the same Petri dish side-by-side (Fig. 1A). Ten days after germination, the Petri dish with plants was frozen to -1°C. After inoculation with ice chips to initiate freezing, the temperature was decreased to -2°C and remained at -2°C overnight. When both the medium and plants reached equilibrium freezing at -2°C, the temperature was decreased to -8°C at 1°C per hour. The Petri dishes remained at -8°C for additional 3 hours. After freezing treatment, plants were recovered at 4°C in the dark overnight and at 22°C for two days under continuous illumination of 150μmol quanta $m^{-2}\cdot s^{-1}$. Survival of plants was scored by visual observation. In the absence of cold acclimation, none of the wild-type plants survived such freezing treatment while all of the *esk1* plants did (Fig. 1A). Mutant plants were capable of continued growth after freezing and completed their life cycle. In the same experiment, two days of cold acclimation at 4°C allowed 100% survival of both wild-type and *esk1* plants after freezing to -8°C (not shown). To determine the extent of freezing tolerance that is constitutively activated in *esk1*, samples of plants grown on agar were frozen to temperatures ranging from -4°C to -16°C (Fig. 1B) following the freezing protocol described above. These experiments indicated that nonacclimated wild-type plants showed 50% survival at -5.5°C. Fully acclimated wild-type plants increased survival to -12.6°C while nonacclimated *esk1* plants showed 50% survival at -10.6°C. These results indicate that *esk1* mutation has instituted approximately 70% of the freezing tolerance generated by full acclimation of wild-type plants. Interestingly, cold acclimation of *esk1* plants increased their tolerance beyond that of acclimated wild-type to -14.8°C, indicating that other signaling pathways, in addition to the one constitutively activated in *esk1* plants are also likely to be involved in cold acclimation.

To determine the genetic basis of the *esk1* mutation, an *esk1-1* mutant was crossed to a wild-type *Arabidopsis* (ecotype Columbia) reciprocally. F1 plants derived from either cross showed a freezing sensitivity similar to wild-type plants (Table 1). Segregation of

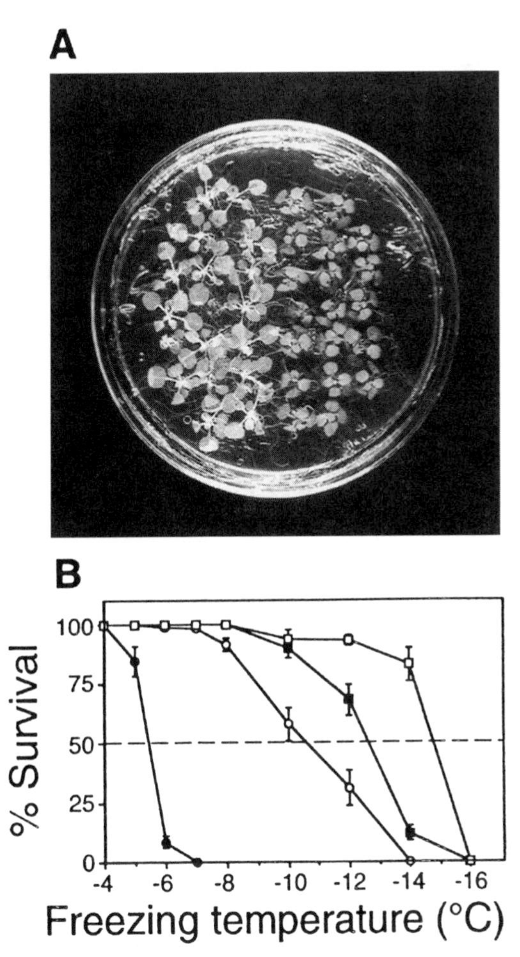

Figure 1. Freezing survival of wild-type and *esk1* plants. (*A*) Nonacclimated wild-type (*Left*) and *esk1* plants were frozen at −8°C. (*B*) Percent survival of nonacclimated and acclimated plants after freezing to different temperatures. Nonacclimated wild-type (●) and *esk1* (○) were frozen in a temperature-controlled chamber. At the temperatures as shown on the X-axis, samples of plants were removed from the chamber, allowed to recover at 4°C over night in the dark and then at 22°C under 150μmol quanta $m^{-2} \cdot s^{-1}$. Percent survival was scored visually. Alternatively, wild-type (■) and *esk1* (□) plants were cold acclimated at 4°C for 2days before being subjected to the same freezing test. The data are means ± SE from three separate experiments (from Xin and Browse, 1998, with permission).

Table 1. The *esk1-1* is a recessive mutation. The *esk1-1* was reciprocally crossed to wild-type (ecotype Columbia) *Arabidopsis*. Freezing tolerance was determined by Petri dish freezing assay as described in the text.

Genotype	Freezing sensitive	Freezing tolerant
WT-Col	32	0
esk1-1	0	32
WT**esk1-1*	43	0
*esk1-1**WT	27	0

Table 2. High proline content phenotype cosegregated with freezing tolerance. Free proline content in leaf sap was measured by amino acid analyzer from the third true leaf of individual F2 plants. Freezing tolerance was determined with F3 progenies derived from individual F2 plants that were used for proline analysis.

Genotype	Number of plants	Proline content µmole/ml	Freezing Tolerant
ESK1/ESK1	29	0.13±0.04	None
ESK1/esk1	52	0.13±0.03	Segregating
esk1/esk1	25	20.52±3.26	All

freezing tolerance was determined with F3 progenies derived from a cross between wild-type and *esk1-1* mutant. Among 106 F3 progenies tested, 29 lines were wild-type, 52 lines were heterozygous, and 25 lines were homozygous for *esk1* (Table 2). This ratio is consistent with a recessive mutation on a single nuclear gene.

3. FREEZING TOLERANCE IN *ESK1* WAS NOT DEPENDENT ON EXPRESSION OF FOUR *COR* GENES

Many genes induced during cold acclimation have been cloned. The most strongly induced of these cold regulated genes in *Arabidopsis* include *COR6.6*, *COR15a*, *RAB18*, *COR47* and *COR78* (Thomashow, 1994). The precise roles of these genes in cold acclimation remain unknown, and several of them are induced by both drought and abscisic acid, as well as by low temperatures (Gilmour and Thomashow, 1991). This suggests that they may help to protect cells during dehydration, which is a major component of freezing stress. Constitutive over expression of the *COR15a* gene alone in transgenic *Arabidopsis* has been shown to provide a moderate increase in freezing tolerance to chloroplasts and protoplasts derived from nonacclimated plants (Artus et al., 1996). A much greater protection has been achieved by overexpressing the transcription factors, *CBF1/DREB1B* or *CBF3/DREB1A*, which activate expression of *COR6.6*, *COR15a*, *RAB18*, *COR47,* and *COR78* genes, as well as a battery of other changes (Jaglo-Ottosen et al., 1998; Kasuga et al., 1999; Gilmour et al., 2000). It appears that these cold inducible changes may act in concert to bring about a greater protection against freezing injury.

To determine whether expression of these cold regulated genes contributes to the constitutive freezing tolerance of *esk1* plants, mRNA levels corresponding to five cold inducible genes were assayed by Northern blot analysis. Only one of the genes, *RAB18*, showed significant constitutive expression in *esk1* plants (Fig. 2). The remaining genes showed very strong induction (ranging from 25- to 100-fold) following cold acclimation of either wild-type or *esk1* plants but transcript levels in nonacclimated *esk1*

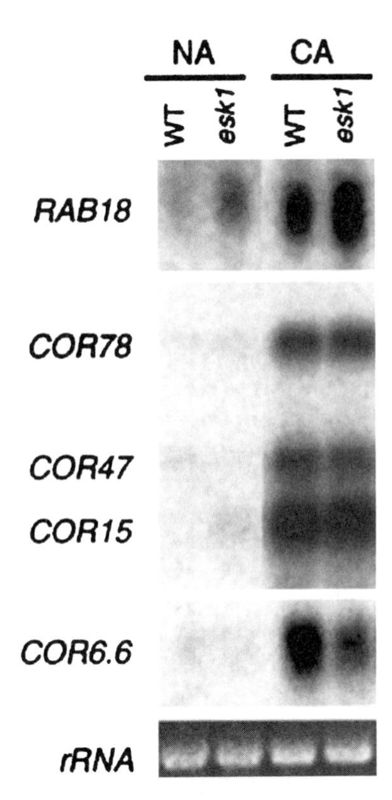

Figure 2. Northern-blot analysis of transcripts of five cold-regulated genes in wild-type (WT) and *esk1* *Arabidopsis*. Plants were grown at 22°C under 150µmol quanta $m^{-2} \cdot s^{-1}$. Total RNA (10µg) isolated from leaf tissue of nonacclimated plants (NA) or from plants cold acclimated at 4°C for 2 days (CA) was separated on a 1.2% agarose-formaldehyde gel and probed successively with cDNAs corresponding to the cold-regulated genes indicated. The ribosomal 25S rRNA was visualized on the nylon membrane by using ethidium bromide to demonstrate equal loading. Quantitative comparisons reported in the text are based on PhosphorImager analyses of the blots. The experiment was repeated four times with similar results (from Xin and Browse, 1998, with permission).

plants were essentially the same as in wild-type controls. Thus, the extensive freezing tolerance achieved in the *esk1* mutants is not dependent on high levels of *COR* gene expression.

4. THE *ESK1* MUTANTS ACCUMULATE FREE PROLINE AND SOLUBLE SUGAR

Proline is one of several compounds that act as compatible osmolytes to ameliorate the effect of dehydration which occurs during freezing and drought stress (Delauney and Verma, 1993). Increases in proline content occur in many plant species during cold acclimation (Koster and Lynch, 1992). In our experiments, cold acclimation of wild-type plants resulted in a ten-fold increase in free proline from 4.3 to 47.0 μmol g^{-1}dry wt. Such data are comparable with those obtained in other studies of cold acclimation (Koster and Lynch, 1992) and drought stress (Peng et al., 1996; Verbruggen et al., 1996). Analyses showed that levels of other amino acids in the mutant remained close to those in wild-type plants, indicating that constitutive accumulation of proline as a compatible solute may indeed be a component of freezing tolerance in the mutant. We also determined the levels of soluble sugars, another class of compatible osmolytes, in wild-type and *esk1* plants. The level found constitutively in leaf tissue of *esk1* was 5.9 mg g^{-1} fresh wt. compared with only 2.3 mg g^{-1} fresh wt. for wild-type controls. To determine if changes in free proline accumulation was indeed caused by *esk1* mutation instead of by closely linked mutations, cosegregation of freezing tolerance and proline accumulation were analyzed in 106 F3 families derived from a cross between *esk1* and wild-type plants (Table 2). Three weeks after planting, the third true leaf was harvested from each individual F2 plant for proline analysis. Free proline content in leaf sap was determined with an amino acid analyzer. After sampling for proline analysis, the plants were grown to maturity to set seeds for determining the genotype at *esk1* locus of each F2 plant through F3 progeny freezing test. As shown in Table 2, accumulation of free proline segregated perfectly with freezing tolerance. Furthermore, proline levels in heterozygous plants were largely the same as those found in wild-type leaves, suggesting that *esk1* mutation is completely recessive. Later in our screening, two additional alleles of *esk1* were identified. These new alleles of *esk1* were named *esk1-2* and *esk1-3* respectively. Both *esk1-2* and *esk1-3* alleles accumulated high levels of free proline and soluble sugars. Thus, accumulation of free proline and soluble sugars must be one of the physiological processes regulated by the *ESK1* gene.

5. *ESK1* GENE ENCODES A NOVEL PROTEIN

The recessive nature of *esk1* mutation and the pleiotropic effects on freezing tolerance and solute accumulation suggest that *ESK1* may serve as a negative regulator of cold acclimation. To understand how an *ESK1* gene product represses cold acclimation, we took a positional cloning strategy to identify the gene since *esk1* mutation was induced chemically by ethyl methane sulfonate. The *esk1-1* mutant (Columbia ecotype) was outcrossed to a wild-type plant of Landsberg ecotype to produce the mapping population. The mutation was fine-mapped to within 20 kb on chromosome III. This 20 kb genomic DNA was amplified and sequenced from both wild-type Columbia and *esk1-*

1 mutant. A novel protein within this region was found to contain a mis-sense mutation that caused a change of amino acid (data not shown). To confirm this novel protein is indeed ESK1, the putative gene plus 500 bp of flanking sequence at each end was sequenced from both *esk1-2* and *esk1-3* mutants. In each mutant line, a distinctive mutation was found in the *ESK1* gene. Thus, the gene we have identified is indeed *ESK1*. Sequence analysis and a database search revealed that *ESK1* encodes a novel protein in a moderate gene family. Homologs of *ESK1* were identified in both dicot and monocot crop plants. *ESK1* may act as a conserved regulator mediating stress adaptation in a wide range of plant species.

6. COLD ACCLIMATION IS REGULATED BY MULTIPLE SIGNALING PATHWAYS

Elucidation of the signaling processes involved in the response of plants to low nonfreezing temperatures is essential to understanding of cold acclimation. Recently, Gilmour *et. al.* showed that transgenic *Arabidopsis* plants overexpressing *CBF3/DREB1A* contain increased proline and soluble sugars as well as exhibiting constitutive expression of the *COR* genes (Gilmour et al., 2000). When the freezing tolerance was examined by an electrolyte leakage assay, the transgenic plants displayed a better constitutive freezing tolerance in the absence of cold acclimation than that found in wild-type plants acclimated for 7 days at 5°C. Based on these results, *CBF3/DREB1A* is proposed to act as a key regulator that integrates the activation of multiple components of cold acclimation responses (Gilmour et al., 2000). It is unclear whether *CBF3/DREB1A* is the sole signaling pathway that controls cold acclimation and freezing tolerance or if there are additional signaling pathways.

Many lines of evidence indicate that parallel or branched signaling pathways activate distinct suites of cold acclimation responses (Browse and Xin, 2001). For example, studies of *COR* gene expression indicate that both ABA-dependent and ABA-independent pathways are involved (Gilmour and Thomashow, 1991; Nordin et al., 1993; Ishitani et al., 1997). Although overexpression of *CBF3/DREB1A* results in a considerable increase in freezing tolerance *in vivo* as well as in ion-leakage assays performed on excised leaves, the transgenic plants can develop much higher levels of freezing tolerance following cold acclimation at 5°C for 7 days (Kasuga et al., 1999; Gilmour et al., 2000). A straightforward explanation for these observations is that *CBF3/DREB1A* activates a subset of the total cold acclimation responses and that a high degree of freezing tolerance is achieved because the *CBF3/DREB1A*-responsive components are hyperactivated in the transgenic plants. Seki *et al.* produced a microarray of 1,300 *Arabidopsis* full length cDNAs and used it to investigate changes in transcript levels in response to dehydration or cold (4°C) treatments (Seki et al., 2001). In addition, a comparison between wild-type *Arabidopsis* and plants overexpressing *DREB1A* (*35S:DREB1A*) was used to identify *CBF3/DREB1A* target genes. Besides the 10 known cold inducible *COR/DRE* genes which are included as internal controls, nine new cold inducible genes were identified. Of these nine new genes, only 2 are identified as *CBF3/DREB1A* target genes. Based on these small numbers, we would infer that a significant number of cold-inducible genes are regulated through signaling pathways other than *CBF3/DREB1A*. Furthermore, the *esk1* mutant of *Arabidopsis* revealed that considerable freezing tolerance could be achieved in the absence of *COR* gene expression

(Xin and Browse, 1998). Support for a model with multiple pathways also comes from the analysis of *sfr* mutants (sensitive to freezing) that are not able to fully acclimate (McKown et al., 1996; Knight et al., 1999). Most of the *sfr* mutants retain over 50% of the wild-type capacity to cold acclimate. The simplest explanation is that each *sfr* mutation blocks one signaling pathway. Therefore, each mutant is still able to partially cold acclimate through signaling pathways that are not disrupted in the mutant plant. Molecular characterization of the *ESK1* gene and cloning of *sfr* loci and genes corresponding to other constitutively freezing tolerant mutants (Xin and Browse, 1998) will undoubtedly contribute to the understanding of the complexities of cold acclimation signaling.

7. ACKNOWLEDGEMENTS

This project was initiated at Dr. John Browse's laboratory at Washington State University. The cloning of the *ESK1* gene and molecular analysis of the mutant alleles was performed while the author was at Cereon Genomics, LLC. I thank Dr Robert Last for his support in cloning *ESK1* gene and Zhaohui Xiong for her technical assistance. I thank Dr. John Burke and Brian Sanderson for helpful discussion and suggestion of the manuscript.

8. DISCLAIMER

Mention of trade names or commercial products in this article is solely for the purpose of providing specific information and does not imply recommendation or endorsement by the U.S. Department of Agriculture.

REFERENCES

Artus, N. N., Uemura, M., Steponkus, P. L., Gilmour, S. J., Lin, C. and Thomashow, M. F., 1996, Constitutive expression of the cold-regulated *Arabidopsis thaliana COR15a* gene affects both chloroplast and protoplast freezing tolerance, *Proc. Natl. Acad. Sci. USA* **93:** 13404-13409.

Browse, J. and Xin, Z., 2001, Temperature sensing and cold acclimation, Curr. Opin. Plant Biol. **4:** 241-246.

Chen, H. -H., Brenner, M. L. and Li, P. H. 1983, Involvement of abscisic acid in potato cold acclimation, *Plant Physiol.* **71:** 362-365.

Delauney, A. J. and Verma, D. P. S., 1993, Proline biosynthesis and osmoregulation in plants, *Plant J.* **4:** 215-223.

Gilmour, S. J., Sebolt, A. M., Salazar, M. P., Everard, J. D. and Thomashow, M. F., 2000, Overexpression of the *Arabidopsis* CBF3 transcriptional activator mimics multiple biochemical changes associated with cold acclimation, *Plant Physiol.* **124:** 1854-1865.

Gilmour, S. J. and Thomashow, M. F., 1991, Cold acclimation and cold-regulated gene expression in ABA mutants of *Arabidopsis thaliana, Plant Mol. Biol.* **17:** 1233-1240.

Ishitani, M., Xiong, L., Stevenson, B., Zhu, J. K., 1997, Genetic analysis of osmotic and cold stress signal transduction in *Arabidopsis*: interactions and convergence of abscisic acid-dependent and abscisic acid-independent pathways, *Plant Cell* **9:** 1935-1949.

Jaglo-Ottosen, K. R., Gilmour, S. J., Zarka, D. G., Schabenberger, O. and Thomashow, M. F., 1998, *Arabidopsis CBF1* overexpression induces *COR* genes and enhances freezing tolerance, Science **280:** 104-106.

Jarillo, J. A., Leyva, A., Salinas, J., Martinez Zapater, J. M., 1993, Low temperature induces the accumulation of alcohol dehydrogenase mRNA in *Arabidopsis thaliana*, a chilling-tolerant plant, *Plant Physiol.* **101:** 833-837.

Kasuga, M., Liu, Q., Miura, S., Yamaguchi-Shinozaki, K. and Shinozaki, K., 1999, Improving plant drought, salt, and freezing tolerance by gene transfer of a single stress-inducible transcription factor, *Nat Biotechnol.* **17:** 287-291.

Knight, H., Veale, E. L., Warren, G. J. and Knight, M. R., 1999, The *sfr6* mutation in *Arabidopsis* suppresses low-temperature induction of genes dependent on the CRT/DRE sequence motif, *Plant Cell* **11:** 875-886.

Koster, K. L. and Lynch, D. V., 1992, Solute accumulation and compartmentation during the cold acclimation of Puma rye, *Plant Physiol.* **98:** 108-113.

Lang, V., Mantyla, E., Welin, B., Sundberg, B. and Palva, E. T., 1994, Alterations in water status, endogenous abscisic acid content, and expression of rab18 gene during the development of freezing tolerance in *Arabidopsis thaliana. Plant Physiol.* **104:** 1341-1349.

Leyva, A., Jarillo, J. A., Salinas, J. and Martinez Zapater, J. M., 1995, Low temperature induces the accumulation of phenylalanine ammonia-Lyase and chalcone synthase mRNAs of *Arabidopsis thaliana* in a light-dependent Manner, *Plant Physiol.* **108:** 39-46.

McKown, R., Kuroki, G. and Warren, G., 1996, Cold responses of *Arabidopsis* mutants impaired in freezing tolerance, *J. Exp. Bot.* **47:** 1919-1925.

Nordin, K., Vahala, T. and Palva, E. T., 1993, Differential expression of two related, low-temperature- induced genes in *Arabidopsis thaliana* (L.) Heynh, *Plant Mol. Biol.* **21:** 641-653.

Peng, Z., Lu, Q. and Verma, D. P., 1996, Reciprocal regulation of delta 1-pyrroline-5-carboxylate synthetase and proline dehydrogenase genes controls proline levels during and after osmotic stress in plants, *Mol. Gene. Genet.* **253:** 334-341.

Ristic, Z. and Ashworth, E. N., 1993, Changes in leaf ultrastructure and carbohydrates in *Arabidopsis thaliana* L. (Heyn) cv. Columbia during rapid cold acclimation, *Protoplasma* **172:** 111-123.

Seki, M., Narusaka, M., Abe, H., Kasuga, M., Yamaguchi-Shinozaki, K., Carninci, P., Hayashizaki, Y. and Shinozaki, K., 2001, Monitoring the expression pattern of 1300 *Arabidopsis* genes under drought and cold stresses by using a full-length cDNA microarray, *Plant Cell* **13:** 61-72.

Steponkus, P. L., Uemura, M., Joseph, R. A., Gilmour, S. J. and Thomashow, M. F., 1998, Mode of action of the *COR15a* gene on the freezing tolerance of *Arabidopsis thaliana*, *Proc. Natl. Acad. Sci. USA* **95:** 14570-14575.

Thomashow, M. F. 1994, *Arabidopsis thaliana* as a model for studying mechanisms of plant cold tolerance, in: *ARABIDOPSIS*, E. M. Meyerowitz, C. R. Somerville, eds, Cold Spring Harbor Laboratory Press, Cold Spring Harbor, pp 807-834.

Thomashow, M. F., 1999, Plant cold acclimation: Freezing tolerance genes and regulatory mechanisms. Annu. Rev. Plant Physiol. *Plant Mol. Biol.* **50:** 571-599.

Uemura, M., Joseph, R. A. and Steponkus, P. L., 1995, Cold acclimation of *Arabidopsis thaliana*. Effect on plasma membrane lipid composition and freeze-induced lesions, *Plant Physiol.* **109:** 15-30.

Verbruggen, N., Hua, X. J., May, M. and Van Montagu, M., 1996, Environmental and developmental signals modulate proline homeostasis: evidence for a negative transcriptional regulator, *Proc. Natl. Acad. Sci. USA* **93:** 8787-8791.

Webb, M. S., Uemura, M. and Steponkus, P. L., 1994, A comparison of freezing injury in oat and rye: two cereals at the extremes of freezing tolerance, *Plant Physiol.* **104:** 467-478.

Xin, Z. and Browse, J., 1997, Constitutive freezing tolerant mutants in *Arabidopsis*, in: *Plant Cold Hardiness*, P. H. Li, T.H. Chen, *eds*, Plenum Press, New York, pp 35-44.

Xin, Z. and Browse, J., 1998, *Eskimo1* mutants of *Arabidopsis* are constitutively freezing-tolerant, Proc. Natl. Acad. Sci. USA **95:** 7799-7804.

Xin, Z. and Browse, J., 2000, Cold comfort farm: the acclimation of plants to freezing temperatures, *Plant Cell Environ.* **23:** 893-902.

EARLY EVENTS DURING LOW TEMPERATURE SIGNALING

Veena Sangwan, Bjorn L. Örvar and Rajinder S. Dhindsa*

1. INTRODUCTION

Environmental factors, such as temperature and water availability, are important determinants of plant growth, development and geographical distribution. Thus crop production is severely limited by stresses due to freezing temperatures and drought. The vulnerability of plants to environmental stresses is primarily due to their sessile growth habit. However, the same growth habit, by enforcing a selection pressure, has also led to the development of sophisticated mechanisms by which plants constantly monitor their environment and activate mechanisms to tolerate these stresses. Plants incapable of mounting such responses succumb to the stressful environment. Thus, freezing tolerant plants can survive sub-zero temperatures for months. They do so by sensing the non-lethal initial decline in temperature as seen in nature at the onset of winter and launching the processes involved in cold acclimation. Cold acclimation is a complex process comprising perception of non-freezing low temperature, transmission of this perception to the nucleus through a cascade of transduction events, and activation of gene transcription; products resulting from this step then confer freezing tolerance on the plant. Cold acclimation is a time-dependent process, the completion of which may take days or weeks. The state of acclimation temporally coincides with the stress and as the latter is relieved, de-acclimation occurs rapidly. Therefore, in order to improve crop production, and to extend geographical range of crop growth, a clear understanding of the processes involved in cold acclimation is essential.

In this chapter, we review our recent studies which suggest that the events during cold signaling follow the temporal sequence: membrane rigidification, cytoskeleton rearrangements, opening of calcium channels, calcium influx, activation of calcium-dependent protein kinases (CDPKs), activation of a MAP kinase cascade, and expression of cold-induced genes.

---

* Veena Sangwan, Bjorn L. Örvar and Rajinder S. Dhindsa, McGill University, Department of Biology, 1205 Avenue Docteur Penfield, Montreal, Quebec, Canada H3A 1B1

Plant Cold Hardiness, edited by Li and Palva
Kluwer Academic/Plenum Publishers, 2002

2. END-POINT MARKERS FOR LOW TEMPERATURE SIGNALING

In order to study low temperature signal transduction (LTST), it is essential that specific and reliable markers be used. The use of more than one marker to study the progress of signal transduction strengthens the hypotheses derived from the experiments. Also, the use of more than one plant species underscores the importance of an event taking place in the signaling pathway. Therefore, we have used suspension cultures of alfalfa (*Medicago sativa* spp. falcata cv. Anik) and seedlings of *Brassica napus* cv. Westar to study the early events of LTST.

In studies of LTST in *Medicago sativa*, we have used the accumulation of transcripts of the cold acclimation-specific (cas) gene *cas*30 as a marker for early events occurring in this pathway. To study the physiological outcome of the LTST pathway, we have used the ability of these cells to acquire freezing tolerance. In studies of LTST in *Brassica napus*, we used transcriptional activation of the promoter from the cold-inducible *Brassica napus* gene *BN115*. To monitor activation of this promoter, we used a transgenic line of *Brassica napus* possessing both the endogenous cold-inducible *BN115* gene and the coding region of the GUS (β-glucuronidase) reporter gene under the control of the *BN115* promoter (White *et al.*, 1994). Thus we monitored the activity of the *BN115* promoter at the transcriptional level by determining the accumulation of the endogenous *BN115* transcripts; and at the translational level by measuring the enzyme activity of GUS histochemically. As a control for cold-induced GUS activity, we used another transgenic line of *Brassica napus* constitutively expressing GUS activity under the control of the tobacco cryptic constitutive promoter *tCUP* (Foster *et al.*, 1999). Finally, in this study, the development of freezing tolerance was monitored as an end-point marker.

3. ROLE OF MEMBRANE FLUIDITY IN LTST

Temperature is known to rapidly and reversibly affect the viscosity or fluidity of biological membranes. Thus, an increase in temperature renders the membranes more fluid, whereas a decrease rigidifies them. Maintenance of proper membrane fluidity is essential for normal cellular metabolism and function. Organisms regulate the fluidity of their membranes by modulating the degree of desaturation (number of double bonds in fatty acids) of membrane phospholipids. Increased desaturation renders the membranes more fluid. Thus increased membrane fluidity at high temperatures is lowered to normal levels by saturating the phospholipids. On the other hand, decreased membrane fluidity at low temperatures is brought up to normal levels by desaturating the phospholipids. Injuries caused by chilling are thought to result from a decrease in membrane fluidity, resulting in reduction in growth rate, electrolyte leakage and leaf chlorosis (Murata, 1989). Since the plasma membrane separates the cell from its environment, it has been considered the site of perception of temperature change (Wada *et al.*, 1990). Murata *et al.* (1992) induced desaturation in the chloroplast membrane, using glycerol-3-phosphate acyltransferase from *Arabidopsis* and squash, (chilling-tolerant and -sensitive plants, respectively), and transformed these genes into tobacco, which is chilling-sensitive. Plants transformed with the squash gene became more sensitive to chilling, whereas those transformed with the *Arabidopsis* gene became less sensitive. The role of desaturation was further investigated by Gombos *et al.* (1992) in a study where strains of the cyanobacterium *Synechocystis* possessing different levels of fatty acid saturation were compared, and the level of saturation was found to inversely correlate with chilling tolerance. Since the disruption of *desA*, a *Synechocystis* cold-induced gene, leads to the absence of diunsaturated lipids, the authors concluded that desaturation of fatty acids plays a role in protection against chilling damage.

The effects of overall saturation of membrane lipids on the induction of *desA* in *Synechocyctis* have been examined using Pd-catalyzed hydrogenation of membrane lipids (Vigh *et al.*, 1993). Saturation of membrane lipids, expected to rigidify the membranes, induced *desA* expression, leading to the conclusion that membrane rigidification activates the low temperature-signaling pathway. Using electron paramagnetic resonance and fatty acid spin probes to assess membrane fluidity, Alonso *et al.* (1997) demonstrated that chilling leads to an increase in membrane rigidity in roots of coffee seedlings. Thus, it is well established that a primary effect of a decrease in external temperature is an increase in the rigidity of cellular membranes, and that at least in *Synechocystis*, mimicking of this cold-induced rigidification leads to activation of the low temperature signaling pathway. However, it has been unclear whether membrane fluidity was causally involved in the activation of *desA* gene. In alfalfa cells, the application of the well known membrane rigidifier, dimethylsulfoxide (DMSO) at 25°C caused membrane rigidification and the frequently used membrane fluidizer benzyl alcohol (BA) prevented the cold-induced membrane rigidification (Örvar *et al.*, 2000). In both alfalfa cells (Örvar *et al.*, 2000) and *Brassica napus* seedlings (Sangwan *et al.*, 2001), DMSO treatment mimics the effects of cold on calcium influx, gene expression and development of freezing tolerance. Clearly, cold-triggered membrane rigidification is causally involved in cold-induced gene expression.

4. ROLE OF THE CYTOSKELETON IN SIGNALING

If cold-triggered membrane rigidification is causally involved in cold-induced calcium influx, gene expression and development of freezing tolerance, what is the nature of events that occur immediately downstream of the changes in membrane fluidity? There is evidence in the literature that cytoskeletal components such as microfilaments and microtubules are attached to the plasma membrane directly or indirectly, and hold it under tension. Furthermore, microfilaments are also known to be attached to ion channels. Recent studies have shown that dynamic interconversions of F-actin (filamentous actin polymers) and G-actin (globular actin monomers) regulate ion channels in the plasma membrane which control osmoregulation (Schwiebert *et al.*, 1994; Tilly *et al.*, 1996), cell polarity (Drubin & Nelson, 1996), cell growth and proliferation, secretion and cell-wall interactions (Grabski *et al.*, 1998). Stabilization of the radial orientation of microfilaments using phalloidin prevents K^+ influx into guard cells and inhibits light-mediated stomatal opening (Kim *et al.*, 1995; Hwang *et al.*, 1997). Treatment with the microfilament destabilizer cytochalasin D causes influx of K^+ into guard cells and led to partial opening of dark-closed stomata (Kim *et al.*, 1995) or promoted the light-mediated opening of the stomata (Hwang *et al.*, 1997). Thus destabilization of microfilaments is essential for stomatal opening. Another study (Liu & Luan, 1998) has shown that microfilaments may be involved in osmosensing via regulation of inward K^+ channels in guard cells. Some of the components involved in actin signaling in animal cells are thought to be Ca^{2+} (Janmey, 1994) and lipids (Ridley & Hall, 1992; Janmey, 1994). Using cell optical density assays, Grabski *et al.* (1998) have shown that in soybean actin tension is regulated by CDPKs (calcium-dependent protein kinases) and calcineurin-type protein phosphatases. The authors have suggested that myosin and/or actin cross-linking proteins may be the targets of these enzymes.

The cytoskeleton is known to be remodelled in association with cell division, growth and differentiation. Various signaling cascades also involve changes in cytoskeletal organization. It is believed that microtubules transmit signals from the receptors to the nucleus, since they span the distance of the cell from the nucleus to the plasma membrane (Gundersen & Cook, 1999). Since microtubules provide a surface area ten times larger than the nuclear envelope, there is ample space for protein-protein interactions on their surface. Microtubules have been shown to play a role in growth orientation in plants (Williamson, 1991; Joshi, 1998). Mathur and Chua (2000), using transgenic plants expressing a fusion product of green fluorescent protein and microtubule associated protein 4, have shown that microtubule stabilization leads to growth reorientation in *Arabidopsis thaliana* trichomes. The role of microtubules in Ca^{2+} channel opening has been suggested by Thion *et al.* (1996). When carrot protoplasts were treated with the microtubule-disrupting drugs colchicine or oryzalin, a 6- to 10-fold increase in Ca^{2+} channel activities was observed. Also, when cold-shocked *Nicotiana plumbagnifolia* protoplasts were treated with oryzalin and cytochalasin D, destabilizers of microtubules and microfilaments respectively, a synergistic increase of Ca^{2+} influx was observed (Mazars *et al.*, 1997). Thus, both microtubules and microfilaments are involved in cold-induced Ca^{2+} influx. However, the role of cytoskeletal disruption in LTST, and its effect, if any, upon the phenotypic consequence of LTST, i.e. freezing tolerance, has been unclear.

In order to investigate the role played by cytoskeletal rearrangements in the LTST pathway, the effect of the microfilament stabilizer, jasplakinolide, on cold acclimation of alfalfa cells have been examined. Jasplakinolide prevents the cold-induced Ca^{2+} influx, accumulation of *cas30* transcripts and the development of freezing tolerance (Örvar *et al.*, 2000). Conversely, when the cells were treated with cytochalasin D, the actin microfilament destabilizer, *cas*30 transcript accumulation and Ca^{2+} influx occur in the absence of cold but freezing tolerance is unaffected. Similarly in *Brassica napus* leaves (Sangwan *et al.*, 2001), jasplakinolide prevents both the cold-induced gene expression and development of freezing tolerance. However, treatment with the microfilament destabilizer latrunculin B, enables the seedlings to bypass the requirement for the cold signal to activate the *BN115* promoter, resulting in the *BN115*-driven expression of GUS activity as well as the accumulation of the endogenous *BN115* transcripts in the absence of the cold signal (Sangwan *et al.*, 2001). Again, freezing tolerance is unaffected. In order to investigate the hypothesis that both microfilaments and microtubules are involved in LTST, the microtubule stabilizer, taxol, was used to determine if it would inhibit cold signaling by preventing the cold-induced microtubule destabilization. Results show that taxol treatment prevents the cold-induced accumulation of the endogenous *BN115* transcripts as well as the GUS activity. On the other hand, oryzalin or colchicine, known to destabilize microtubule cytoskeleton, induces the LTST pathway independent of the cold signal.

An intriguing observation on the effects of cytoskeleton destabilizing drugs is that although in both alfalfa and *Brassica napus they* mimic the effects of cold in triggering calcium influx and specific gene expression, freezing tolerance is either unaffected or even reduced (Örvar *et al.*, 2000; Sangwan *et al.*, 2001). We propose that low temperature induced rearrangements of the cytoskeleton are required for the development of freezing tolerance. Such rearrangements would consist of depolymerization followed by repolymerization in an altered pattern. We suggest that in the continued presence of depolymerizing/destabilizing drugs, cytoskeleton rearrangement in a new pattern does not take place, thus preventing the acquisition of freezing tolerance. It is noteworthy that only the final stages of cold acclimation, i.e. development of freezing tolerance, are prevented by the presence of cytoskeleton destabilizers. Therefore, reorganization of the cytoskeleton is required for the completion of the cold acclimation process but not for cold signaling and cold-induced gene expression.

5. ROLE OF CA^{2+} IN LTST

Given the importance of Ca^{2+} in signaling pathways, it is not surprising that Ca^{2+} influx plays a key role in cold signaling in plants. This requirement for Ca^{2+} influx is transient, since blocking Ca^{2+} influx after 4 hours of exposure to low temperature does not inhibit LTST (Monroy & Dhindsa, 1995). Several recent studies, using end-point markers such as cold-induced gene expression, have provided us with valuable insights into how Ca^{2+} influx fits into these signal transduction pathways (Monroy *et al.*, 1993b; Monroy & Dhindsa, 1995; Knight *et al.*, 1996; Tähtiharju *et al.*, 1997).

Ca^{2+} influx occurs rapidly upon exposing the organism to lower temperatures, with maximum influx occurring at ca. 4-7°C in most plants. Induction of both *Arabidopsis* and alfalfa cold-regulated genes requires Ca^{2+} influx, since the use of Ca^{2+} chelators and calcium channel blockers prevent their upregulation by cold (Monroy *et al.*, 1993b; Monroy & Dhindsa, 1995; Knight *et al.*, 1996; Tähtiharju *et al.*, 1997). Treatment of alfalfa cells with the Ca^{2+} ionophore A23187 or the Ca^{2+} channel agonist Bay K 8644 leads to the induction of *cas15*, a cold-acclimation-specific gene in alfalfa, at 25°C (Monroy *et al.*, 1993b). Interestingly, chemicals known to block intracellular Ca^{2+} release, such as ruthenium red, have little effect on cold-induction of alfalfa *cas* genes, but partially inhibit the induction of the cold-induced *kin* genes in *Arabidopsis* (Tähtiharju *et al.*, 1997). Ca^{2+} influx has been correlated with both the up-regulation of cold-regulated genes and an increase in freezing tolerance in alfalfa (Monroy & Dhindsa, 1995; Örvar *et al.*, 2000) and *Arabidopsis* (Tähtiharju *et al.*, 1997).

Cold-triggered Ca^{2+} influx is known to occur in several plants including *Arabidopsis* and alfalfa, and our recent studies show that it plays a role in the transduction of the cold signal in *Brassica napus* (Sangwan *et al.*, 2001). Blocking Ca^{2+} influx with the calcium chelators BAPTA or EGTA, or with the calcium channel blockers lanthanum and gadolinium prevents both the cold-induced activation of the endogenous *BN115* gene and *BN115*-promoter driven GUS activity. However, chemical induction of calcium influx using the calcium ionophore, A23187 or cADPR (known to cause release of Ca^{2+} from intracellular stores (Allen *et al.*, 1995)) is able to activate *BN115* promoter in the absence of cold. Solute leakage studies indicate that Ca^{2+} influx is essential for the development of freezing tolerance, since prevention of Ca^{2+} influx prevents cold-induced development of freezing tolerance, whereas chemically-induced Ca^{2+} influx enhances it. Interestingly, Gd^{3+}, a mechanosensitive channel blocker, causes complete inhibition of cold-induced GUS activity as well as *BN115* transcript accumulation, indicating that intracellular Ca^{2+} is not required for LTST in *Brassica napus*. However, ruthenium red, a chemical which blocks intracellular Ca^{2+} influx, was able to completely prevent GUS protein accumulation, and the development of freezing tolerance, but only partially decrease endogenous *BN115* transcript levels. These data suggest that both intra- and extracellular Ca^{2+} influx is required for the acquisition of freezing tolerance in *Brassica napus*. It is also possible that calcium-induced calcium release is involved.

It has been proposed (Monroy & Dhindsa, 1995) that cold-induced membrane rigidification and Ca^{2+} channel opening are coupled. However, the nature of such coupling mechanisms remains unclear. How do changes in membrane fluidity lead to Ca^{2+} channel opening? There is evidence in the literature that cold-triggered Ca^{2+} channels are mechanosensitive in nature, suggesting that tensile forces may be involved in triggering their activity. Changes in membrane fluidity are expected to alter these tensile forces. As mentioned earlier, both the microtubules and microfilaments are attached to ion channels which hold them, like the rest of the membrane, under tension in the closed state. It is likely, therefore, that changed membrane fluidity would dissipate these forces, causing the channels to open (Örvar *et al.*, 2000). This hypothesis does not rule out the possibility that temperature may have direct effects on the stability of the microfilament and microtubule cytoskeletal networks.

What roles may Ca^{2+} play during cold signaling? It is now well established that Ca^{2+} influx is required for cold acclimation (Monroy *et al.*, 1993). There are numerous reports in the literature that Ca^{2+} signaling is coupled with altered patterns of protein phosphorylation. Thus it has been suggested (Monroy *et al.* (1993b) that Ca^{2+} influx is followed by changes in protein phosphorylation, since treatment of cells with the Ca^{2+} chelator EGTA, the Ca^{2+} channel blocker La^{3+}, or the calcium-dependent protein kinase (CDPK) inhibitor W7, inhibits cold-induced protein phosphorylation, gene expression and induction of freezing tolerance. Thus it is important to identify the events in cold signaling downstream of Ca^{2+} influx that lead to the activation of cold-inducible gene expression.

6. PROTEIN KINASES AND PHOSPHATASES IN LTST

Transduction of the low temperature signals by changes in protein phosphorylation, leading to the expression of cold-induced genes, has been reported in alfalfa (Monroy *et al.*, 1993b, Monroy *et al.*, 1998) and *Arabidopsis* (Tähtiharju *et al.*, 1997). In *Arabidopsis*, use of kinase inhibitors led to a decrease in cold-induction of *kin1* (Tähtiharju *et al.*, 1997). Using two-dimensional gel electrophoresis to compare phosphorylation levels of cold-treated and untreated alfalfa cells, Monroy *et al.* (1993b) observed that some proteins are hyperphosphorylated in cold, whilst others are hypophosphorylated. It has been argued that because temperature is a pervasive thermodynamic factor, it should be sensed independently at many places in the cell, including different organelles (Dhindsa *et al.*, 1998). Therefore the effects of cold on phosphorylation levels of proteins in nuclei isolated from un-acclimated control and cold-acclimated plants of frost-sensitive and frost-tolerant cultivars of alfalfa were compared (Kawczynski & Dhindsa, 1996). Interestingly, nuclei from the freezing-tolerant cultivar constitutively contained larger number of cold-responsive phosphoproteins than the nuclei from freezing-sensitive cultivar. Also, the nuclei cold acclimated plants possess additional phosphoproteins compared with nuclei from plants maintained at 25°C. This suggests that some phosphoproteins are imported into the nucleus during cold acclimation. These phosphoproteins may be involved in cold acclimation as transcription factors activating the expression of cold-inducible genes. The relative roles played by protein kinases and phosphatases in LTST have been explored using such inhibitors as staurosporine, a general protein kinase inhibitor, which decreased the cold-induction of *cas15* in alfalfa, and okadaic acid, a protein phosphatase inhibitor which induces *cas15* expression at 25°C (Monroy *et al.*, 1998).

The importance of the role played by protein phosphorylation is further underscored by the results of our studies on *Brassica napus* plants (Sangwan *et al.*, 2001). Specific inhibitors of several types of protein kinases such as staurosporine and K252a (general kinase inhibitors) and tyrosine kinase inhibitors such as genestein prevent the cold-induced activation of *BN115* promoter and the development of freezing tolerance. Interestingly, the phosphoinositide kinase inhibitor, wortmannin, was also able to prevent LTST. Conversely, the inhibitors of protein phosphatase type 1 and 2A, okadaic acid and calyculin A, mimic the effects of cold and cause the activation of *BN115* promoter as well as the acquisition of freezing tolerance. These results underscore the complexity of LTST in that it involves the activities of several different types of protein kinases and phosphatases.

Mitogen-activated protein kinase (MAPK) cascades are ubiquitously involved in eukaryotic signaling in response to many stimuli (Moriguchi *et al.*, 1996). A MAPK cascade consists of sequential phosphorylation-mediated activation of three protein kinases functionally linked as a cassette. A MAPK is activated through phosphorylation by a MAPK kinase (MAPKK) and the latter is activated through phosphorylation by a MAPKK kinase (MAPKKK). MAPKKK is activated through phosphorylation by another kinase. A cold activated MAPK named SAMK (Stress Activated MAP Kinase) has been identified (Jonak *et al.*, 1996). The results of recent studies in our laboratory show that SAMK and the recently discovered heat shock-activated MAP kinase, HAMK, are activated by opposite changes in membrane fluidity. For example, cold activation of SAMK is mediated by cold-triggered rigidification of the membranes whereas heat shock-triggered activation of HAMK is mediated by membrane fluidization (Sangwan *et al.*, 2002). The activation of both SAMK and HAMK by respective temperature changes requires destabilization of the cytoskeleton, Ca^{2+} influx and the action of CDPKs. It is not known at present whether SAMK and HAMK are causally involved in mediating the cold-specific or heat-specific gene expression. However, because SAMK activation requires the same changes as the cold-induced gene expression and development of freezing tolerance, it is likely that SAMK is involved in cold acclimation.

What is the role of phosphatases in cold signaling? The phosphorylation level of a protein is determined by the relative activities of its kinase and phosphatase. Therefore, an increase in the phosphorylation level is due to a relative increase in kinase activity over phosphatase activity. A decline in temperature inhibits the rate of all biochemical reactions. Thus it is expected that on exposure to cold, activities of both protein kinase and phosphatase would be inhibited. Indeed, Monroy *et al.* (1997) have demonstrated that as alfalfa cells are transferred from 25°C to 4°C, total cellular phosphatase activity is decreased by 95%. However, the proportional decrease in activities of casein and histone kinases is much less. Such differential inhibition of a protein kinase and the related protein phosphatase would change the equilibrium between the phosphorylation/ dephosphorylation reactions. For example, the phosphorylation level of one of the proteins examined is doubled with every 4°C decrease in temperature below 12°C (Monroy *et al.*, 1997). It has been suggested that for cold acclimation to occur, stimulation of protein kinases and inhibition of protein phosphatases are required (Dhindsa *et al.*, 1998; Monroy *et al.*, 1998). The inhibitors of protein phosphatases type 1 and 2A, okadaic acid and calyculin A activate the expression of cold-inducible genes at 25°C (Monroy *et al.*, 1998). Evidence from mutation genetics has revealed that protein phosphatases negatively regulate the expression of stress-inducible genes and act as attenuators of signaling mechnisms. Duration of MAPK activity plays a significant role in the stimulus-specific response (Marshall, 1995), the activation of MAPK cascades is generally transient. Thus cold activation of SAMK and heat-activation of HAMK are transient declining after about an hour. It is quite likely, therefore, that protein phosphatases act as regulators or attenuators of MAPK activities.

The inhibition of cold activation of the *Brassica napus BN115* promoter by wortmannin, an inhibitor of phosphoinositide kinase, suggests the involvement of phosphoinositide signaling in cold acclimation. It is relevant to mention here that phospholipase C, a membrane-anchored mechanosensitive enzyme involved in phosphoinositide signaling is upregulated by cold (Hirayama *et al.,* 1995). Phospholipase C catalyzes the production of inositol triphosphate (IP_3) and diacylglycerol (DAG) from phosphatidyl-inositol triphosphate. Whereas IP_3 is known to cause release of Ca^{2+} from intracellular stores, DAG is known to activate protein kinase C. Interestingly, cold activation of the *BN115* gene and cold acclimation in *Brassica napus* require Ca^{2+} influx from extracellular as well as from intracellular stores. Furthermore, activation of SAMK by cold and of HAMK by heat in alfalfa cells is inhibited by H7, an inhibitor of protein kinase C. Protein kinase C is not well characterized in plants and its role in cell function is unclear. It has been reported that some CDPKs mimic the activity of protein kinase C (Farmer & Choi, 1999) but whether such CDPKs are also inhibited by H7 is not known.

7. TEMPORAL SEQUENCE OF EVENTS INVOLVED IN LTST

Since Ca^{2+} influx occurs very rapidly in response to cold, it has been suggested that it is a very early step in cold signaling. Changes in membrane fluidity and rearrangements of the cytoskeleton are also known to cause influx of Ca^{2+} into the cytosol. Since all of these events are thought to occur in the early phase of signaling response, we studied the temporal sequence of events taking place upon perception of a cold signal. Using alfalfa suspension cultures treated with DMSO, which had earlier been shown to chemically induce membrane rigidification, we studied the induction of *cas*30. Our results indicate that chemically induced membrane rigidification, which resulted in the *cas*30 transcript accumulation in the absence of the cold signal, could be prevented by pretreatment of the cells with the calcium chelator BAPTA. Also, cytochalasin D-induced cytoskeletal destabilization, which induces *cas*30 transcript accumulation in the absence of low temperature, could be prevented by pretreatment of cells with BAPTA. These data suggest that Ca^{2+} influx occurs after membrane rigidification and cytoskeletal rearrangements have taken place. The observation that BAPTA can prevent Bay K 8644-induced *cas*30 transcript accumulation, but jasplakinolide is unable to do so, provides further support for this temporal sequence of events. Since the cold activation of SAMK is inhibited by Ca^{2+} chelators and channel blockers, as well as by W7, an inhibitor of CDPKs, it appears that any role of MAPK cascades in cold signaling is downstream of the changes in membrane fluidity, cytoskeleton organization, Ca^{2+} influx and the activity of CDPKs. A proposed pathway for low temperature signal transduction based on our studies of cold acclimation in alfalfa and *Brassica napus* is shown in Figure 1. It is necessarily incomplete and does not show cross-talk and overlap with other signaling pathways involved in cold acclimation such as those initiated by the phytohormone abscisic acid or by other stresses.

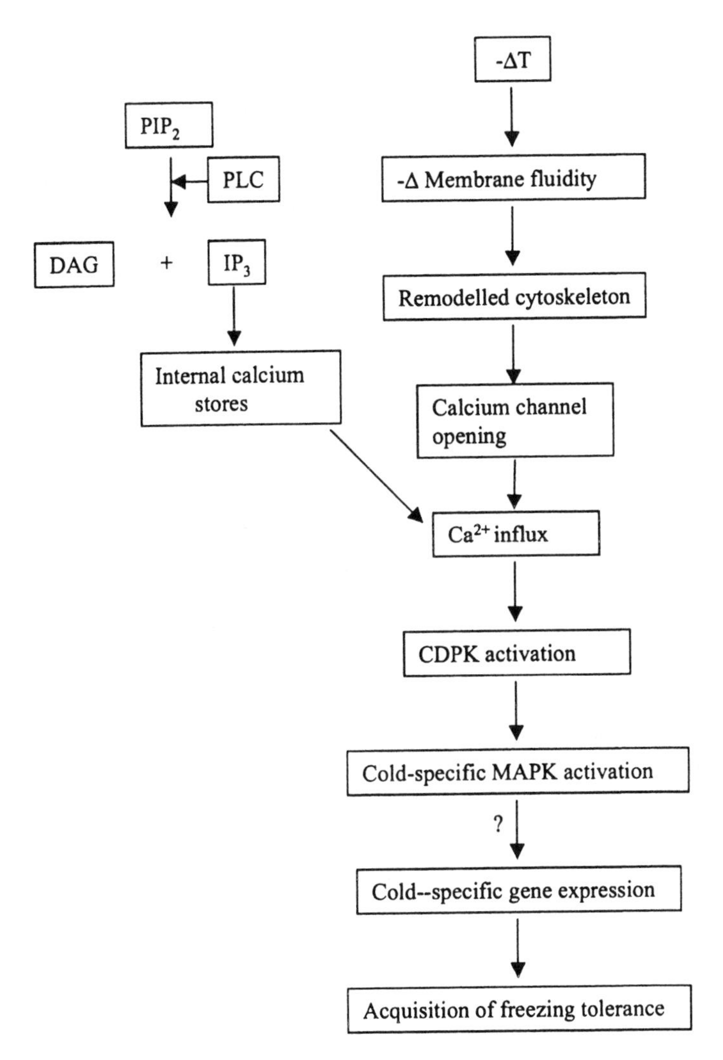

Figure 1. Proposed model for low temperature signaling. PIP_2, phosphatidyl inositol bisphosphate; IP_3, Inositol triphosphate; DAG, Diacylglycerol.

REFERENCES

Allen, G.J., Muir, S.R. and Sanders, D., 1995, Release of Ca^{2+} from individual plant vacuoles by both $InsP_3$ and cyclic ADP-ribose, *Science* **268**: 735-737.

Alonso, A., Queiroz, C.S. and Magalhaes, A.C., 1997, Chilling stress leads to increased membrane rigidity in roots of coffee (*Coffea arabica* L.) seedlings, *Biochim. Biophys. Acta* **1323**: 75-84.

Dhindsa, R.S., Monroy, A.F., Sangwan, V., Kawczynski, W. and Labbe, E., 1998, Low temperature signal transduction during cold acclimation of alfalfa, in: *Plant Cold Hardiness, Molecular Biology, Biochemistry and Physiology,* P.H. Li and T.H.H. Chen, eds., Plenum Press, New York, pp 15-28.

Drubin, D.G. & Nelson, W.J., 1996, Origins of cell polarity, *Cell* **84**: 335-344.

Farmer, P.K. & Choi, J.H., 1999, Calcium and phospholipid activation of a recombinant calcium-dependent protein kinase (DcCPK1) from carrot (*Daucus carota* L.), *Biochim. Biophys. Acta* **1434**: 6-17.

Foster, E., Hattori, J., Labbe, H., Ouellet, T., Fobert, P.R., James, L.E., Iyer, V. and Miki, B.L., 1999, A tobacco cryptic constitutive promoter, tCUP, revealed by T-DNA tagging, *Plant Mol. Biol.* **41**: 45-55.

Gombos, Z., Wada, H. and Murata, N., 1992, Unsaturation of fatty acids in membrane lipids enhances tolerance of the cyanobacterium Synechocystis PCC6803 to low-temperature photoinhibition, *Proc. Natl. Acad. Sci. USA.* **89**: 9959-9963.

Grabski, S., Arnoys, E., Busch, B. and Schindler, M., 1998, Regulation of actin tension in plant cells by kinases and phosphatases, *Plant Physiol.* **116**: 279-290.

Gundersen, G.G. and Cook, T.A., 1999, Microtubules and signal transduction, *Curr.Opin. Cell Biol.* **11**: 81-94

Hirayama, T., Ohto, C., Mizoguchi, T., Shinozaki, K., 1995, A gene encoding a phosphatidylinositol-specific phospholipase C is induced by dehydration and salt stress in *Arabidopsis thaliana, Proc. Natl. Acad. Sci. USA* **92**: 3903-3907.

Horváth, I., Glatz, A., Varvasovszki, V., Torok, Z., Pali, T., Balogh, G., Kovacs, E., Nadasdi, L., Benko, S., Joo, F. and Vigh, L., 1998, Membrane physical state controls the signaling mechanism of the heat shock response in Synechocystis PCC 6803: identification of *hsp17* as a "fluidity gene", *Proc. Natl. Acad. Sci. USA* **95**: 3513-3518

Hwang, J-U, Suh, S., Yi, H., Kim, J. and Lee, Y., 1997, Actin filaments modulate both stomatal opening and inward K^+ -channel activities in guard cells of *Vicia faba* L, *Plant Physiol.* **115**: 335-342.

Janmey, P.A. (1994) Phosphoinositides and calcium as regulators of cellular actin assembly and disassembly. *Ann. Rev. Physiol.* **56**: 169-191.

Joshi, H.C., 1998, Microtubule dynamics in living cells, *Curr. Opin. Cell Biol.* **10**: 35-44.

Jonak, C., Kiegerl, S., Ligterink, W., Barker, P. J., Huskisson, N. S. and Hirt, H., 1996, Stress signaling in plants: a mitogen-activated protein kinase pathway is activated by cold and drought, *Proc. Natl. Acad. Sci. USA* **93**: 11274-11279.

Kawczynski, W. and Dhindsa, R.S., 1996, Alfalfa nuclei contain cold-responsive phophoproteins and accumulate heat-stable proteins during cold treatment of seedlings, *Plant Cell Physiol.* **37**:1204-1210.

Kim, M., Hepler, P.K., Eun, S-O, Ha, K.S. and Lee, Y., 1995, Actin filaments in mature guard cells are radially distributed and involved in stomatal movement, *Plant Physiol.* **109**: 1077-1084.

Knight, H., Trewavas, A.J. and Knight, M.R., 1996, Cold calcium signaling in Arabidopsis involves two cellular pools and a change in calcium signature after acclimation, *Plant Cell* **8**: 489-503.

Liu, K. and Luan, S., 1998, Voltage-dependent K^+ channels as targets of osmosensing in guard cells. *Plant Cell* **10**: 1957-1970.

Mathur, J. and Chua, N.H., 2000, Microtubule stabilization leads to growth reorientation in Arabidopsis trichomes, *Plant Cell* 12: 465-477.

Marshall, C.J., 1995, Specificity of receptor tyrosine kinase signaling: Transient versus sustained extracellular signal-regulated kinase activation, *Cell* **80**: 179-185.

Mazars, C., Thion, L., Thuleau, P., Graziana, A., Knight, M.R., Moreau, M. and Ranjeva, R., 1997, Organization of cytoskeleton controls the changes in cytosolic calcium of cold-shocked *Nicotiana plumbaginifolia* protoplasts, *Cell Calcium* **22**: 413-420.

Monroy, A.F., Sarhan, F. and Dhindsa, R.S., 1993b, Cold-induced changes in freezing tolerance, protein phosphorylation, and gene expression. Evidence for a role of calcium, *Plant Physiol.* **102**: 1227-1235.

Monroy, A.F. and Dhindsa, R.S., 1995, Low-temperature signal transduction: induction of cold acclimation-specific genes of alfalfa by calcium at 25 °C, *Plant Cell* **7**: 321-331.

Monroy, A.F., Labbe, E. and Dhindsa, R.S., 1997, Low temperature perception in plants: effects of cold on protein phosphorylation in cell-free extracts, *FEBS Lett.* **410**: 206-209.

Monroy, A.F., Sangwan, V. and Dhindsa, R.S., 1998, Low temperature signal transduction during cold acclimation: protein phosphatase 2A as an early target for cold-inactivation, *Plant J.* **13**: 653-660.

Moriguchi, T., Gotoh, Y. and Nishida, E., 1996, Roles of the MAP kinase cascade in vertebrates, *Advances in Pharmacology* **36**: 121-137.
Murata, N., 1989, Low-temperature effects on cyanobacterial membranes, *J. Bioenerg. Biomembr.* **21**: 61-75.
Murata, N., Ishizaki-Nishizawa, O., Higashi, S., Hayashi, H., Tasaka, Y. and Nishida, I., 1992, Genetically engineered alteration in the chilling sensitivity of plants, *Nature* **356**: 710-713.
Örvar, B.L., Sangwan, V., Omann, F. and Dhindsa, R.S., 2000, Early steps in cold sensing by plant cells: the role of actin cytoskeleton and membrane fluidity, *Plant J.* **23**: 785-794.
Ridley, A.J. and Hall, A., 1992, The small GTP-binding protein regulates the assembly of focal adhesions and actin stress fibers in response to growth factors, *Cell* **70**: 389-399.
Sangwan, V., Örvar, B.L., Beyerly, J., Hirt, H., and Dhindsa, R.S., 2002, Opposite changes in membrane fluidity mimic cold and heat stress activation of distinct plant MAP kinase pathways, (submitted).
Sangwan, V., Foulds, I., Singh J. and Dhindsa, R.S., 2001, Cold-activation of *Brassica napus* BN115 promoter is mediated by structural changes in the membrane and cytoskeleton, and requires Ca^{2+} influx, *Plant J.* **27**: 1-12.
Schwiebert, E.M., Mills, J.W. and Stanton, B.A., 1994, Actin-based cytoskeleton regulates a chloride channel and cell volume in a renal cortical collecting duct cell line, *J. Biol. Chem.* **269**: 7081-7089.
Tähtiharju, S., Sangwan, V., Monroy, A.F., Dhindsa, R.S. & Borg, M., 1997, The induction of *kin* genes in cold-acclimating *Arabidopsis thaliana*. Evidence of a role for calcium, *Planta* **203**: 442-447.
Thion, L., Mazars, C., Thuleau, P., Graziana, A., Rossignol, M., Moreau, M. and Ranjeva, R., 1996, Activation of plasma membrane voltage-dependent calcium-permeable channels by disruption of microtubules in carrot cells, *FEBS Lett.* **393**: 13-18
Tilly, B.C., Edixhoven, M.J., Tertoolen, L.G., Morii, N., Saitoh, Y., Narumiya, S. and de Jonge, H.R., 1996, Activation of the osmo-sensitive chloride conductance involves P21ρ and is accompanied by a transient reorganization of the F-actin cytoskeleton, *Mol. Biol. Cell* **7**: 1419-1427.
Vigh, L., Los, D.A., Horváth, I. and Murata, N., 1993, The primary signal in the biological perception of temperature: Pd-catalyzed hydrogenation of membrane lipids stimulated the expression of the *desA* gene in *Synechocystis* PCC6803, *Proc. Natl. Acad. Sci. USA* **90**: 9090-9094.
Wada, H., Gombos, Z. & Murata, N., 1990, Enhancement of chilling tolerance of a cyanobacterium by genetic manipulation of fatty acid desaturation, *Nature* **347**: 200-203.
White, T.C., Simmonds, D., Donaldson, P. and Singh, J., 1994, Regulation of BN115, a low-temperature-responsive gene from winter *Brassica napus, Plant Physiol.* **106**: 917-928.
Williamson, R.E., 1991, Orientation of cortical microtubules in interphase plant cells, *Int. Rev. Cytol.* **129**: 135-208.

ATPP2CA NEGATIVELY REGULATES ABA RESPONSES DURING COLD ACCLIMATION AND INTERACTS WITH THE POTASSIUM CHANNEL AKT3

Sari Tähtiharju[1], Pekka Heino[2], and E. Tapio Palva[1,2]

1. INTRODUCTION

The phytohormone abscisic acid (ABA) regulates diverse developmental and physiological responses, including seed maturation, dormancy and germination as well as guard cell closure. ABA also mediates adaptive responses to abiotic environmental stresses such as drought (Leung and Giraudat, 1998). The role of ABA in cold acclimation has been the center of much debate. However, several lines of evidence suggest that the ABA may have an important role in the cold acclimation process. First, application of ABA at normal growth temperatures can induce an increase in freezing tolerance in a wide range of plants, including *Arabidopsis* (Guy, 1990; Lång et al., 1989). Furthermore, endogenous ABA levels increase transiently in response to low temperature (Lång et al., 1994). In addition, the ABA-deficient mutants of *Arabidopsis*, *aba1* and *aba4*, are severely impaired in their ability to cold-acclimate (Heino et al., 1990; Gilmour and Thomashow, 1991). However, application of ABA could suppress the impaired cold-acclimation phenotype (Heino et al., 1990). In addition to mutants in ABA biosynthesis, also *Arabidopsis* mutants defective in ABA responsiveness appear to affect cold acclimation. The ABA-insensitive mutant *abi1* is impaired in development of freezing tolerance (Mäntylä et al., 1995) as well as in the cold-induced expression of several cold-responsive genes (Lång and Palva, 1992; Nordin et al., 1993). Identification of ABI1 as a protein phosphatase 2C (Leung et al., 1994; Meyer et al., 1994) suggested that protein dephosphorylation might be involved in cold signal transduction.

[1] Institute of Biotechnology, P.O. Box 56, FIN-00014 University of Helsinki, Finland. For correspondence: Sari.Tahtiharju@helsinki.fi. Tel. +358-9-191 58758, fax +358-9-191 59366.
[2] Department of Biosciences, Division of Genetics, P.O. Box 56, FIN-00014 University of Helsinki, Finland.

Plant Cold Hardiness, edited by Li and Palva
Kluwer Academic/Plenum Publishers, 2002

2. PROTEIN PHOSPHATASES DURING COLD ACCLIMATION

A common way cells relay molecular messages is reversible protein phosphorylation and dephosphorylation catalysed by protein kinases and protein phosphatases, respectively. Several genes encoding protein phosphatases have been isolated from a number of plant species (Smith and Walker, 1996) and based on their substrate specificity and structural differences they have been classified into two major groups; serine/threonine and tyrosine phosphatases (Luan, 1998).

2.1. Protein Tyrosine Phosphatases

The protein tyrosine phosphatases (PTPases) are further divided into three subgroups: receptor-like PTPases; intracellular PTPases; and dual-specificity PTPases (Luan, 1998). Recently, the first plant cDNA encoding putative PTPase, AtPTP1, was isolated from *Arabidopsis*. Remarkably, the expression of the *AtPTP1* gene is transiently down regulated by cold temperature. Such down regulation by stress factors has not been reported for PTPases in any other organism before (Xu et al., 1998) implicating AtPTP1 in a unique mechanism for plant response to cold stress. However, the function of AtPTP1 as well as other PTPases in signalling pathways remains to be elucidated in plants.

2.2. Serine/Threonine Protein Phosphatases

The serine/threonine protein phosphatases (PPases), which are categorized into four types; 1, 2A, 2B, or 2C (also known as PP1, PP2A, PP2B and PP2C) (Luan, 1998). These types represent more than 80% of protein phosphatase activity in plant cells (Smith and Walker, 1996). Numerous genes encoding the various PPases have been isolated from different plant species, including *Arabidopsis* (Kudla et al., 1999; Smith and Walker, 1996). Although little is known about the roles of these PPs in plants, recent studies indicate that these PPases are involved in various signalling cascades including those for ABA, pathogen and stress responses as well as developmental processes (Luan, 1998). However, their role in cold signal transduction has not been characterised in any detail.

PP2B has been shown to act as primary sensors of Ca^{2+} signals (Sanders et al., 1999). The expression of an *Arabidopsis* AtCBL1 gene encoding PP2B (Ca^{2+} binding calcineurin-B-like protein) is highly up-regulated in addition to drought and salt also by cold (Kudla et al., 1999). Calcineurin B-like proteins have been demonstrated to play a role in salt stress signalling (Hasegawa et al., 2000). Whether these Ca^{2+} regulated protein phosphatases play roles also in cold signal transduction remains to be elucidated.

PP2A has been suggested to be an early target for cold-inactivation in low temperature signal transduction, since low temperature caused a rapid and dramatic decrease in PP2A activity (Monroy et al., 1998). Moreover, unspecific inhibition of type 1 and 2A PP activity induced expression of cold-responsive genes (Monroy et al., 1998; Sangwan et al., 2001) as well as caused an increase in freezing tolerance at normal growth temperature (Sangwan et al., 2001) suggesting a role for these type of PPs in cold acclimation.

The *abi1* mutation has been shown to affect both cold-inducible gene expression as well as development of freezing tolerance (Lång and Palva, 1992; Nordin et al., 1991; 1993; Mäntylä et al., 1995). Both *ABI1* and homologous *ABI2* encode PP2Cs, which act as negative regulators of ABA signalling (Gosti et al., 1999; Leung et al., 1994; Leung et al., 1997; Merlot et al., 2001; Meyer et al., 1994; Sheen, 1998). The expression of *ABI1* and *ABI2* is upregulated by ABA and water stress (Leung et al., 1997). Notably, ABI1 acts as a negative regulator of ABA-mediated responses to drought (Gosti et al., 1999). ABI1 transcript levels increase also in response to cold (Tähtiharju and Palva, 2001). Another *Arabidopsis* PP2C gene, *AtPP2CA* (Kumori and Yamamoto, 1994) is also upregulated by these same treatments (Tähtiharju and Palva, 2001). AtPP2CA is also a negative regulator of ABA signalling inhibiting activity of barley *HVA* gene promoter (Sheen, 1996), which is both ABA and cold inducible (Straub et al., 1994). Studies with unspecific inhibitor of PP1 and PP2A also suggest a role for these PPs in regulation of ABA signalling and expression of ABA and cold-responsive genes in *Arabidopsis* (Wu et al., 1997). However, role for these various regulators of ABA signalling during cold acclimation waits to be revealed.

3. ANTISENSE INHIBITION OF AtPP2CA ACCELERATES COLD ACCLIMATION IN *ARABIDOPSIS THALIANA*

Recently, the function of AtPP2CA in cold acclimation and freezing tolerance was characterized by silencing the corresponding gene by antisense inhibition. Exposure to both low temperature and exogenous ABA resulted in clearly accelerated development of freezing tolerance in transgenic *AtPP2CA* antisense plants compared to wild-type plants (Figures 1a and 1b) suggesting a role for AtPP2CA in cold and ABA responses. Enhanced ABA sensitivity of transgenic *AtPP2CA* antisense plants during cold acclimation (Figure 1b) also indicated that AtPP2CA might regulate ABA signalling (Tähtiharju and Palva, 2001).

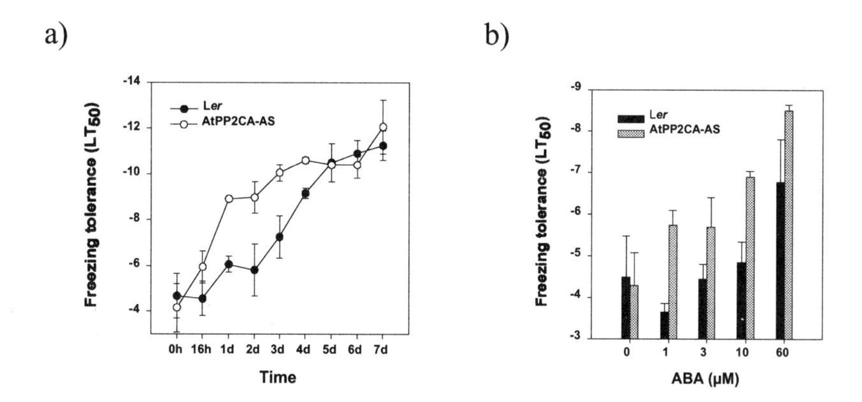

Figure 1. Effect of cold acclimation and exogenous ABA on freezing tolerance of wild-type and transgenic *AtPP2CA* antisense plants. (a) Freezing tolerance of 2-week-old wild type (L*er*) and *AtPP2CA* antisense (*AtPP2CA*-AS) after cold acclimation for times indicated. Control plants were grown at 20°C. (b) Freezing tolerance of 2-week-old wild type (L*er*) and *AtPP2CA* antisense plants (*AtPP2CA*-AS) after exposure to 0, 1, 3, 10 and 60 μM ABA for 24h. Data is a mean ± SD of three to four independent experiments.

4. AtPP2CA NEGATIVELY REGULATES EXPRESSION OF COLD AND ABA-INDUCIBLE GENES

AtPP2CA was also shown to regulate the expression of several cold- and ABA-inducible genes (Tähtiharju and Palva, 2001). Low temperature-induced expression of *LTI78* (Nordin et al., 1991), *RAB18* (Lång and Palva, 1992) and *RCI2A/LTI6* (Capel et al., 1997; Nylander et al., 2001) as well as ABA-induced expression of *RAB18* was enhanced in transgenic *AtPP2CA* antisense plants suggesting that at the onset of cold acclimation AtPP2CA modulates ABA signalling and thus regulates the expression of ABA-responsive genes. However, there seems to be differences in this regulation due to the differences in ABA-dependency (Tähtiharju and Palva, 2001). Although all of the above genes have putative ABA-responsive elements in their promoters and are induced by ABA, some of them can also be triggered through an ABA-independent pathway (Lång and Palva, 1992; Nordin et al., 1993; Nylander et al., 2001). Notably, neither expression of *CBF/DREB1* (Gilmour et al., 1998; Liu et al., 1998; Stockinger et al., 1997) nor *DREB2A* genes (Liu et al., 1998), encoding transcriptional activators involved in cold- and dehydration-induced gene expression, was affected by antisense inhibition of *AtPP2CA*. This indicates that AtPP2CA is a component of a CBF/DREB-independent signal pathway and suggests a role for ABRE-binding factors (Choi et al., 2000; Uno et al., 2000) in this process (Tähtiharju and Palva, 2001). The hypothetical model of AtPP2CA regulated signalling during cold acclimation is presented in Figure 2.

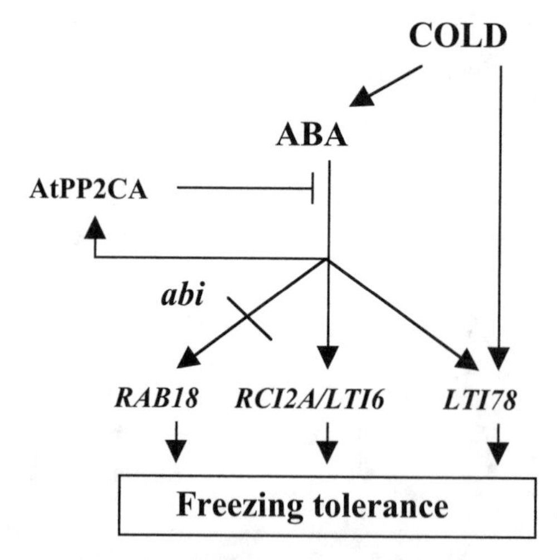

Figure 2. A hypothetical model for AtPP2CA-mediated cold stress signal transduction. At the onset of cold acclimation the ABA signalling goes through the AtPP2CA controlled pathway. Perception of cold signal triggers an increase in endogenous ABA, which in turn triggers signal cascades leading to an expression of *AtPP2CA* and cold-inducible target genes downstream of *AtPP2CA*. Induction of *AtPP2CA* allows feedback regulation of ABA signalling and hence controlled induction of the cold-inducible genes.

5. AtPP2CA SPECIFICALLY INTERACTS WITH AN UNIQUE TYPE OF K^+ CHANNEL

The results obtained from yeast two-hybrid screen show that AtPP2CA specifically interacts with the potassium channel AKT3 (Vranová et al., 2001), also known as AKT2 (Lacombe et al., 2000). Based on RNA blots and promoter activity studies, AKT3 was shown to be predominately expressed in leaves (Dennison et al., 2001; Lacombe et al., 2000; Marten et al., 1999). Since AKT3 was shown to be expressed also in phloem and it was capable of mediating K^+ fluxes, AKT3 was proposed to be involved in K^+ transport in phloem tissues and thus play a role in sugar translocation (Lacombe et al., 2000; Marten et al., 1999). As a consequence, an impairment of AKT3 activity would be expected to affect growth. Yet, studies with plants containing a T-DNA insertion mutation in *AKT3* showed that AKT3 does not contribute to seedling growth rate (Dennison et al., 2001). Thus, the function of AKT3 is still unknown.

Based on the amino acid sequence homology to previously identified plant genes encoding K^+ channels, AKT3 was classified as an inward rectifying K^+ (K^+_{in}) channel (Cao et al., 1995; Ketchum and Slayman, 1996). K^+_{in} channels have been implicated in specific physiological processes such as stomatal movements, which require both action of ABA and K^+ (Maathuis et al., 1997), suggesting that AKT3 might also be involved in such a response. Stomatal closure is partially achieved through ABA-induced increase in pH, which inhibits K^+_{in} channels and activates K^+_{out} channels leading to K^+ efflux from the guard cells and the consequent decline in turgor (Zimmermann et al., 1999). However, recent studies have revealed that AKT3 has unique functional features that are not displayed by any other K^+ channel studied thus far. In addition to inward K^+ currents AKT3 also mediates K^+ efflux. Most strikingly, both extracellular and intracellular acidification decreases the activity of AKT3 (Lacombe et al., 2000; Marten et al., 1999) whereas the other K^+_{in} channels are activated by a decrease in pH (Zimmermann et al., 1999). Thus, it is evident that AKT3 is a unique type of K^+ channel.

6. HYPOTHETICAL ROLE OF AKT3 IN ABA-MEDIATED COLD SIGNAL TRANSDCUTION

Since cytosolic pH is strictly regulated, it has been proposed to serve as a second messenger. However, it is not known how this signal is sensed. It has been put forward that either pH sensitivity of the channels themselves or a membrane-delimited signalling pathway mediates pH sensing. ABA is known to evoke a considerable alkalisation of the cytoplasm (Zimmermann et al., 1999). The activity of PP2C ABI1 is highly sensitive to pH *in vitro* (Leube et al., 1998) suggesting a role for PP2C in pH mediated ABA responses. Notably, ABI1, a negative regulator of ABA responses to drought, was shown to regulate K^+ channel activity in guard cells despite of a pH shift in the cytosol (Armstrong et al., 1995) suggesting a role for another signalling component in this response. AtPP2CA, which interacts with AKT3, is also negative regulator of ABA signalling. Whether pH regulates the activity of AtPP2CA is unknown. However, the unique response of AKT3 to pH (Lacombe et al., 2000; Marten et al., 1999) as well as ABA-induced upregulation of *AKT3* gene expression (Lacombe et al., 2000) suggests a key role for the AKT3 in pH-mediated ABA responses.

7. CONCLUDING REMARKS

The results demonstrate that AtPP2CA is an important modulator of the ABA action during cold stress (Tähtiharju and Palva, 2001). Induction of AtPP2CA may allow feedback regulation of ABA signalling and hence controlled induction of the target genes and development of freezing tolerance. In this way AtPP2CA could continuously modulates ABA signalling during the cold acclimation process. Interestingly ERA1, another negative regulator of ABA signalling, appears to repress the expression of AtPP2CA during cold acclimation (Tähtiharju and Palva, 2001). Indeed, recent studies suggest the presence of several negative regulators of ABA signalling in plants (Cutler et al., 1996; Foster and Chua, 1999; Gosti et al., 1999; 2001; Sheen, 1998), which could reflect the complexity of ABA-controlled responses in plants. It is likely that these regulators of ABA signalling monitor and reset the ABA-signalling network allowing the cell to continuously monitor the presence or absence of ABA as suggested by Gosti et al. (1999).

The study with AtPP2CA also supports the role of ABA in cold signal transduction (Tähtiharju and Palva, 2001). Furthermore, it shows that genetic engineering of signalling components other than transcriptional activators (Jaglo-Ottosen et al., 1998; Liu et al., 1998) can be employed to increase stress tolerance in plants.

REFERENCES

Armstrong, F., Leung, J., Grabov, A., Brearley, J., Giraudat, J. and Blatt, M., 1995, Sensitivity to abscisic acid of guard-cell K^+ channels is suppressed by *abi1-1*, a mutant *Arabidopsis* gene encoding a putative protein phosphatase, *Proc. Natl. Acad. Sci. USA*, **92**: 9520-9524.

Cao, Y., Ward, J.M., Kelly, W.B., Ichida, A.M., Gaber, R.F., Anderson, J.A., Uozumi, N., Schroeder, J.I. and Crawford, N.M., 1995, Multiple genes, tissue specificity, and expression-dependent modulation contribute to the functional diversity of potassium channels in Arabidopsis thaliana, *Plant Physiol.* **109**: 1093-1106.

Capel, J., Jarillo, J.A., Salinas, J. and Martinez-Zapater, J.M., 1997, Two homologous low-temperature-inducible genes from *Arabidopsis* encode highly hydrophobic proteins, *Plant Physiol.* **115**: 569-576.

Choi, H-I., Hong, J-H., Ha, J-O., Kang, J-Y. and Kim, S.Y., 2000, ABFs, a family of ABA-responsive element binding factors, *J. Biol. Chem.* **275**: 1723-1730.

Cutler, S., Ghassemian, M., Bonetta, D., Cooney S., and McCourt, P., 1996, A protein farnesyl transferase involved in abscisic acid signal transduction in *Arabidopsis*, *Science*, **273**: 1239-1240.

Dennison, K.L., Robertson, W.R., Lewis, B.D., Hirsch, R.E., Sussman, M.R. and Spalding, E.P., 2001, Functions of AKT1 and AKT2 potassium channels determined by studies of single and double mutants of Arabidopsis, *Plant Physiol.* **127**: 1012-1019.

Foster, R., and Chua, N.-H., 1999, An *Arabidopsis* mutant with deregulated ABA gene expression: implications for negative regulator function, *Plant J.* **17:** 363-372.

Gilmour, S.J. and Thomashow, M.F., 1991, Cold acclimation and cold regulated gene expression in ABA mutants of *Arabidopsis thaliana*, *Plant Mol. Biol.* **17**: 1233-1240.

Gilmour, S.J., Zarka, D.G., Stockinger, E.J., Salazar, M.P., Houghton, J.M.. and Thomashow, M.F., 1998, Low temperature regulation of the *Arabidopsis* CBF family of AP2 transcriptional activators as an aerly step in cold-induced COR gene expression, *Plant J.* **16**: 433-442.

Gosti, F., Beaudoin, N., Serizet, C., Webb, A.A.R., Vartanian, N. and Giraudat, J., 1999, ABI1 protein phosphatase 2C is a negative regulator of abscisic acid signaling, *Plant Cell* **11**: 1897-1909.

Guy, C.L., 1990, Cold acclimation and freezing stress tolerance: role of protein metabolism. Annu. Rev. Plant Physiol. *Plant Mol. Biol.* **41**, 187-223.

Hasegawa, P.M., Bressan, R.A., Zhu, J-K. and Bohnert, H.J., 2000, Plant cellular and molecular responses to high salinity, *Annu. Rev. Plant Physiol. Plant Mol. Biol.* **51**: 463-499.

Heino, P., Sandman, G., Lång, V., Nordin, K. and Palva, E.T., 1990, Abscisic acid deficiency prevents development of freezing tolerance in *Arabidopsis thaliana* (L.) Heynh, *Theor. Appl. Genet.* **79**: 801-806.

Jaglo-Ottosen, K.R., Gilmour, S.J., Zarka, D.G., Schabenberger, O. and Thomashow, M., 1998, *Arabidopsis* CBF1 overexpression induces COR genes and enhances freezing tolerance, *Science* **280**: 104-106.

Ketchum, K.A. and Slayman, C.W., 1996, Isolation of an ion channel gene from Arabidopsis thaliana using the H5 signature sequence from voltage-dependent K^+ channels, *FEBS Lett.* **378**: 19-26.

Kudla, J., Xu, Q., Harter, K., Gruissem, W. and Luan, S., 1999, Genes for calcineurin B-like proteins in *Arabidopsis* are differentially regulated by stress signals, *Proc. Natl. Acad. Sci. USA* **96**: 4718-4723.

Kumori, T., and Yamamoto, M., 1994, Cloning of cDNAs from *Arabidopsis thaliana* that encode putative protein phosphatase 2C and human Dr-1-like protein by transformation of a fission yeast mutant, *Nuc. Acid Res.* **22**: 5296-5301.

Lacombe, B., Pilot, G., Michard, E., Gaymard, F., Sentenac, H. and Thibaud, J-B., 2000, A shaker-like K^+ channel with weak rectifications is expressed in both source and sink phloem tissues of *Arabidopsis, Plant Cell* **12**: 837-851.

Leube, M.P., Grill, E. and Amrheim, N., 1998, ABI1 of *Arabidopsis* is a protein serine/threonine phosphatse highly regulated by the proton and magnesium ion concentration, *FEBS Lett.* **424**: 100-104.

Leung, J., Bouvier-Durand, M., Morris, P-C., Guerrier, D., Chefdor, F. and Giraudat, J., 1994, *Arabidopsis* ABA response gene *ABI1*: features of a calcium-modulated protein phosphatase, *Science* **264**: 1448-1452.

Leung, J. and Giraudat, J., 1998, Abscisic acid signal transduction, *Annu Rev. Plant Physiol. Plant Mol Biol.* **49**: 199-222.

Leung, J., Merlot, S. and Giraudat, J., 1997, The Arabidopsis *ABSCISIC ACID-INSENSITIVE2* (*ABI2*) and *ABI1* genes encode homologous protein phosphatases 2C involved in abscisic acid signal transduction, *Plant Cell* **9**: 759-771.

Liu, Q., Kasuga, M., Sakuma, Y., Abe, H., Miura, S., Yamaguchi-Shinozaki, K. and Shinozaki, K., 1998, Two transcription factors, DREB1 and DREB2, with an EREBP/AP2 DNA binding domain separate two cellular signal transduction pathways in drought- and low temperature-responsive gene expression, respectively, in Arabidopsis, *Plant Cell* **10**: 1391-1406.

Luan, S., 1998, Protein phosphatases and signaling cascades in higher plants, *Trends Plant Sci.* **3**: 271-275.

Lång, V., Heino, P. and Palva, E.T., 1989, Low temperature acclimation and treatment with exogenous abscisic acid induce common polypeptides in *Arabidopsis thaliana* (L.) Heynh, *Theor. Appl. Genet.* **77**: 729-734.

Lång, V., Mäntylä, E., Welin, B., Sundberg, B. and Palva, E.T., 1994, Alterations in water status, endogenous abscisic acid content, and expression of *rab18* gene during the development of freezing tolerance in *Arabidopsis thaliana, Plant Physiol.* **104**: 1341-1349.

Lång, V. and Palva, E.T., 1992, The expression of a rab-related gene, *rab18*, is induced by abscisic acid during the cold acclimation process of *Arabidopsis thaliana* (L.) Heynh, *Plant Mol. Biol.* **20**: 951-962.

Maathuis, F.J.M., Ichida, A.M., Sanders, D. and Schroeder, J.I., 1997, Roles of higher plant K^+ channels, *Plant Physiol.* **114**: 1141-1149.

Marten, I., Hoth, S., Deeken, R., Ache, P., Ketchum, K.A., Hoshi, T. and Hedrich, R., 1999, AKT3, a phloem-localized K^+ channel, is blocked by protons, *Proc. Natl. Acad. Sci. USA* **96**: 7581-7586.

Merlot, S., Gosti, F., Guerrier, D., Vavasseur, A. and Giraudat, J., 2001, The ABI1 and ABI2 protein phosphatases 2C act in a negative feedback regulatory loop of the abscisic acid signalling pathway, *Plant J.* **25**: 315-324.

Meyer, K., Leube, M.P. and Grill, E., 1994, A protein phosphatase 2C involved in ABA signal transduction in *Arabidopsis thaliana, Science* **264**: 1452-1455.

Monroy, A.F., Sangwan, V. and Dhindsa, R.S., 1998, Low temperature signal transduction during cold acclimation: protein phosphatase 2A as an early target for cold-inactivation, *Plant J.* **13**: 653-660.

Mäntylä, E., Lång, V. and Palva, E.T., 1995, Role of abscisic acid in drought-induced freezing tolerance, cold acclimation, and accumulation of LTI78 and RAB18 proteins in *Arabidopsis thaliana*, *Plant Physiol.* **107**: 141-148.

Nordin, K., Heino, P. and Palva, E.T., 1991, Separate signal pathways regulate the expression of a low-temperature-induced gene in *Arabidopsis thaliana* (L.) Heynh, *Plant Mol. Biol.* **16**: 1061-1071.

Nordin, K., Vahala, T. and Palva, E.T., 1993, Differential expression of two related, low-temperature-induced genes in *Arabidopsis thaliana* (L.) Heynh, *Plant Mol Biol.* **21**: 641-653.

Nylander, M., Heino, P., Helenius, E., Palva, E.T., Ronne, H. and Welin, B.V., 2001, The low-temperature- and salt-induced *RCI2A* gene of *Arabidopsis* complements the sodium sensitivity caused by a deletion of the homologous yeast gene *SNA1*, *Plant Mol. Biol.* **45**: 341-352.

Sanders, D., Brownlee, C. and Harper, J.F., 1999, Communicating with calcium, *Plant Cell* **11**: 691-706.

Sangwan, V., Foulds, I., Singh, J. and Dhindsa, R.S., 2001, Cold-activation *of Brassica napus BN115* promoter is mediated by structural changes in membranes and cytoskeleton, and requires Ca^{2+} influx, *Plant J.* **27**: 1-12.

Sheen, J., 1996, Ca^{2+}-dependent protein kinases and stress signal transduction in plants, *Science* **274**: 1900-1902.

Sheen, J., 1998, Mutational analysis of protein phosphatase 2C involved in abscisic acid signal transduction in higher plants, *Proc. Natl. Acad. Sci. USA* **95**: 975-980.

Smith, R.D. and Walker, J.C., 1996, Plant Protein phosphatases, *Annu. Rev. Plant Physiol. Plant Mol. Biol.* **47**: 101-125.

Stockinger, E.J., Gilmour, S.J. and Thomashow, M., 1997, *Arabidopsis thaliana* CBF1 encodes an AP2 domain-containing transcriptional activator that binds to the C-repeat/DRE, a cis-acting DNA regulatory element that stimulates transcription in response to low temperature and water deficit, *Proc. Natl. Acad. Sci. USA* **94**: 1035-1040.

Straub, P.F., Shen, Q. and Ho, T-H.D., 1994, Structure and promoter analysis of an ABA- and stress-regulated barley gene, *HVA1*, *Plant Mol. Biol.* **26**: 617-630.

Tähtiharju, S. and Palva, T., 2001, Antisense inhibition of protein phosphatse 2C accelerates cold acclimation in *Arabidopsis thaliana*, *Plant J.* **26**: 461-470.

Uno, Y., Furihata, T., Abe, H., Yoshida, R., Shinozaki, K. and Yamaguchi-Shinozaki, K., 2000, *Arabidopsis* basic leucine zipper transcription factors involved in an abscisic acid-dependent signal transduction pathways under drought and high-salinity conditions, *Proc. Natl. Acad. Sci. USA* **97**: 11632-11637.

Vranová, E., Tähtiharju, S., Sriprang, R., Willekens, H., Heino, P., Palva, E.T., Inzé, D., and Van Camp, W., 2001, The AKT3 potassium channel protein interacts with the AtPP2CA protein phosphatase 2C, *J. Exp. Bot.* **52**:181-182.

Wu, Y., Kuzma, J., Maréchal, E., Graeff, R., Lee, H.C., Foster, R. and Chua, N.-H., 1997, Abscisic acid signaling through cyclic ADP-ribose in plants, *Science* **278**: 2126-2130.

Xu, Q., Fu, H-H., Gupta, R. and Luan, S., 1998, Molecular characterization of a tyrosine-specific protein phosphatase encoded by a stess-responsive gene in *Arabidopsis, Plant Cell* **10**: 849-857.

Zimmermann, S., Ehrhardt, T., Plesch, G. and Müller-Röber, B., 1999, Ion channels in plant signaling, *Cell. Mol. Life Sci.* **55**: 183-203.

Part II

Physiological Aspects of Plant Cold Hardiness

6

PHYSIOLOGICAL ASPECTS OF COLD HARDINESS IN NORTHERN DECIDUOUS TREE SPECIES

Olavi Junttila, Annikki Welling, Chunyang Li, Berhany A. Tsegay, E.Tapio Palva*

1. INTRODUCTION

Tree species growing at northern latitudes are characterised by significant annual variation in their cold hardiness. Due to their height, trees are not protected by insulating snow cover and, in order to survive, they must be able to develop high level of cold hardiness. Lack of capacity for sufficient cold hardiness is, however, often not a limiting factor for winter survival of deciduous and evergreen tree species growing in boreal and subarctic regions. Their survival is primarily dependent on a proper timing of cold hardening and dehardening. This is a well-known fact and has been demonstrated in numerous studies, and the basic aspects of the environmental regulation of these processes have been studied comprehensively. These studies have provided strong evidence for the central role of photoperiod as a cue for timing of hardening in northern tree species, and for interaction between photoperiod and temperature. Due to the decisive effect of photoperiod on cold hardening, adaptation to photoperiodic conditions appears to be a prominent feature in these tree species.

Physiological basis of hardening and dehardening processes in trees has been studied exensively and some general features have emerged, but at the molecular level almost next to nothing is known. This is in contrast with the progress on frost hardiness research in herbaceous plants, particularly with the model plant *Arabidopsis* (Thomashow, 2001). However, studies with trees have been, and still are, hampered by the lack of characterised mutants and, in comparison with herbaceous plants, of long generation time, often several years. Recently transgene technology has provided possibilities for new approaches even for tree species, but also this type of studies are tedious, time consuming and, consequently, expensive. So far, hybrid aspen (*Populus tremula* L. x *P. tremuloides* Michx) is the only tree species where transgenic plants have been used to study frost hardening (Junttila et al., 1998). However, currently ongoing research, among others with

* Olavi Junttila, Annikki Welling, Chunyang Li, E. Tapio Palva, University of Helsinki, Helsinki, Finland, FIN-00014. Berhany A. Tsegay, University of Tromsö, Tromsö, Norway, N-9037.

Plant Cold Hardiness, edited by Li and Palva
Kluwer Academic/Plenum Publishers, 2002

Silver birch (*Betula pendula* Roth), is expected to result in new transgenic lines of particular interest for cold hardiness studies.

In this paper we give a short overview on the environmental regulation of cold harde-ning in deciduous trees and present some more recent data on changes in dehydrins and abscisic acid (ABA) in relation to hardening, based on experiments with hybrid aspen and Silver birch.

2. ENVIRONMENTAL REGULATION OF COLD HARDINESS

From an evolutionary point of view, it is understandable that those environmental conditions, which normally precede the cold stress, are the main factors affecting cold hardening. At the northern latitudes those conditions are declining daylength (short photoperiod) and declining temperature. Similarly, dehardening is promoted by conditions that normally precede the new growth period, mainly increasing temperature. In trees hardening induced by short photoperiod and low temperature is generally rather slow process and will therefore not provide protection for episodic, unexpected frost periods occurring outside the normal, annual cycle. There are, however, also examples of rapid changes, within hours or a day, in hardiness of citrus (*Poncirus trifoliate*; Tignor et al., 1997) and *Nothofagus* (Meza-Basso et al., 1986) leaves. Such changes could be caused by modulations of tissue water content. Reduced water content is commonly connected with increased hardiness, and reduced availability of water, mild drought, is known to accelerate cold hardiness even in woody plants. In northern deciduous trees, relatively rapid modulation of leaf hardening might be of some advantage for the plants; it would provide a possibility to protect assimilating leaves against late spring or/and early autumn frosts and thus extend the period for photosynthesis. As shown below, deciduous leaves of Silver birch are capable of rather extensive hardening and also show ecotypic differences.

2.1. Photoperiod

Short photoperiod as an important signal for cold hardening in deciduous woody plants has been known for a long time (for references, see Weiser, 1970). This applies also to Silver birch and hybrid aspen (Li et al., unpublished results; Junttila et al., 1998). Research on transgenic hybrid aspen has shown that, in this species, the main effect of short photoperiod on hardening can be understood as an indirect effect that is dependent on the induction of apical growth cessation. Hybrid aspen plants over-expressing the oat phytochrome A did not stop elongation growth under short photoperiod, and they were unable to develop cold hardiness, even after a long exposure to low temperature (Junttila et al., 1998; Olsen et al., 1997). Thus, in hybrid aspen, and probably in other northern tree species with a similar growth pattern as well, cessation of apical shoot elongation (bud set) appears to be a prerequisite for cold hardening. Shoot elongation in plants over-expressing the phytochrome A ceased, when they were transferred to low temperature (6-

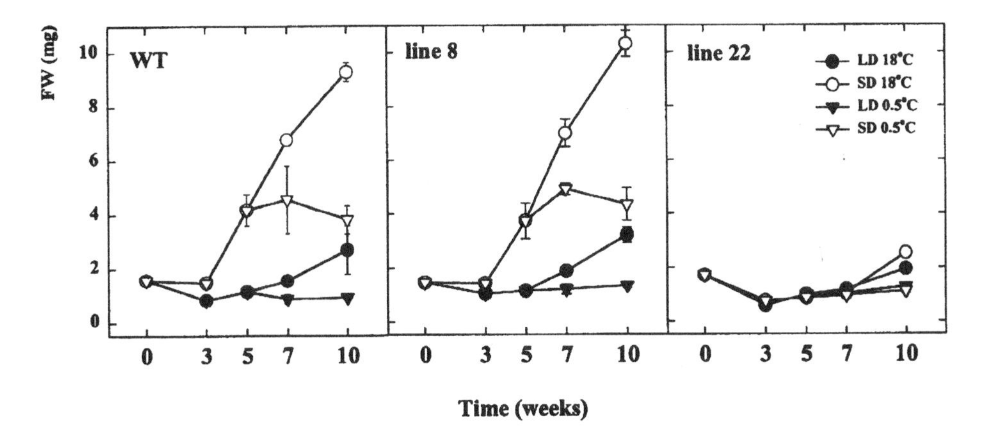

Figure 1. Effects of photoperiod and temperature on fresh weight of lateral buds in hybrid aspen. WT: wild type; line 8: transgenic line with wild type phenotype; line 22: transgenic line over-expressing the oat phytochrome A. LD: 16 h; SD: 10 h. Vertical bars show the double standard deviation.

0.5°C), but no apical bud set or growth of lateral buds occurred. It should be mentioned that in some species, including Silver birch, older plants do not form apical bud, but the whole apical meristem is aborted (see f. ex. Junttila, 1976). However, this abscission of the apical meristem is accompanied by a rapid development of lateral buds, comparable to seedlings forming an apical bud.

Also the normal senescence processes, yellowing and abscission of leaves, were impaired in the transgenic hybrid aspen line 22, which did not respond to short photoperiod. This indicates that all the major developmental processes that normally take place during the fall, are blocked as long as the apical meristem remains in an active or a growing phase. Physiological nature of this correlative action of growing/non-growing (bud) meristem on cold hardening of the shoot is unknown. It could be understood as an inhibitory action induced by an active meristem, or as an enhancing effect induced by inactive meristem (bud), or that the regulation includes both a negative and a positive influence.

In hybrid aspen, cessation of apical elongation growth initiates a rapid growth of lateral buds and development of bud dormancy. In wild type and line no. 8, which has similar phenotype as wild type, elongation growth stops after about 3 weeks in short photoperiod (10-12 h). As shown in Figure 1, fresh weight of lateral buds increased linearly after the cessation of elongation growth as long as plants were kept at 18°C and in short photoperiod. No significant changes in lateral bud size were observed in phytochrome A over-expressing line 22 which is insensitive to short photoperiod (Figure 1). Dormancy in lateral buds, measured as an inability to burst under long day conditions without chilling treatment, develops gradually during an exposure to short photoperiod. After five weeks under 10-h photoperiod, lateral buds in wild type plants had developed dormancy and did not initiate new growth when the plants were transferred to long photoperiod. However, no development of dormancy was observed in the transgenic line no. 22 (Figure 2). So far, no correlation between the dormancy and level of abscisic acid (ABA) in the lateral buds of hybrid aspen has been observed (Welling et al., unpublished).

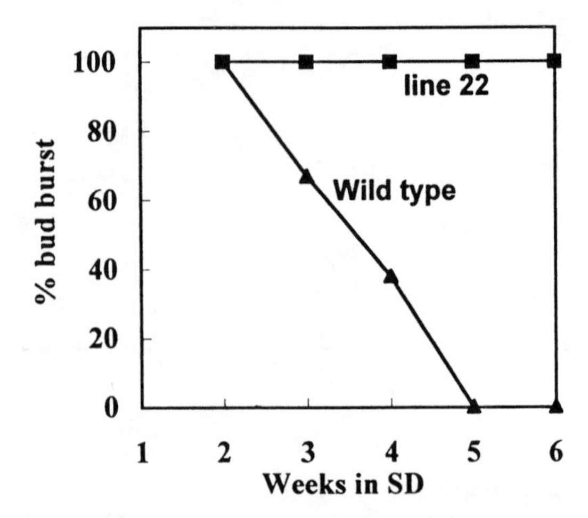

Figure 2. Development of dormancy in lateral buds of hybrid aspen during an exposure to a 10-h photoperiod at 18°C. At indicated time points, plants were transferred to 24-h photoperiod at 18°C and decapitated. Dormancy is expressed as per cent bud burst in three topmost lateral buds after 14 days. Results are means of five plants.

Although the main effect of photoperiod on cold hardening in hybrid aspen and Silver birch is mediated through regulation of the apical meristem, photoperiod may induce cold hardening apparently also more directly. Recent experiments with Silver birch have shown that an exposure to short photoperiod induces a rapid cold hardening in mature leaves (Li et al., unpublished). Leaves also responded to low temperature and developed significant cold hardiness (Figure 3). In this experiment, a mild water stress had a minor effect only on hardiness of the leaves. When the treatments used in this experiment were combined, an additive effect was observed. Results in Figure 3 are after seven days of hardening, but low temperature treatment resulted in a significant increase in frost hardi-ness already after one day (24 h). Also the effect of short photoperiod was visible after 24 h. Leaves of Silver birch are annual and there are no studies indicating that cold hardiness of birch leaves could be of some adaptive significance. We can, however, speculate that some capacity for hardening in leaves during late summer and early fall would prolong the period for photosynthesis at the end of the season and thereby enhance possibilities for a successful overwintering. In addition, cold acclimation capacity of birch leaves may have importance as a protective system for cold spell during spring season. To our know-ledge, these questions have not been studied in northern deciduous tree species to any ex-tent.

Another aspect of interest is the question of interaction between short day induced cessation of apical shoot elongation and short day induced frost hardening in the leaves.

In the present experiments, whole plants were exposed to short photoperiod and the possibility that the observed hardening is a response to a signal from the apical meristem or stem can not be ruled out. If so, translocation of the signal is initiated rapidly and affects leaf hardiness before any visible changes in growth behaviour of the apex can be observed. On the other hand, the photoperiodic signal is supposed to be perceived by the leaves and also the frost hardening in the leaves could be a direct response to photoperiod. Experiments with detached leaves could provide some answers to these questions. Low temperature, most probably, is acting directly on the leaves, and there are no indica-

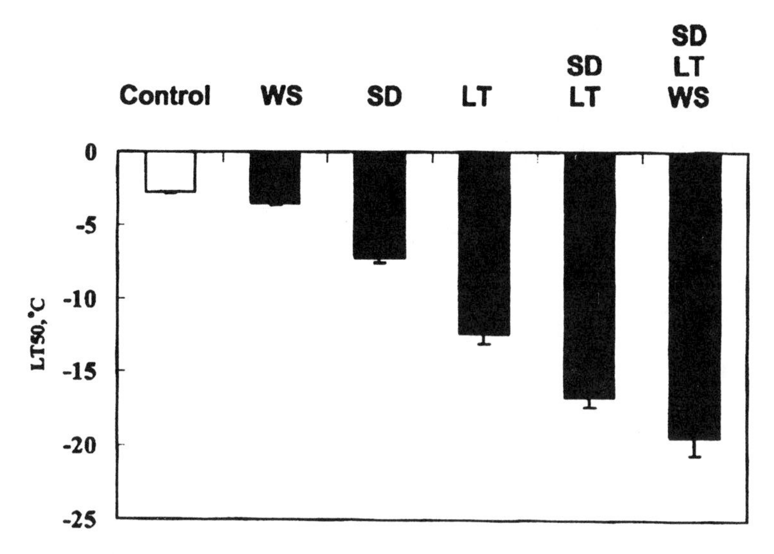

Figure 3. Effect of mild water stress (WS), short photoperiod (SD, 12 h) and low temperature (LT, +4°C) on frost hardiness (LT50) of Silver birch leaves. Treatments were given for seven days. In combined treatments WS+SD, WS+LT, WS was given for 7 days both before and during the SD or LT treatments. Similarly, in SD+LT treatment, SD was given for 7 days before and during the LT treatment, and in WS+SD+LT treatment both WS and SD were applied for 7 days before and during the LT treatments.

tions of translocation of a low temperature induced frost hardening signal from one tissue to another.

Within the boreal zone, the critical photoperiod for cessation of growth in most deciduous tree species changes as a cline over the latitudes (for references, see Junttila and Nilsen, 1993). Similarly, timing of cold hardening in latitudinal ecotypes of Scots pine (*Pinus sylvestris* L.) changes in a clinal pattern (Hurme et al., 1997). Presence of such clines suggests that the traits are under a strong selection (Aitken and Hannerz, 2001),

and demonstrates genetic adaptation to the local light climate. Thus, latitudinal origins of such species can be easily distinguished from each other due to their photoperiodic re-sponses. In addition to the differences in critical photoperiod, latitudinal ecotypes also appear to differ in their rate of response to photoperiod. Under a given photoperiod, time course of bud set is inversely related to the critical photoperiod, those with the longest critical photoperiod (the most northern ecotypes) set bud earliest, and those with the shortest (the most southern ecotypes) set bud latest (Junttila, 1982). The physiological mechanisms regulating these two responses, the critical photoperiod and the rate of re-sponse to photoperiod, can, however, be different (Howe et al., 1995).

Ecotypic differences in timing of frost hardening in perennial tissues of northern tree species are easy to understand because of the decisive importance of this timing for winter survival of the genotype. As mentioned above, also deciduous leaves of Silver birch have a capacity for frost hardening. Further studies have shown that also hardening of the leaves show similar differences between latitudinal ecotypes as observed for cessation of elongation growth and following hardening of bud and stem tissues. In a study with Silver birch (Li et al. unpublished), leaves of a northern ecotype, originating from the latitude of 67°44'N, hardened significantly faster than leaves of a southern ecotype (58°10'N), while the rate of hardening in leaves of an ecotype from 63°14'N was intermediate. After two weeks, leaves of all three ecotypes had reached the same level of cold hardiness. Hardening of the leaves was faster than hardening of stems and, as discussed above, deciduous leaves may provide interesting information on the initial phases of frost hardening in woody plants.

2.2. Temperature

Although short photoperiod may induce a significant level of cold hardiness even at normal growing temperature (Junttila and Kaurin, 1990), low temperature is generally needed for development of maximal hardiness. During the years, extensive research has been carried out to study interactions and quantitative aspects of photoperiod and temperature for induction of cold hardiness. Based on such studies, mathematic models for hardening and dehardening in trees have been developed (e.g. Leinonen, 1996).

The maximum level of cold hardiness is not necessarily different in contrasting latitu-dinal ecotypes, as illustrated for *Salix pentandra* (Junttila and Kaurin, 1990). However, the rate of hardening, both in response to photoperiod as discussed above, and in response to low temperature is generally higher in northern than southern ecotype, as demonstrated for example for Scots pine by Repo et al. (2000). Also the northern ecotype of Silver birch acclimated faster than the southern ecotype when exposed to 6°C and 0.5°C in short photoperiod (Li et al., unpublished). Results from that study also demonstrated that during low temperature induced hardening, frost hardiness in Silver birch was separated from bud dormancy, defined as ability for bud burst after a transfer to long photoperiod and high temperature. Dormancy induced by short photoperiod at 18°C was rapidly bro-ken when plants were transferred to low temperature, but frost tolerance continued to increase during the 6-week period at low temperature. Temperature effects on frost har-dening in trees like Silver birch have not been analysed quantitatively in any detail, and we do not know, whether contrasting ecotypes may differ, for example, in respect to the temperature requirements for hardening or dehardening.

In Silver birch, as well as in hybrid aspen, also deciduous leaves are capable of cold hardening in response to low temperature (Figure 3). In Silver birch, LT_{50} value of leaves of a northern (67°44'N) and a southern ecotype (58°10'N) changed from –2°C to –14°C and to -11°C, respectively. Even leaves of transgenic hybrid aspen over-expressing phytochrome A, responded to low temperature (+4°C), after a week LT_{50} was -8°C compared to -4.5°C in control (long photoperiod, 18°C). Effect of 12-h photoperiod at 18°C on hardening of transgenic leaves was not significant (LT_{50} -5.5°C). On the other hand, both short photoperiod (LT_{50} -5°C) and low temperature (LT_{50} -6°C) enhanced frost hardiness of leaves of wild type (LT_{50} in control –3.5°C). Taken together, these results indicate that photoperiod and low temperature regulate hardening in leaves through different mechanisms (Welling et al., unpublished).

3. PHYSIOLOGICAL ASPECTS OF COLD HARDINESS

Development of cold hardiness is based on complex metabolic and structural changes in plant cells involving regulation at various levels. For example, in *Arabidopsis*, more than hundred cold induced genes are known, but functions for almost all of those genes are not yet known (Hiilovaara-Teijo and Palva, 1999). The metabolic changes include al-terations in plasma membrane lipids, modulation of photosynthesis, carbohydrate accu-mulation, altered protein synthesis, etc. Perhaps due to this complexity, transgenic ap-proaches to improve frost hardiness in herbaceous plants have resulted, so far, in mode-rate results only. Among the complex metabolic changes, altered carbohydrate meta-bolism and enhanced production of certain soluble carbohydrates during hardening ap-pears to be among responses that are often observed in most herbaceous and woody spe-cies. One possible function of such compounds is to protect cell membranes and vital macromolecules against freeze induced dehydration. Also dehydrins may have similar functions. Hardening of stem and bud tissue in deciduous trees is strongly dependent on regulation of growth and control mechanisms for this involves hormonal regulation, particularly gibberellin (GA) and auxins. There is strong evidence for the role of abscisic acid (ABA) in hardening of herbaceous plants (Hiilovaara-Teijo and Palva, 1999), but the function of ABA in woody plants is less clear.

3.1. Water Content

Like in herbaceous plants, also in deciduous species of trees the increase in frost hardiness is often correlated with decreased water content. This partial dehydration takes place both in stem and bud tissues. Simultaneously, dry matter content increases due to an increased synthesis of soluble carbohydrates. In hybrid aspen, decrease in water content seems to be closely connected to the cessation of apical shoot elongation and following bud set. Thus, in transgenic plants over-expressing phytochrome A, no changes in tissue water content could be detected during a 10-week period in short photoperiod compared to long photoperiod, bud water content was above 70% in both treatments. In wild type, on the other hand, bud water content had declined to about 55% after 10 weeks in short photoperiod

3.2. Dehydrins

Dehydrins, encoded by a well-conserved gene family, are rapidly accumulating when plant cells are exposed to decreased water content and potential. Although the mode of action of these proteins is not known in detail, they appear to have significant protective functions in plant cells against dehydration induced stresses (Campbell and Close, 1999; Ismail et al., 1999). Dehydrins accumulate during cold acclimation correlating with freezing tolerance in woody plants as well (Wisniewski et al., 1996, 1999; Welling et al., 1997; Rinne et al., 1998). According to Rinne et al. (1999) dehydrins in birch may rescue hydrolytic enzyme functions during dehydration of the cells.

In hybrid aspen, dehydrin gene expression has been shown in bud, stem and apex tissues (Welling et al., unpublished). In wild type of hybrid aspen, transcription and protein levels of dehydrins are enhanced significantly during frost hardening, both during a short day treatment at high temperature and during an exposure to low temperature treatment. Dehydrin-related proteins were studied using a polyclonal antibody. It recognized, among others, a 30 kDa protein that correlated with dehydrin gene expression (Figure 4). On the other hand, short photoperiod, applied at 18°C, had no effect on dehydrin gene expression in the transgenic line of hybrid aspen over-expressing the oat phytochrome A, and no signal was detected in western blot. However, when plants were exposed to low temperature, either under long or short photoperiod, both wild type and the transgenic

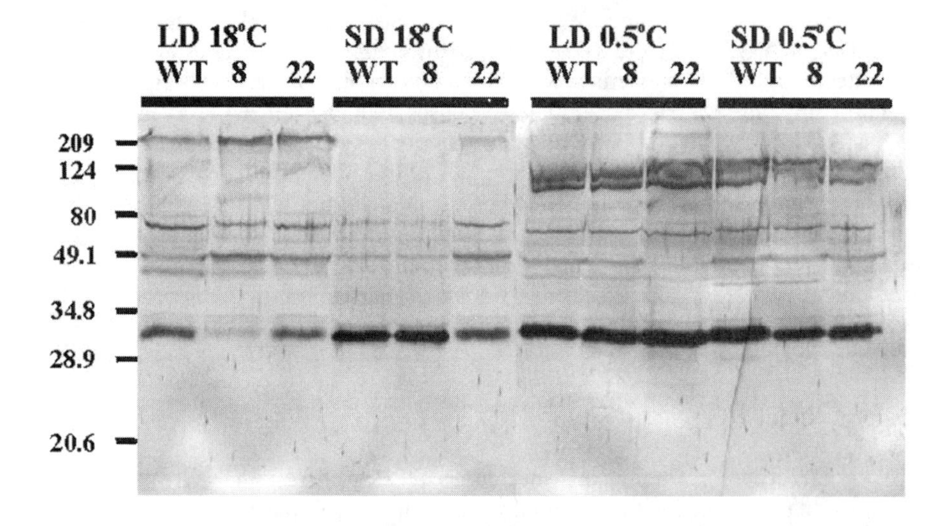

Figure 4. Western-blot analysis of the dsp 16-like proteins from stems of wild type (WT) and two transgenic lines (8, 22) of hybrid aspen. For the transgenic lines, see Figure 1. Treatments: SD, 10-h photoperiod; LD, 16-h photoperiod. At 18°C samples were collected after 10. Plants exposed to 0.5°C were first kept for 5 weeks at 18°C before they were transferred to 0.5°C for 5 weeks. The polyclonal antibody was developed against desiccation-inducible proteins (dsp 16) of *Craterostigma plantagineum* (Schneider et al., 1993).

line produced increased levels of dehydrin transcript and protein (Figure 4). Similar responses were observed in buds (Welling et al., unpublished). These results suggest that photoperiodic control of dehydrin production in hybrid aspen can be mediated through phytochrome A. They also suggest that low temperature and photoperiod regulate dehydrin levels through different mechanisms. In wild type of hybrid aspen, expression of dehydrins correlated with freezing tolerance, but accumulation of dehydrin in stem and bud tissue of the transgenic line at low temperature was not accompanied by increased tolerance. As mentioned above, low temperature induced some hardening in leaves of this transgenic line, but dehydrins in leaves have not yet been investigated. Clearly, further studies are needed on dehydrins in tree species. It would be particularly interesting to look at regulation of dehydrine genes in contrasting photoperiodic ecotypes of northern tree species, both in respect to photoperiodic responses and development of freezing tolerance.

3.3. Gibberellins and ABA

As discussed above, cessation of apical shoot elongation, induced by short photoperiod in hybrid aspen and Silver birch, is a prerequisite for cold hardening. Of this reason, physiological studies on these trees have to a great extent been focused on the mechanisms of growth regulation. Research in several species has resulted in a strong line of evidence for the role of gibberellins (GA) in control of shoot elongation in trees (for references, see Eriksson 2000 and Eriksson et al., 2000). These studies have included application experiments, chemical control of GA biosynthesis, metabolic studies, analysis of endogenous GAs, studies of gene expression, and use of transgenic plants modified for GA levels. In Silver birch and hybrid aspen, active GA is needed for normal shoot elongation. GA_1 is the main active GA in these species, as well as in most species of higher plants studied so far. Identification of GA_1 as the active compound was first concluded from studies of GA-deficient mutants in maize (Phinney, 1984), in woody plants the main evidence is obtained from studies with inhibitors of GA biosynthesis (Junttila et al., 1991). In respect to shoot elongation, the subapical meristem can be the site of action of GA_1, and GA_1 is required for cell division (Hansen et al., 1999).

In hybrid aspen, like in willow, short photoperiod appears to block certain steps in biosynthesis of GA_1, and this effect can be mediated through phytochrome A (Olsen et al. 1995; Olsen et al., 1997; Eriksson, 2000). Similar results have been obtained with Silver birch as well (Tsegay et al., unpublished). Inhibition of GA biosynthesis as a response to short photoperiod is an important step in induction of growth cessation and following dormancy development, leaf abscission and frost hardening. In the transgenic line of hybrid aspen expressing the *PHYA* gene, where all these processes are impaired, level of GA_1 was not reduced by an exposure to short photoperiod (Olsen et al., 1997).

Very little is known about the physiological and molecular basis of the action of photoperiod and phytochrome on GA biosynthesis. Studies with herbaceous plants and with seed suggest transcriptional control of GA C20-oxidase and/or 3β-hydroxylase (for references, see Kamiya and Garcia-Martinez, 1999). The former enzyme catalyses multiple oxidation steps to GA_{20}, the immediate precursor of GA_1, while 3β-hydroxylase converts GA_{20} to the bioactive GA_1. Eriksson et al. (2000) have demonstrated that, in hybrid aspen, expression of C20-oxidase is reduced under short photoperiod. Besides these enzymes, photoperiod may also control enzymes acting at early steps of GA biosynthesis (Zeevaart and Gage, 1993).

Already about 40 years ago, Irving and Lanphear (1967) proposed that long day leaf could be a source of cold hardiness inhibitor. GA could be such inhibitor, acting partly indirectly through stimulation of shoot elongation. Functions of GA have been investigated primarily in respect to regulation of shoot growth, and almost nothing is known about the effects of GA on cold hardening in non-growing systems, for example in mature leaves.

If GA can be described as a negative modulator of cold hardening, ABA is thought to be one of the key regulators enhancing cold hardening (Heino et al., 1990; Lång et al., 1994, Tamminen et al., 2001). For example, ABA-deficient mutants of *Arabidopsis* are impaired for cold hardening. There is also some evidence for involvement of ABA in cold hardening of woody plants as well. For example, both low temperature and short photoperiod have been shown to result in transient increase in ABA levels (Welling et al., 1997; Rinne et al., 1998). ABA application has been shown to increase freezing tolerance in some woody taxa (Irving, 1969; Welling et al., 1997; Rinne et al., 1998). We have analysed ABA levels in several series of samples from Silver birch and results from those studies demonstrate in general correlation between ABA levels and freezing tolerance (Li et al., unpublished). In those studies ABA levels were measured in samples from apical tissues. Birch leaves, with their capacity for hardening, will provide an interesting system for further studies on the relationship between ABA and cold hardening in tree species.

We also have carried out some preliminary studies on ABA levels in wild type and the *PHYA* expressing line of hybrid aspen (Welling et al., unpublished). In the wild type, ABA levels in lateral buds, calculated per dry weight, were significantly lower under short than under long photoperiod. However, in the transgenic line no significant differences were observed between the photoperiod, showing that also in this respect the *PHYA* over-expression disturbed the photoperiodic responses. Exposure to 0.5°C resulted in a significant increase in ABA levels both in the wild type and the transgenic line, irrespective of the photoperiod. These levels declined during an extended treatment at low temperature. Thus, genetic manipulation of phytochrome A in hybrid aspen did not prevent the effect of low temperature on ABA levels.

4. CONCLUDING REMARKS

As mentioned in the introduction, progress in cold hardiness research on trees has been relatively slow compared with herbaceous plants. Most probably many of the basic processes leading to frost hardening are similar in herbaceous and woody plants. This may obviously be the case with hardening of leaf tissue. Results presented above suggest that differences in timing of cold hardening between latitudinal ecotypes of Silver birch are present not only in perennial tissues but also in leaf tissues, where they are expressed rather rapidly. Leaves may provide an interesting and relatively fast system for examination of ecotypic differences at a molecular level. Further, if detached leaves are able to show similar responses, they could be used to study cold hardening independent of growth and dormancy regulation. On the other hand, regulation of growth cessation and influence of this process on all the various developmental processes initiated after bud set, remains as a highly interesting area of research.

Further progress in cold hardiness research at the whole tree level can be expected when transgenic approaches are used more extensively and systematically to test hypothesis developed from traditional physiological experiments. In addition, new and powerful technologies within functional genomics and proteomics will be used in research on trees as well. Combination of such different approaches will, hopefully, give us a more complete understanding of the various regulatory mechanisms in trees than what we have today.

5. ACKNOWLEDGEMENTS

This study was supported in part by grants from the Norwegian Research Council (NFR), Academy of Finland, Biocentrum Helsinki, Helsinki University Foundation, SNS and Technology Development Centre of Finland (TEKES).

6. REFERENCES

Aitken, S. N., and Hannerz, M., 2001, Genecology and gene resource management strategies for conifer cold hardiness, in: *Conifer Cold Hardiness*, F. J. Bigras and S. J. Colombo, eds., Kluwer Academic Publ., Dordrecht, pp. 23-54.

Campbell, S. A., and Close, T. J., 1997, Dehydrins: genes, proteins, and associations with phenotypic traits, *New Phytol.* **137**:61-74.

Eriksson, M., 2000, The role of phytochrome A and gibberellins in growth under long and short day conditions. Studies in hybrid aspen, Doctoral Thesis, Umeå, *Acta Univ. Agric. Suecica, Silvestria* **164**:1-55.

Eriksson, M. E., Israelsson, M., Olsson, O., and Moritz, T., 2000, Increased gibberellin biosynthesis in transgenic trees promotes growth, biomass production and xylem fiber length, *Nat. Biotechnol.* **18**:784-788.

Hansen, E., Olsen, J. E. and Junttila, O. 1999. Gibberellins and subapical cell division in relation to bud set and bud break in *Salix pentandra. J. Plant Growth Regul.* **18**:167-170.

Heino, P., Sandman, G., Lång, V., Nordin, K., and Palva, E. T., 1990, Abscisic acid deficiency prevents development of freezing tolerance in *Arabidopsis thaliana* (L.) Heynh. *Theor. Appl. Genet.* **79**:801-806.

Hiilovaara-Teijo, M., and Palva, E. T., 1999, Molecular responses in cold-adapted plants, in: *Cold-Adapted Organisms: Ecology, Physiology, Enzymology, and Molecular Biology*, R. Margesin and F. Schinner, eds., Springer-Verlag, Berlin Heidelberg, pp. 349-384.

Howe, G. T., Hackett, W. P., Furnier, G. R., and Klevorn, R. E., 1995, Photoperiodic responses of a northern and southern ecotype of black cottonwood, *Physiol. Plant.* **93**:695-708.

Hurme, P., Repo, T.,Savolainen, O., and Pääkkönen, T., 1997, Climatic adaptation of bud set and frost hardiness in Scots pine (*Pinus sylvestris* L.), *Can. J. For. Res.* **27**:716-723.

Irving, R. M., 1969, Characterization and role of an endogenous inhibitor in the induction of cold hardiness in *Acer negundo, Plant Physiol.* **44**:801-805.

Irving, R. M., and Lanphear, F. O., 1967, The long day leaf as a source of cold hardiness inhibitors, *Plant Physiol.* **42**:1191-1196.

Ismail, A. M., Hall, A. E., and Close, T. J., 1999, Purification and partial characterization of a dehydrin involved in chilling tolerance during seedling emergence of cowpea, *Plant Physiol.* **120**:237-244.

Junttila, O., 1976, Apical growth cessation and shoot tip abscission in *Salix*, *Physiol. Plant.* **38**:278-286.

Junttila, O., 1982, Cessation of apical growth in latitudinal ecotypes and ecotype crosses of *Salix pentandra* L., *J. Exp. Bot.* **33**:1021-1029.

Junttila, O. and Kaurin, Å., 1990, Environmental control of cold acclimation in *Salix pentandra*, *Scand. J. Forest Res.* **5**:195-204.

Junttila, O,. and Nilsen, J., 1993, Growth and development of northern forest trees as affected by temperature and light, in: *Forest Tree Development in Cold Climates*, Alden, J. N., Ødum, S., and Mastrantonio, J. L., eds., Plenum Publishing Corporation, New York, pp. 43-57

Junttila, O., Jensen, E., and Ernstsen, A.,1991, Effects of prohexadione (BX-112) and gibberellins on shoot growth in seedlings of *Salix pentandra*, *Physiol. Plant.* **83**:17-21.

Junttila, O., Olsen, J. E., Nilsen, J., Martinussen, I., Moritz, T., Eriksson, M., Olsson, O., and Sandberg, G., 1998, Phytochrome A overexpression and cold hardiness in transgenic *Populus*, in: *Plant Cold Hardiness*, P. H. Li and T. H. H. Chen, eds., Plenum Publishing Corporation, New York, pp. 245-256.

Kamiya, Y., and Garcia-Martinez, J. L., 1999, Regulation of gibberellin biosynthesis by light, *Curr. Opin. Plant Biol.* **2**:398-403.

Leinonen, I., 1996, A simulation model for the annual frost hardiness and freeze damage of Scots pine, *Ann. Bot.* **78**:687-693.

Lång, V., Mäntylä, E., Welin, B., Sundberg, B., and Palva, E. T., 1994, Alterations in water status, endogenous abscisic acid content, and expression of *rab18* gene during the development of freezing tolerance in *Arabidopsis thaliana*, *Plant Physiol.*. **104**:1341-1349.

Meza-Basso, L., Guarda, P., Rios, D., and Alberdi, M., 1986, Changes in free amino acid content and frost resistance in *Notofagus dombey* leaves, *Phytochemistry* **25**:1843-1846.

Olsen, J. E., Junttila, O., and Moritz, T., 1995, A localised decrease of GA_1 in the elongation zone of *Salix pentandra* seedlings precedes cessation of shoot elongation induced by short photoperiod, *Physiol. Plant.* **95**:627-632.

Olsen, J. E., Junttila, O., Nilsen, J., Martinussen, I., Olsson, O., Sandberg, G., and Thomas, M., 1997, Ectopic expression of phytochrome A in hybrid aspen changes critical daylength for growth and prevents cold acclimation. *Plant J.* **12**:1339-1350.

Phinney, B. O., 1984, Gibberellin A1, dwarfism and the control of shoot elongation in higher plants, in: *The Biosynthesis and Methabolism of Plant Hormones*, A. Crozier and T. R. Hillman, eds., Cambridge University Press, Cambridge, pp. 17-41.

Repo, T., Zhang, G., Ryyppö, A., Rikala, R. and Vuorinen, M., 2000, The relationship between growth cessation and frost hardening in Scots pines of different origins, *Trees* **14**:456-464.

Rinne, P., Welling, A., and Kaikuranta, P., 1998, Onset of freezing tolerance in birch (*Betula pubescens* Ehrh.) involves LEA proteins and osmoregulation and is impaired in an ABA-deficient genotype, *Plant Cell Environ.* **21**:601-611.

Rinne, P.L.H., Kaikuranta, P.L.M., van der Plas, L.H.W., and van der Schoot, C., 1999, Dehydrins in cold-acclimated apices of birch (*Betula pubescens* Ehrh.). production, localization and potential role in rescuing enzyme function during dehydration, *Planta* **209**:377-388.

Schneider, K., Wells, B., Schmelzer, E., Salamini, F., and Bartels, D., 1993, Desiccation leads to the rapid accumulation of both cytosolic and chloroplastic proteins in the resurrection plant *Craterostigma plantagineum* Hochst., *Planta* **189**:120-131.

Tamminen, I., Mäkelä, P., Heino, P., and Palva, E. T.., 2001, Ectopic expression of ABI3 gene enhances freezing tolerance in response to abscisic acid and low temperature in *Arabidopsis thaliana*,. *Plant J.* **25**:1-8.

Thomashow, M. F., 2001, So what's new in the field of plant cold acclimation? Lots, *Plant Physiol*. **125**:89-93.

Tignor, M. E., Davies, F. S., and Sherman, W. B., 1997, Rapid freeze acclimation of *Poncirus trifoliata* seedlings exposed to 10°C and long days, *HortScience*. **32**:854-857.

Welling, A., Kaikuranta, P., and Rinne, P. (1997). Photoperiodic induction of dormancy and freezing tolerance in *Betula pubescens*. Involvement of ABA and dehydrins, *Physiol. Plant.* **100**:119-125.

Wisniewski, M., Close, T.J., Artlip, T., and Arora, R., 1996, Seasonal patterns of dehydrins and 70-kDa heat-shock proteins in bark tissues of eight species of woody plants. *Physiol. Plant.* **96**:496-505.

Wisniewski, M., Webb, R., Balsamo, R., Close, T.J., Yu, X.-M., and Griffith, M., 1999, Purification, immunolocalization, cryoprotective, and antifreeze activity of PCA60: a dehydrin from peach (*Prunus persica*), *Physiol. Plant.* **105**:600-608.

Weiser, C. J. 1970. Cold resistance and injury in woody plants, *Science* **169**:1269-1278.

Zeevaart, J. D. A., and Gage, D. A., 1993, Ent-kaurene biosynthesis is enhanced by long-day in long-day plants Spinacia oleracia L. and Agrostemma githago L. Plant Physiol. 101:25-29.

COLD ACCLIMATION IN RHODODENDRON
A genetic and physiological study

Rajeev Arora*

1. INTRODUCTION

Woody landscape plants residing in temperate zone experience adverse winter conditions during their annual cycle and, therefore, must develop sufficient cold hardiness in the fall of each year (cold acclimation) in order to prepare for over-wintering and surviving low mid-winter temperatures. Lack of adequate cold acclimation (CA), a genetically determined trait, often limits where many landscape plants can be grown successfully. Besides affecting their geographic distribution, low temperatures often cause winter-injury to many landscape plantings thereby causing substantial economic losses. To be able to expand the range of landscape plants that could be grown in areas with cold winters, one must first acquire a basic understanding of cold acclimation process, however, few attempts have been made to study the genetic, physiological or molecular mechanism(s) of CA in woody perennials.

Woody plants have several physiological traits that complicate cold hardiness (CH) and cold acclimation research. For example, a juvenile period in woody perennials raises the possibility of differences in freeze-tolerance (FT) and cold acclimation ability between the juvenile vs. mature (flowering) phases of development (Lim et al., 1998 b). In addition, tissues within an over-wintering plant can exhibit different FT mechanisms — for example, supercooling in xylem parenchyma and bud tissues in contrast to equilibrium freezing in leaf and bark tissues (Wisniewski and Arora, 1993). Furthermore, some woody plant tissues (e.g. buds) undergo dormancy and CA transitions simultaneously, making it difficult to associate physiological changes with one or the other phenological event (Arora et al. 1992).

*Rajeev Arora, Department of Horticulture, Iowa State University, Ames, Iowa 50011.

Plant Cold Hardiness, edited by Li and Palva
Kluwer Academic/Plenum Publishers, 2002

2. USEFULNESS OF RHODODENDRON IN COLD HARDINESS RESEARCH

We are using species of *Rhododendron* to study CA and FT in a woody plant. This genus has a number of attributes that make it amenable to cold hardiness research. There are over 800 species of *Rhododendron* distributed throughout the Northern Hemisphere, ranging from tropical to polar climates and varying widely in FT. Sakai et al. (1986) found that many species in the *Ponticum* subsection, such as *R. maximum* and *R. brachycarpum* are leaf-hardy to –60 °C and bud-hardy to –30 °C, whereas cold-tender species such as *R. griersonianum* and *R. barbatum* showed both leaf and bud damage at temperatures approaching –18 °C® hardiness correlated with provenance (altitude and latitude), suggesting that the trait has evolved through natural selection on existing genetic variability. Wide range of FT among species and relative ease of cross hybridization in this genus makes it possible the generation of progenies segregating for CH to be used in genetic studies. Moreover, availability of 'super-hardy' species in this genus allows one to study physiological and genetic control of "extreme freezing tolerance" displayed by certain plants.

Some *Rhododendron* species, notably those in section *Ponticum*, are also broad-leaved evergreens and allow the use of leaves, year-round, for CH studies. By using leaves to estimate FT, the problem of dormancy transitions in buds is avoided. In addition, FT is conferred without supercooling in *Rhododendron* leaves (Sakai et al. 1986), enabling the use of freeze-thaw experiments and ion-leakage assays for the determination of leaf freeze-tolerance (LFT). We have previously demonstrated that this method of estimating FT provided LFT rankings in an array of evergreen *Rhododendron* cultivars that were consistent with USDA hardiness zone rankings (Lim et al. 1998 a). Advantages of using leaves notwithstanding, it may be argued, however, that since the nursery industry in colder climates desires new *Rhododendron* hybrids and cultivars with improved bud-hardiness, use of leaves (as opposed to buds) may not be appropriate for cold hardiness research. In this regard, the data on the relationship between leaf and bud hardiness of about 35 *Rhododendron* species indicate that the estimates of FT based on leaf performance can be used as a general measure of flower bud hardiness (Fig. 1), although cold-hardened leaf tissues are typically hardier than floral buds®maximum hardiness of the latter being limited by the supercooling of bud primordia.

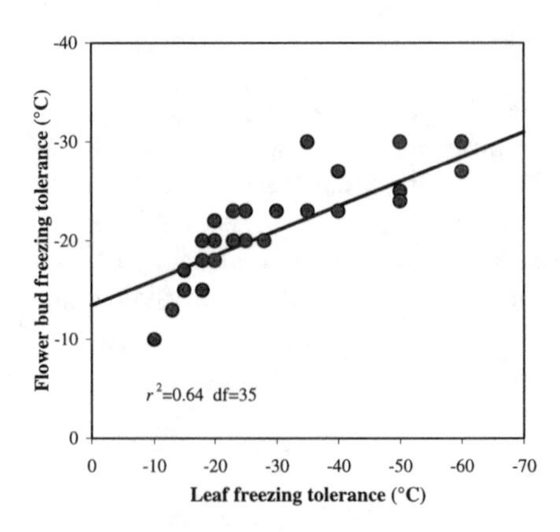

Figure 1. Regressions of floral bud freeze-tolerance on leaf freeze-tolerance (LFT) in *Rhododendron* species accessions (37 plants total); regression analysis of data presented by Sakai et al. 1986. (*From Lim et al., 1998 b*)

3. GENETIC AND PHYSIOLOGICAL STUDY OF COLD ACCLIMATION

The primary goal of most rhododendron breeders working in cold climates has been to combine the ornamental diversity found in less hardy Asian taxa with the super cold hardiness characteristic of several North American and Asian species. This goal has been achieved in many instances, but the process has often been slow and inefficient. Part of the reason for this has been methodological. The conventional approach of pedigree breeding and single site evaluation is not an optimal method for breeding a multigenic and physiologically complex trait. Breeding efficiency has been improved in programs where mass selection and multiple site testing are used to increase cold hardiness (Uosukainen and Tigerstedt, 1988).

A limited understanding of genes controlling freeze tolerance has also hampered efforts in CH breeding. It would be very useful, for example, to be able to predict the outcomes of cold hardy *x* cold tender crosses and to have a sense of how much variability in hardiness occurs among progeny. Prediction is important because it influences population size, number of generations, and breeding strategy needed to arrive at the desired trait. We know, anecdotally, that segregation for the trait exists among advanced generation progeny (e.g. cultivar grexes in which siblings differ by few °C in bud hardiness) but the full variation in the discarded populations has been largely ignored. The efficiency of breeding programs is further restricted by the difficulty in obtaining early and reliable estimates of a plant's cold hardiness. Currently, seedlings are grown to maturity in order to assess winter damage to floral buds. This is a slow (5+ years) and, sometimes, an unreliable process – bud damage on a single plant can often range from 0 to 100% in one season (Gilkey, 1996). Year to year variation in field estimates of CH can also occur due to uncontrollable factors such as snow cover, wind, fluctuating temperatures, and desiccation. Horticultural evaluations would benefit from having a more accurate and consistent method of measuring CH, and breeding efficiency would improve dramatically by using a seedling freeze test that predicts mature plant performance. In addition, identification of a genetic marker, such as, a protein or the gene encoding it, that is associated with cold hardiness would clearly benefit breeders in their screenings and improve our basic understanding of cold acclimation in *Rhododendron.*

We addressed several of these issues in our research program which centered on developing a laboratory protocol that simulates natural freeze-thaw stress under controlled conditions, and using this method to assess CH variation, using electrolyte leakage in segregating progeny arrays derived from a super cold-hardy (*R. catawbiense*) x moderately cold-hardy (*R. fortunei*) cross. The progeny distributions were then used to understand genetic control of CH in *Rhododendron*— number of genes that control FT and the type of gene action involved. Leaf tissues of these progenies were also used to determine whether the presence and / or quantitative accumulation of a plant stress protein, "dehydrin" could serve as genetic marker for cold hardiness in *Rhododendron.*

4. GENETIC STUDY

We selfed a single F_1 plant, 'Ceylon' (*R. catawbiense x R. fortunei*), to create an F_2 population. In addition, two reciprocal backcrosses were made; *R. fortunei x 'Ceylon'* and *'Ceylon' x R. fortunei.* Nonacclimated and cold acclimated plants were evaluated for LFT by subjecting leaf-discs to laboratory based controlled freeze-thaw protocol and determining freeze-injury using electrolyte leakage assay (Lim et al. 1998 a). Much of the experimental variability commonly associated with field trials is significantly reduced by the controlled freeze method and the somewhat subjective method of estimating CH by visual assessment of leaf injury is replaced with a more quantitative procedure (ion-leakage). Ion leakage over a range of freezing temperatures typically displayed a sigmoidal response in *Rhododendron* leaves (Fig 2). A number of statistical procedures have been used to fit this response curve to known arithmetic functions to help determine the rates of freezing injury (Lim et al. 1998 a). For this study, we fitted percent-injury data to a Gompertz function and used the parameter T_{max}, defined as the temperature causing the maximum rate of injury, as an estimator of CH (Fig. 2). Our adaptation of

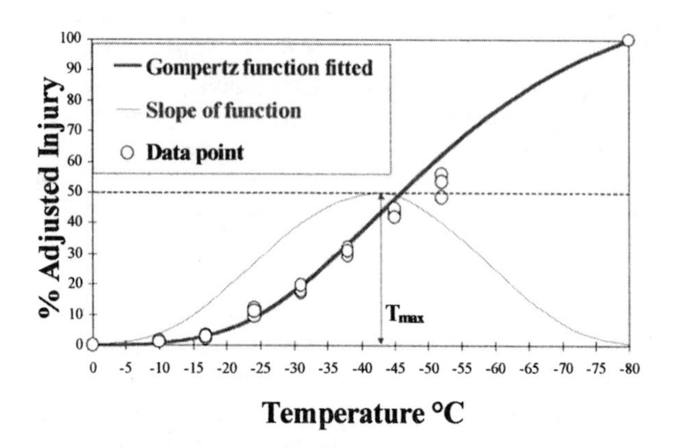

Figure 2. Percent-injury as a function of freezing temperatures in *R.* 'Ceylon' leaf tissues. Data are fitted to a Gompertz function where the slope of the curve (d Ad Injury/d Temp) is the rate of injury. Tmax, inflection point of the slope, is the temperature causing maximum injury rate and defined as leaf freezing tolerance (*From Lim et al., 1998 b*)

this method to rhododendrons — the determination of a diagnostic range of treatment temperatures and the statistical curve fitting of percent injury data in order to estimate Tmax—appears to be reliable, correlating well (r=0.79) with visual estimates of freeze injury based on leaf necrosis and browning (Lim *et al.*, 1998 a).

4.1. Gene Action of Cold Hardiness in *Rhododendron*

Our data (Table 1) showed significant differences in parental LFTs – *R. catawbiense* had a Tmax of –52 °C and the Tmax of *R. fortunei* was –31 °C. The hybrid 'Ceylon' was intermediate, leaf hardy to about –43 °C. These measurements of the parents were made on clones 30-40 years old. LFTs of 47 F_2 and 62 each of the two BC progenies were evaluated as 3-year-old seedlings maintained in outdoor ground frames (Table 1). The F_2 and BC populations of rhododendrons exhibited continuous and normal (bell-shaped) distributions of progeny CH values (Fig. 3). A continuous pattern, as opposed to discrete groupings of hardiness types, suggests that genetic control at this level of freezing tolerance is multigenic, an interpretation consistent with reports from other genera (Pellett, 1998). In order to make inferences about the type of gene action involved

Table 1. Leaf freeze-tolerance in various *Rhododendron* populations (*From Lim et al., 1998 b*)

Parents	**N** [z]	**T_{max}(°C) ± SE** [y]
R. catawbiense	1	-51.2 ± 0.2[a]
R. fortunei	1	-32.4 ± 0.9[c]
R. 'Ceylon', F_1	1	-43.2 ± 1.3[b]
Populations		
BC_F progeny, *R. fortunei* x *R.* 'Ceylon'	62	-20.3 ± 0.6[e]
BC_C progeny, *R.* 'Ceylon' x *R. fortunei*	62	-15.1 ± 0.4[f]
F_2 progeny, *R.* 'Ceylon' selfed	47	-27.6 ± 0.7[d]
Physiological- aged		
R. maximum, juvenile seedlings	7	-38.5 ± 2.0
R. maximum, mature plants	3	-50.2 ± 0.2

[z]N= number of plants evaluated. Three replicate discs were measured at each temperature. Parents and F_1 were included as standards in all population screenings. [y]Mean separation in column by multiple *t* test, $P = 0.05$.

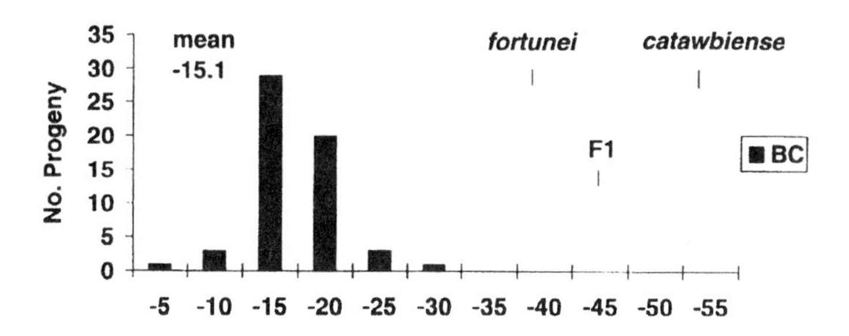

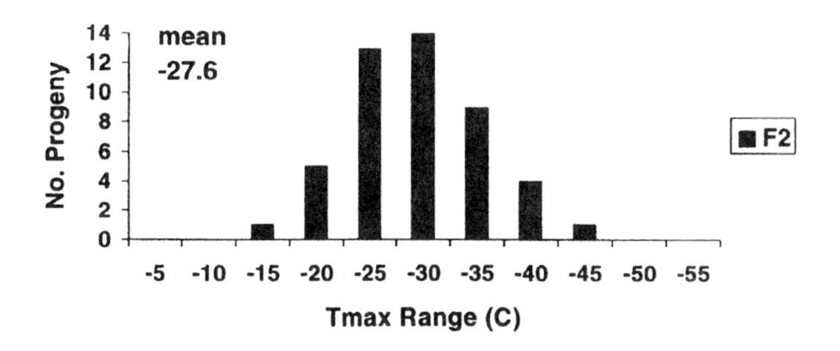

Figure 3. Segregation for leaf freezing tolerance in backcross and F_2 populations derived from a *R. catawbiense* x *R. fortunei* cross.

(dominant, recessive, or additive), geneticists often measure parent offspring resemblance. If the average progeny resembles one parent more strongly than the other, then dominant or recessive gene action is implied, depending on which parent has the greater effect on offspring. If most offspring are intermediate in character to both parents, then the type of gene action is termed 'additive', referring to genes with incremental positive or negative effects on a metric trait such as CH.

4.1.1. Age-dependent LFTs of mature parents and juvenile progenies confound parent-offspring comparisons

A major difficulty in our work results from the determination that parent-offspring comparisons are confounded by age-dependent freezing tolerances. For example, the freeze-tolerance values for the BC and F_2 three-year-old seedling populations were substantially lower than those of their 30 to 40-year-old parents (Table 1), indicating that the majority of the progeny from a cold-hardy by cold-tender cross were more sensitive (or less hardy) than the tender parent. This is an unexpected result, and we sought to explain it by looking at the effect of physiological age on CH. To this end, we conducted a two-year study of different-aged seedlings from a wild *R. maximum* population and found that rhododendrons can gain about 10°C or more in freezing tolerance as they age physiologically from the pre-flowering juvenile stage (1 to ~4 years) to maturity (Table 1). Therefore, inferences on gene action in our study are best made on comparisons of averages of CH values from BC and F_2 populations consisting of similar-aged juvenile plants. The average freeze-tolerance of reciprocal backcrosses ('Ceylon' crossed with *R. fortunei*) was 26 and 45% lower than the F_2 values (Table 1). The nuclear genetic background of the F_2 is 50:50 (*fortunei:catawbiense*) whereas in both BCs it is approximately75:25 (*fortunei:catawbiense*). A significant additive component to CH genes in these populations is inferred from the reasonably close correspondence between the 25 - 45% reduction in BC mean FT and the 50% reduction in genetic contribution from the hardier *R. catawbiense* parent.

Another method of comparing parent-offspring performance in these populations is to predict the 'mature' F_2 or BC hardiness distributions by adding a 'correction factor' of ~10°C (the gain in freezing tolerance from juvenility to maturity as evidenced in different-aged *R. maximum* populations). For example, the predicted mature F_2 CH distribution would range from -25°C to -51°C (instead of observed range of -15°C to -41°C), well within the temperature interval defined by the parent species. By this method, there could be at least one segregant in this population of 47 plants that recovers the hardiness level of the *R. catawbiense* parent as evident form CH distributions. Given the probability that $(1/4)^n$ F_2 segregants will resemble either parent value (where n = number of gene pairs) and assuming additive gene effects, it is possible that as few as 3 genes are responsible for the phenotypic variation observed in these populations $[(1/47) \approx (1/4)^3]$.

4.2. Nonacclimated CH and Cold Acclimation Ability are Independently Controlled

Midwinter CH status is really the outcome of two properties in an individual plant – its freeze tolerance in the non-acclimated state and its acclimating ability. The cold acclimation process is triggered by environmental cues such as shorter day length and cooler temperatures beginning in late summer and extending into late fall (Fuchigami et al., 1971), and acclimating plants undergo physiological and structural changes that condition them to low midwinter temperatures.

The acclimating ability of a plant in our research was defined as the LFT in the cold-acclimated condition minus the LFT in the non-acclimated (NA) condition. As expected, NA LFTs from summer-collected progenies in BC and F_2 populations were much lower than CA LFTs determined for the same individuals during the previous winter. There was, however, significant variation in NA LFTs (-2 to -9 C in the F_2), and we were interested in determining whether these freeze tolerance differences in the NA state had an impact on a plant's midwinter hardiness level. Scatter-plots of these CH components

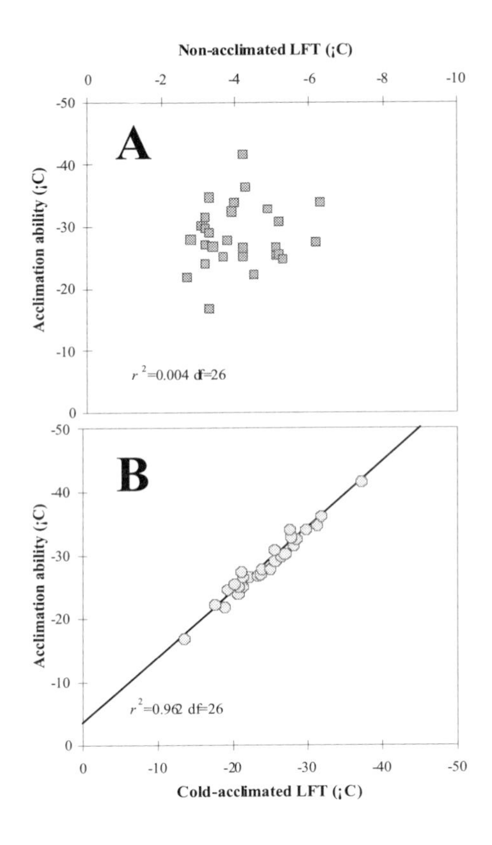

Figure 4. Relationships between components of freezing tolerance in F_2 progenies. (A) Non-acclimated LFT vs. Cold acclimation ability. (B) Cold-acclimated LFT vs. Cold acclimation ability. (*From Lim et al., 1998 b*)

indicate that NA-LFT is not correlated with cold acclimation ability (Fig. 4A), whereas CA-LFT was highly correlated with acclimation ability (Fig. 4B). These results suggest that the physiological processes involved in NA hardiness and acclimation ability are independent of each other and that these two components of mid-winter hardiness are under independent genetic control, a conclusion also drawn from research on potatoes (Stone *et al.*, 1993). The relationships also indicate that, of the two components, it is the variation in acclimating ability (–13.6 to –37.3 C in the F2 progenies) rather than NA hardiness that accounts for the observed differences in Tmax in rhododendrons. This is the meaningful variation from a CH breeding perspective, and from a more basic point of view, it is quite possible that our fairly simple genetic model based on Tmax segregation involves a few genes with major effects on acclimation ability.

5. FREEZING TOLERANCE OF JUVENILE SEEDLINGS AS A PREDICTOR OF FIELD PERFORMANCE BY MATURE PLANTS?

Although floral bud survival is conventionally used to describe CH in rhododendrons and certain other ornamental plants, the use of leaves for the same purpose is more desirable in research and breeding programs, because one can assess juvenile populations for CH without waiting 5+ years for flowering to occur. The applicability of this method to 'real life' conditions involves two factors – 1) how well leaf CH assay predict floral bud hardiness and 2) the degree to which freezing tolerance of juvenile seedlings predicts mature plant performance. As mentioned earlier, evidence suggests that, for at least *Rhododendron*, estimates of CH based on leaf performance can be used as a general measure of flower bud hardiness (Fig 1). A regression analysis of data presented by Sakai et al. (1986), indicated that leaf and floral bud hardiness in *Rhododendron* species were significantly correlated ($r^2 = 0.64$) (Lim et al., 1998 b).

Surveys at the juvenile seedling stage will probably underestimate LFT levels determined at physiological maturity (flowering). As we indicated above, juvenile plants in our experiments have displayed significantly lower LFT than did mature plants. The ability to predict mature plant performance by ranking cold hardiness in juvenile *Rhododendron* populations based on T_{max} distributions would be of great value in breeding programs, and therefore, it is important to know whether or not relative LFT rankings *per se* among genotypes change with physiological age.

We investigated the yearly LFT changes in *Rhododendron* seedling populations and examined the relationship between LFT and aging. Data from one of the backcross (n=20) populations derived from *R. catawbiense* and *R. fortunei* parents indicated a significant increase in mean LFT of ~4-6 °C/year as the seedlings aged from 2-to 5-years old (unpublished results). By the 5th year, most progenies flowered and essentially no further increase in mean LFT was noted during the 6th year. We also observed that although mean LFT of BC populations increased yearly, the annual increment of cold hardiness in juvenile progenies was not always uniform. In other words, there were varying degrees of LFT increment in individual progenies from year to year; some exhibited high while others exhibited low yearly increase in CH, irrespective of their relatively "tender" or "hardy" rankings (unpublished results). LFT of mature/ parental plants (*R. catawbiense*, *R. fortunei*, *R.* 'Ceylon' and *R. maximum*), on the other hand, did not change significantly over 5 years of CH evaluation (Lim et al. 1998b, Lim et al. 1999). These observations clearly indicate that non-flowering, juvenile rhododendron seedlings incrementally gain cold hardiness with physiological ageing. But, the absolute level of CH of a mature rhododendron plant in the landscape cannot be predicted by the CH status at its juvenile stage. Moreover, data on variable, yearly fluctuations in the CH of individual progenies was also somewhat puzzling. Future studies conducted under controlled environmental regimes may reveal whether these fluctuations stem from genetic differences or due to microclimatic variations to which these plants may have been exposed.

6. PHYSIOLOGICAL STUDY

6.1. Dehydrins and Cold Acclimation

Dehydrins (also known as group 2 LEA family of proteins) are hydrophilic, heat-stable proteins that are induced in response to plant stresses possessing a dehydration component, such as salt, water, or freezing stress (Close 1996). Accumulation of dehydrin protein and transcripts during CA has been amply documented in a number of herbaceous species (Close 1997), where FT of cold-acclimated tissues typically does not exceed –15 °C. Investigations of dehydrin expression and its association with CA in woody perennials, which exhibit significantly higher freezing tolerance in cold acclimated state (up to –50 °C or lower) than herbaceous plants, are comparatively scarce. Thus far, studies of woody perennials have used deciduous species to document CA and dehydrin profiles in over wintering tissues such as xylem, bark, and floral buds (Arora and Wisniewski 1994; Muthalif and Rowland 1994; Salzman et al. 1996; Artlip et al. 1997; Rinne et al., 1998). The results of these studies are in general agreement with findings from herbaceous plants, indicating that dehydrins accumulate during the acclimation period.

A functional role for dehydrins in freezing-tolerance (FT) is suggested, in part, by their *in vitro* cryoprotectant properties (Lin and Thomashow 1992; Wisniewski et al. 1999). A direct relationship between dehydrins and FT was demonstrated by the ability of constitutively regulated *cor* proteins (some of which are dehydrins) in *Arabidopsis* to confer FT without prior acclimation (Jaglo-Ottosen et al. 1998). It has also been postulated that dehydrins may act as ion-sequesters (Palva and Heino 1998) or as molecular chaperones (Campbell and Close 1997; Close 1997) under stressful conditions — stabilizing proteins and membranes.

We conducted experiments to determine whether dehydrin profiles differ qualitatively and quantitatively among progeny from a population segregating for CH. This line of inquiry was derived from earlier work (*see above*), which documented that F_2 segregants from a cross between moderately cold-hardy and super cold-hardy *Rhododendron* species varied from -18 to -48°C in LFT (Lim et al. 1998 b). Remnant leaf tissue from progeny already evaluated for LFT was used for dehydrin analysis — in order to determine the relationship between biochemical phenotypes (dehydrin accumulation) and CH phenotypes in the F_2 population. Additionally, age-dependent CH in rhododendrons and parallel changes in dehydrin profiles were studied by comparing juvenile and mature (flowering) plants in natural populations of *R. maximum*.

6.2. A 25 kD Dehydrin as Genetic Marker for Freeze-Tolerance in *Rhododendron*

Comparisons of LFT and dehydrin profiles among F_2 segregants included a super cold-hardy parent (*R. catawbiense*, ~40-year-old), a moderately-hardy parent (*R. fortunei*, ~40-year-old), the F_1 hybrid cultivar 'Ceylon' (~30-year-old) derived from the cross *R. catawbiense* x *R. fortunei*, and F_2 seedlings resulting from the self-pollination of

'Ceylon'. Remnant leaf tissue (left over after LFT study) from selected progeny was used for dehydrin analysis. Initially, the progeny were classified by hardiness phenotype into "low", "medium", and "high" LFT groups. Each group differed significantly from the adjacent group by a mean leaf T_{max} of ~10°C (Table 2). Leaves from 5 random progeny in each group were pooled equally on a fresh weight basis, and the three phenotypic "bulks" were extracted and analyzed for differences in dehydrin profiles. Once bulk differences in dehydrin expression were evident, leaves from 3 individuals in each F_2 bulk were extracted and evaluated separately.

Coomassie-stained SDS-PAGE profiles of total proteins and their anti-dehydrin immunoblots for the two parents, F_1 and F_2 "bulks" are presented in Fig. 5 A & B. The highest optical density (O.D.) value was derived from a 25 kD dehydrin (through densitometry scans of immunoblots) that was present in the super-hardy *R. catawbiense* parent, absent in an equal loading of protein from the moderately-hardy *R. fortunei* parent, and present at intermediate levels in the F_1 hybrid 'Ceylon', which also displayed an intermediate LFT (Table 2).

Table 2. Leaf freezing-tolerance (T_{max}) and corresponding levels of a 25kD dehydrin in *Rhododendron* population segregating for mid-winter leaf freezing tolerance. *(Modified from Lim et al., 1999)*

Plant Groups	N [z]	Mean T_{max} (°C) ± S.E. [y]	Mean O.D. of 25 kD dehydrin ± S.E. [x]
Parents			
R. catawbiense, P_1	1	-52.0±1.3 [a]	0.95±0.03 [a]
R. fortunei, P_2	1	-31.4±1.5 [i]	0.00±0.00 [r]
R. 'Ceylon', P_1 x P_2, F_1	1	-43.4±3.4 [cdefgh]	0.63±0.04 [bcd]
Bulked & individual F_2 progenies			
F_2-Low group	5	-20.5±1.0 [k]	0.40±0.05 [eghikmnq]
F_2-Medium group	5	-32.1±0.2 [i]	0.57±0.04 [cfg]
F_2-High group	5	-43.7±1.7 [defg]	0.76±0.04 [b]
L_1	1	-21.3±0.5 [k]	0.00±0.00 [r]
L_2	1	-22.5±0.7 [jk]	0.26±0.07 [hikmnpq]
L_3	1	-22.5±0.8 [jk]	0.00±0.00 [r]
M_1	1	-31.6±0.7 [I]	0.37±0.08 [deghikmnpq]
M_2	1	-31.8±1.2 [i]	0.00±0.00 [r]
M_3	1	-32.6±1.9 [i]	0.58±0.03 [ce]
H_1	1	-39.3±1.2 [h]	0.32±0.02 [lm]
H_2	1	-40.9±1.3 [efgh]	0.75±0.03 [b]
H_3	1	-47.1±1.7 [bd]	0.45±0.01 [fh]

[z] N = number of plants. Three replicate discs were measured at each treatment temperature. [y] Estimated by using Gompertz function fitted to % adjusted injury data, mean & std. errors estimated by replicates (24 leaf discs) using the Jackknife method (Lim et al., 1998a). Mean separation in column by multiple *t*-test, significant at $P < 0.05$. [x] Using Optimas System, mean & std. errors estimated from 3 separate gray values assigned. Mean separation in column by multiple *t*-test, significant at $P < 0.05$

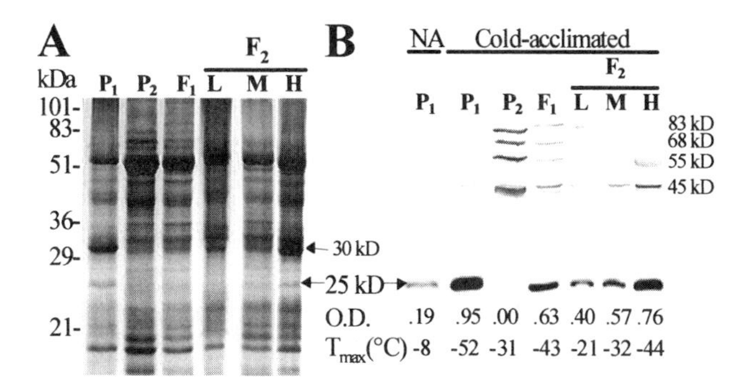

Figure 5. (A) SDS-PAGE profiles of total soluble proteins from cold-acclimated leaves (30 µg per lane). P_1 = *R. catawbiense*, P_2 = *R. fortunei*, F_1 = *R. catawbiense* x *R. fortunei* = 'Ceylon', F_2 = 'Ceylon' selfed. L, M, and H correspond to the "low", "medium", and "high" freeze-tolerant F_2 bulks, (B) Anti-dehydrin immunoblots of parents, F_1 and bulks of F_2 progenies. Protein (15 µg) was loaded in each lane. NA = non-acclimated. O.D. = optical densities. T_{max} = measure of leaf freezing-tolerance. *(From Lim et al., 1999)*

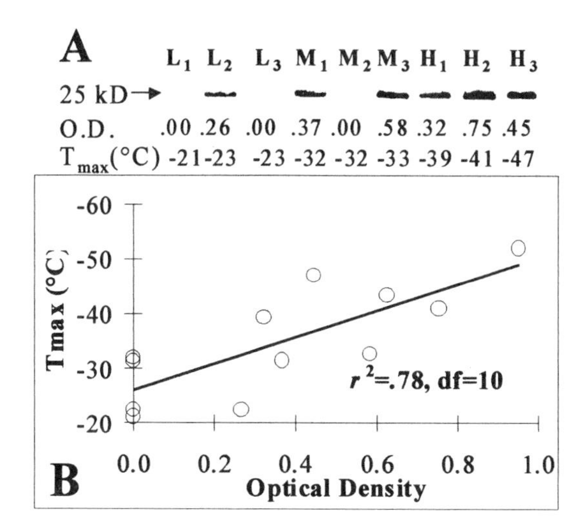

Figure 6. (A) Anti-dehydrin immunoblots of nine individual F_2 progenies. Protein (15 µg) was loaded in each lane, (B) Regression analysis of LFT on dehydrin O.D. in the population comprised of parents, F_1, and 9 F_2 progenies. O.D. = optical densities. T_{max} = quantitative measure of leaf freezing-tolerance. *(From Lim et al., 1999)*

Levels of the 25 kD dehydrin in cold-acclimated leaves from *R. catawbiense* were about 5-fold higher than in non-acclimated leaves from the same source, however, the relatively abundant 25 kD dehydrin on Western blots appeared as a faint band on the corresponding protein gel (Figs. 5 A & B). The reason for apparent lower dehydrin intensities on SDS-PAGE is not clear. Among F_2 seedlings, the 25 kD dehydrin was the only protein clearly associated with differences in CH. Comparisons of the F_2 tissue bulks grouped by phenotypic class — "low", "medium", and "high" LFT — indicated a 50 to 100% increase in the 25 kD dehydrin level as the CH status increased (Table 2; Fig. 5B). When individual progeny from these F_2 bulks were evaluated, most but not all of the offspring displayed increased dehydrin accumulation at higher levels of LFT (Table 2).

6.2.1. Genetic interpretation of dehydrin profiles

Because the immunoblots are detecting temperature-induced proteins, the parental differences in dehydrin profiles (presence vs. absence of a specific molecular weight protein) could be attributed to regulatory genes rather than dehydrin-encoding structural genes. For example, *R. catawbiense* and *R. fortunei* could share identical dehydrin genes under the control of regulatory genes, which respond differently to low temperatures. However, studies with cold-responsive regulatory elements (transcriptional factors) suggest that they promote coordinate expression of a suite of cold-regulated genes, some of which also encode dehydrin or dehydrin-like proteins (Jaglo-Ottosen et al. 1998). Moreover, there has been no evidence in the literature, to date, supporting the scenario whereby an identical dehydrin gene common to two cold-acclimating species is differentially induced by cold in only one of them. Therefore, it appeared likely, then, that the dehydrin profiles observed in acclimated leaf-tissue from each parent, in our study, represented the expression of the full set of cold-regulated dehydrins, and that the presence vs. absence (+/–) of the 25 kD dehydrin was due to structural gene differences rather than regulatory genes. Moreover, accumulations of several dehydrins in the F_1 (based on O.D. or visual estimates) were intermediate to parental levels, suggesting a gene dosage effect, which could result from codominant expression of 'presence' and 'absence' alleles at the corresponding loci. Based on above scenario, we had suggested that the presence or absence of 25 kDa dehydrin could serve as genetic marker to distinguish between super-hardy and less hardy *Rhododendron* species (Lim et al, 1999). However, with improved protein extraction protocols developed in our laboratory, recently we were able to detect the 25 kDa dehydrin in the cold acclimated leaves of *R. fortunei* also, although at a significantly lower levels than in *R. catawbiense* (unpublished results). Therefore, it now appears that differential accumulation of 25 kDa dehydrin (and perhaps differential hardiness) in *R.* catawbiense and *R. fortunei* may be due to differential gene regulation of 25 kDa dehydrin rather than structural gene differences as previously thought. Recent work in our laboratory, with several hardy and tender *Rhododendron* species, further indicated that the *relative, cold-inducibility* of 25 kDa dehydrin, rather than its *absolute quantitative accumulation*, was highly correlated with the cold acclimation ability in these species (unpublished results).

In addition to 25 kDa dehydrin, a 30 kD non-dehydrin protein was visualized at lower levels in non-acclimated conditions compared to cold-acclimated conditions. It was observed at higher levels in *R. catawbiense* compared to *R. fortunei*, and appeared to be one of the most abundant proteins in the "hardiest" F_2 bulk. This undetermined protein also displayed a close quantitative association with age-dependent LFT changes in *R. maximum* (a closely related species to *R. catawbiense* with comparable FT). We plan to look more closely at the association of this 30 kD protein with CH and possibly characterize it in the future.

6.3. Effect of Age on Cold Hardiness and Dehydrin Accumulation

Woody perennials typically have lengthy juvenile phases, which terminate upon flowering (maturation). Between these developmental phases, plants undergo morphological changes (Hartmann et al. 1997) as well as genetic, physiological, or biochemical changes (Hackett et al. 1990); a process known as "phase-change" or "maturation" (Brink 1962). However, very little is known about the effect of phase change on adaptive traits such as CH. We observed that, mature plants (~30-year-old) were more cold-hardy than the juvenile seedlings (~3-year-old) by an average of ~ 10°C T_{max} (Table I). Parallel differences were observed in the relative abundance of the 25 kD dehydrin in leaves from juvenile and mature plants, in that, immunoblots of leaf proteins indicated a ~2.4-fold increase in levels of the 25 kD dehydrin in mature plant leaves relative to juvenile plants (Fig. 7). Phase-related, differential expression of proteins (qualitative and quantitative) has been reported for several woody species (Hand et al. 1996 and references within). The present study is the first to report phase-dependent

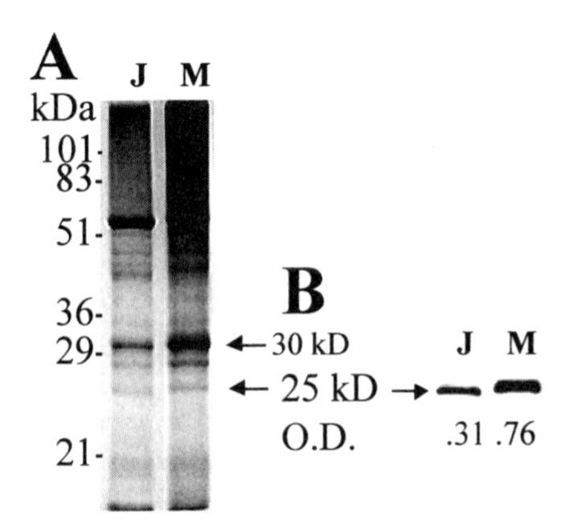

Figure 7. (A) SDS-PAGE profiles of total soluble leaf proteins extracted from cold acclimated *R. maximum* plants. Thirty μg protein was loaded in each lane. (B) Anti-dehydrin immunonblots with 15 μg protein/lane. J = juvenile seedlings (approx. 2 –3 years old); M = mature plants (approx. 30 years old); O.D = optical density. (*From Lim et al., 1999*)

accumulation of a dehydrin and its correlation with CH. The underlying mechanisms for this age-dependent effect are unknown, but may involve regulatory elements (e.g. signal transduction linked with dehydrin expression) or phase changes in nutrient assimilation.

7. ACKNOWLEDGMENTS

This research was supported, in part, by grants from American Rhododendron Society and Horticulture Research Institute.

REFERENCES

Arora, R., Wisniewski, M. E. and Scorza, R., 1992, Cold acclimation in genetically related (sibling) deciduous and evergreen peach (*Prunus persica* [L.] Batsch). I. Seasonal changes in cold hardiness and polypeptides of bark and xylem tissues, *Plant Physiol.* **99**:1562-1568.

Arora, R. and Wisniewski, M. E., 1994, Cold acclimation in genetically related (sibling) deciduous and evergreen peach (*Prunus persica* [L.] Batsch). II. A 60-kilodalton bark protein in cold-acclimated tissues of peach is heat stable and related to the dehydrin family of proteins, *Plant Physiol.* **105**:95-101.

Artlip, T. S., Callahan, A. M. and Bassett, C. L. and Wisniewski, M. E., 1997, Seasonal expression of a dehydrin gene in sibling deciduous and evergreen genotypes of peach (*Prunus persica* [L.] Batsch), *Plant Mol. Biol.* **33**:61-70.

Brink, R. A., 1962, Phase change in higher plants and somatic cell heredity, *Q. Rev. Biol.* **37**:1-22.

Campbell, S. A. and Close, T. J., 1997, Dehydrins: Genes, proteins, and associations with phenotypic traits, *New Phytologist* **137**:61-74.

Close, T. J., 1996, Dehydrins: emergence of a biochemical role of a family of plant dehydration proteins, *Physiol. Plant.* **97**:795-803.

Close, T. J., 1997, Dehydrins: a commonality in the response of plants to dehydration and low temperature, *Physiol. Plant.* **100**:291-296.

Fuchigami, L. H., Weiser, C. J. and Evert, D. R., 1971, Induction of cold acclimation in *Cornus stolonifera* Michx, *Plant Physiol.* **47**: 98-103.

Gilkey, R., 1996, Cold hardiness rankings of rhododendrons by means of flower damage, *J. Amer. Rhodo. Soc.* **50**: 100-102.

Hackett, W. P., Murray, J. R., Woo, H. H., Stapfer, R. E. and Geneve, R., 1990, Cellular, biochemical and molecular characteristics related to maturation and rejuvenation in woody species, *NATO ASI Ser A: Life Sci.* **186**:147-152.

Hand, P., Besford, R. T., Richardson, C. M. and Peppitt, S. D., 1996, Antibodies to phase related proteins in juvenile and mature *Prunus avium*, *Plant Growth Regulat.* **20**:25-29.

Hartmann, H. T., Kester, D. E., Davies, F. T. Jr. and Geneve, R. L., 1997, Plant Propagation: Principles and Practices 6th ed. Prentice Hall, Upper Saddle River, New Jersey.

Jaglo-Ottosen, K. R., Gilmour, S. J., Zarka, D. G., Schabenberger, O. and Thomashow, M. F., 1998, *Arabidopsis* CBF1 overexpression induces *cor* genes and enhances freezing tolerance, *Science* **280**:104-106.

Lim, C. C., Arora, R. and Townsend, E. D., 1998a, Comparing Gompertz and Richards functions to estimate freezing injury in *Rhododendron* using electrolyte leakage, *J. Am. Soc. Hort. Sci.* **123**:246-252.

Lim, C. C., Krebs, S. L. and Arora, R., 1998b, Genetic study of freezing tolerance in *Rhododendron* populations: Implications for cold hardiness breeding, *J. Am. Rhododendron Soc.* **52**:143-148.

Lim, C. C., Krebs, S. L. and Arora, R., 1999, A 25 kD dehydrin associated with with genotype- and age-dependent leaf freezing tolerance in Rhododendron: a genetic marker for cold hardiness? *Theor. Appl. Genet.* **99**: 912-920.

Lin, C. and Thomashow, M. F., 1992, A cold-regulated *Arabidopsis* gene encodes a polypeptide having potent cryoprotective activity, *Biochem. Biophy. Res. Commun.* **183**:1103-1108.

Muthalif M. M. and Rowland, L. J., 1994, Identification of dehydrin-like proteins responsive to chilling in floral buds of blueberry (*Vaccinium*, section *Cyanococcus*), *Plant Physiol.* **104**:1439-1447.

Palva, E. T., Heino, P., 1998, Molecular mechanisms of plant cold acclimation and freezing tolerance, in: *Plant Cold Hardiness: Molecular biology, Biochemistry, and Physiology*, P. H. Li and T. H. H. Chen, eds., Plenum Press, New York, pp. 1-14.

Pellet, H., 1998, Breeding of cold hardy woody landscape plants, in: *Plant Cold Hardiness: Molecular biology, Biochemistry, and Physiology*, P. H. Li and T. H. H. Chen, eds., Plenum Press, New York, pp. 317-324.

Rinne, P., Welling, A. and Kaikuranta, P., 1998, Onset of freezing tolerance in birch (*Betula pubescens* Her.) involves LEA proteins and osmoregulation and is impaired in an ABA-deficient genotype, *Plant Cell Environ.* **21**: 601-611.

Sakai, A., Fuchigami, L. and Weiser, C. J., 1986, Cold hardiness in the genus *Rhododendron, J. Amer. Soc. Hort. Sci.* **111**:273-80.

Salzman, R. A., Bressan, R. A., Hasegawa, P. M., Ashworth, E. N. and Bordelon, B. P., 1996, Programmed accumulation of LEA-like proteins during desiccation and cold acclimation of overwintering grape buds, *Plant Cell Environ.* **19**:713-720.

Stone, J. M., Palta, J. P., Bamberg, J. B., Weiss, L. S. and Harbage, J. F., 1993, Inheritance of freezing resistance in tuber-bearing *Solanum* species: evidence for independent genetic control of nonacclimated freezing tolerance and cold acclimation capacity, *Proc. Natl. Acad. Sci. USA* **90**:7869-7873.

Uosukainen, M. and Tigerstedt, P. M. A., 1988, Breeding of frosthardy rhododendrons, *J. Agric. Sci. Finland* **60**:235-254.

Wisniewski, M. and Arora, R., 1993, Adaptation and response of fruit trees to freezing temperatures, in: *Cytology, Histology and Histochemistry of Fruit Tree Diseases*, A. R. Biggs, ed., CRC Press, Boca Raton, Florida, pp. 299-320

Wisniewski, M., Webb, R., Balsamo, R., Close, T. J., Yu, X.-M. and Griffith, M., 1999, Purification, immunolocalization, cryoprotective, and antifreeze activity of PCA60: a dehydrin from peach (*Prunus persica*), *Physiol. Plant.* **105**:600-608.

8

EARLY ACCLIMATION RESPONSE IN GRAPES (*VITIS*)

Anne Fennell and Katherine Mathiason*

1. INTRODUCTION

Acclimation in response to a decreasing daylength prior to the onset of low temperatures can be critical to the winter survival and productivity of many woody plants. This is particularly true for temperate fruit crops such as grapes. Grapes are native to the northern hemisphere, however they have a very large production range extending from the subtropics to 50 N and S latitude. Acclimation and freezing tolerance characteristics of this high value crop are very important as many production regions incur damaging freezes in fall, winter or spring. Typically production regions are recommended on the basis of the midwinter low temperatures; however, in many North American growing areas, injury during unseasonable low temperatures in the fall when the canes are maturing and acclimating can be as damaging as low midwinter temperatures (Bordelon et al., 1997, Clore et al., 1974, Seyedbagheri and Esmaeil, 1994). Therefore, a better understanding of the mechanisms involved in the onset of acclimation and freezing tolerance in grapes is needed to improve cultivar selection and cultural practices to minimize losses due to freezing injury.

In many temperate woody plants, growth cessation and the induction of dormancy are necessary for acclimation to occur and a decreasing photoperiod is the environmental signal promoting the inductive processes. In grapes, growth cessation is not considered necessary for acclimation; however, some of the more cold hardy genotypes appear to cease growth and acclimate in response to decreasing photoperiods (Fennell and Hoover, 1991, Schnabel and Wample, 1987, Wolpert and Howell, 1986). Therefore we developed a genetic model system to characterize and separate cold acclimation and dormancy responses in grapes. Physiological, biophysical and molecular studies are being used to characterize early acclimation responses and identify traits that can be used in breeding and selecting grapes for northern climates.

---

*Anne Fennell and Katherine Mathiason, Horticulture, Forestry, Landscape and Parks Dept., South Dakota State University, Brookings, SD 57006.

Plant Cold Hardiness, edited by Li and Palva
Kluwer Academic/Plenum Publishers, 2002

2. MATERIALS AND METHODS

2.1. Plant Materials & Environmental Conditions

V. riparia (Manitoba clone), *V.* spp. Seyval Blanc, and an F1 (*V. riparia* x Seyval Blanc) were used for these studies. Three to five-year-old, spur-pruned vines were grown in 15 L containers with media consisting of a 1:2:2 (v/v) mixture of soil, peat and perlite. Three to four shoots (one on each spur) were trained vertically on each plant and all flower clusters were removed (Fennell et al., 1996). The plants were grown under a long photoperiod (PP, 15 h) at 25/20 $\pm$ 3 °C day/night temperatures (D/N) with 600 to 1400 ϑmol∀m^{-2}∀s^{-1} photosynthetic photon flux density in climate-controlled glass greenhouses from May through July in Brookings, South Dakota, USA (42°N). All plants were watered daily and fertilized weekly with a complete nutrient solution.

After 45 days of growth, uniform plants with shoots of 12 or more nodes were randomly assigned to either a long photoperiod (LD, 15 h PP, endodormancy inhibiting) or short photoperiod (SD, 8 h PP, endodormancy inducing) at 25/20 $\pm$ 3 °C D/N. White covered black cloth was used daily to impose SD treatments (Fennell and Line, 2001).

2.2 Physiological/Morphological Characterization

The rate of shoot growth was measured weekly and is presented for 2, 4, and 6 weeks of treatment. Shoot apical meristem (tip) abscission was monitored throughout the study and is expressed as a percentage of shoots measured (n=16) (Wake and Fennell, 2001).

The relative depth of endodormancy was determined in whole plants at 2, 4, and 6 weeks of photoperiod treatment (n=8). Leaves were removed and plants were spur pruned leaving the first 2 basal nodes (Fig. 1) of each shoot (Fennell et al., 1996). The spur-pruned plants were placed in LD growing conditions. Budbreak was monitored for each genotype and photoperiod treatment as previously described (Wake and Fennell, 2000). Number of days to budbreak in the LD-treated vines was used as a basis for determining delay or failure to break bud in SD-treated vines. Delay in budbreak is expressed as the difference between the mean number of days to budbreak in SD-treated plants and the mean number of days to budbreak for LD-treated plants (Fennell et al., 1996). To determine uniformity of bud response, budbreak was also monitored in single node sections of one pruned shoot (Fig. 1, nodes 3 through 12 from the shoot base origin) taken from each of the plants harvested at 2, 4, and 6 weeks of LD or SD treatment. The single-node sections were placed in individual tubes with 5 ml water and then placed in growth chambers with a 15 h PP at 25/20 $\pm$ 1 °C D/N to determine depth of endodormancy by recording the number of days to budbreak.

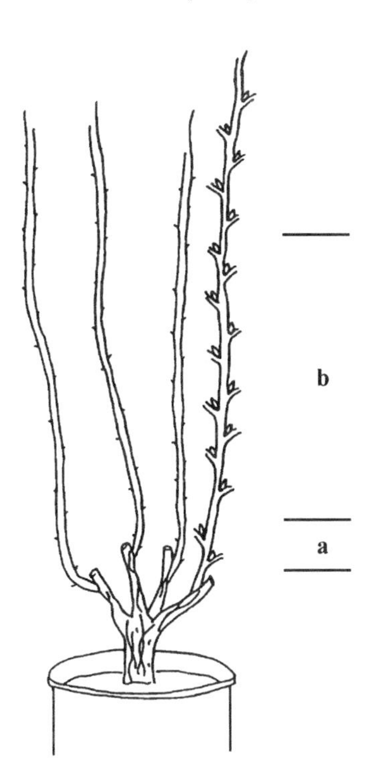

Figure 1. Schematic of treated plant after leaves were removed. a. The two basal nodes remain on the plant for whole plant dormancy determination. b. Nodes 3 to 12 were used for single node dormancy assays or RNA extraction, nodes 3 to 7 were used for freezing tolerance analysis, and node 5 was used for MRI analysis. (Reproduced from Fennell et al. 1996, J. Amer. Soc. Hort. Sci. 121:1112-1116).

Killing temperature of the primary buds was determined for nodes 3 to 7 in five plants for each photoperiod as described by Fennell and Hoover (1991). Single node sections 5.0 cm in length were nucleated and equilibrated at -2°C overnight. The next day samples were frozen at 2°C/h, and samples from each plant removed at 2°C intervals. Samples were thawed at 4°C and placed at 25°C for five days and then examined for tissue browning. Results are presented as mean killing temperature from freezing test conducted in three separate years.

2.3. Biophysical Characterization

Nuclear magnetic resonance imaging (MRI) was used to collect anatomical and biophysical information to determine differential tissue responses during the induction of dormancy and acclimation. MRI analysis was conducted on the fifth node of *V. riparia* of nodes collected at 2, 4 and 6 weeks of LD or SD treatment (Fig. 1). T_2 images and analysis of T_2 relaxation times were used to evaluate whether changes in state of water in different node tissues were related with endodormancy induction or tissue maturation. Node sections were placed in a custom 10mm MRI tube for imaging with a Bruker MSL,

400 MHz (9.4 Tesla) microimaging instrument (Bruker Instr. Co. Bellerica, Mass.), (Fennell and Line, 2001). Sixteen 0.5 mm serial image slices taken perpendicular to the stem axis were made to identify the area containing the center of the primary bud. The slice through the center of the primary bud was used for T_2 imaging and T_2 times were calculated by nonlinear least square fit to the magnitude images after mean filtering as described by Liu et al. (1993) and Parmentier et al. (1998). T_2 times were calculated for bud, gap tissue subjacent to the bud, and the rest of the stem (n=7).

Individual node sections of all photoperiod treatments were also imaged during freezing to determine differential tissue freezing responses during early acclimation. Samples were nucleated with ice chips at -1°C and then frozen at 2°C/h. Samples were imaged at 4°C intervals. Comparison of the T_2 images at each freezing temperature was used to develop a composite image that mapped the freezing temperatures of various tissues (Millard et al., 1995).

2.4. Molecular Characterization

Differential display of mRNA was used to identify and isolate SD specific cDNA fragments as described by Liang and Pardee (1992). RNA was isolated from *V. riparia* buds at 2, 4, and 6 weeks of SD or LD treatment and analyzed by PCR based differential display. Differentially expressed fragments were used to isolate full length cDNAs from *V. riparia*, SD-cDNA libraries for further analysis. A dormancy/early acclimation specific cDNA SD495 was identified and further characterized by northern analysis using bud RNA isolated at 2, 4, and 6 weeks of LD and SD treated *V. riparia*, *V.* spp. Seyval Blanc and the F1 genotype. Southern analysis using genomic DNA used to determine presence and copy number of SD495 in all genotypes.

3. RESULTS

3. 1. Physiological/Morphological Characterization

Shoot growth of LD-treated plants in each genotype decreased slightly but they continued growing at 13 to 15 cm/week throughout the 6 weeks of treatment (Fig. 2). *V. riparia* and F_1 shoot growth significantly decreased after 2 weeks of SD. Shoot growth was minimal in *V. riparia* and F1 by 4 weeks and 6 weeks of SD respectively. Shoot growth in SD-treated Seyval Blanc was not significantly different than LD plants throughout the study. Shoot growth continued through 10 weeks of SD treatment (data not shown).

There was no apical meristem (tip) abscission in the LD plants of all genotypes (data not shown). Shoot tip abscission began at 4 weeks of SD in *V. riparia* and was almost complete by the 6th week of SD (Fig. 3, 4). Tip abscission in the F1 began at 5 weeks and was significantly less than that of *V. riparia* (Fig. 4). No tip abscission occurred in Seyval Blanc.

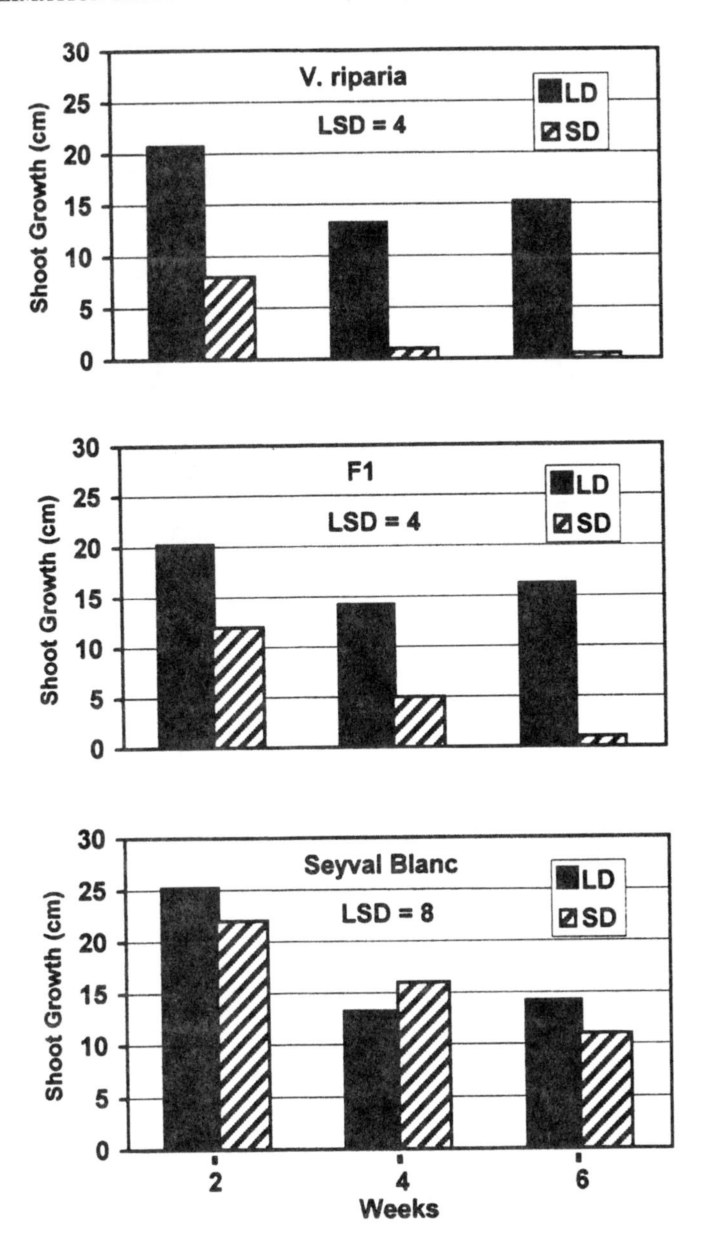

Figure 2. Influence of photoperiod on shoot growth. After 45 days of growth in long photoperiod (LD, 14 h) plants were subjected to LD or short photoperiod (SD, 8 h). Growth rate was determined in *V. riparia*, F1 and Seyval Blanc at 2, 4, and 6 weeks of LD or SD treatment. Student Newman Kuels LSD at $P \leq 0.05$, n=8. (Reproduced with permission from Physiologia Plantarum 109:203-210)

Figure 3. Shoot meristem (tip) abscission in *V. riparia* exposed to SD treatment. Arrow indicates dried abscising shoot meristem.

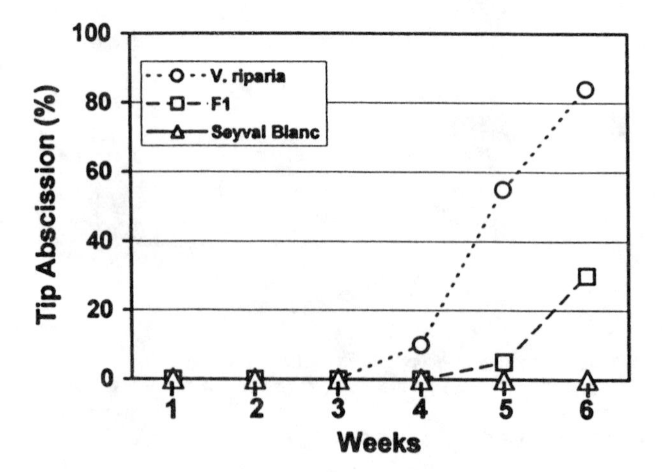

Figure 4. SD induced shoot meristem abscission in *V. riparia*, F1 and Seyval Blanc. Values are expressed as percent of 16 buds monitored. There was no tip abscission in LD treatments. (Reproduced with permission from Physiologia Plantarum 109:203-210).

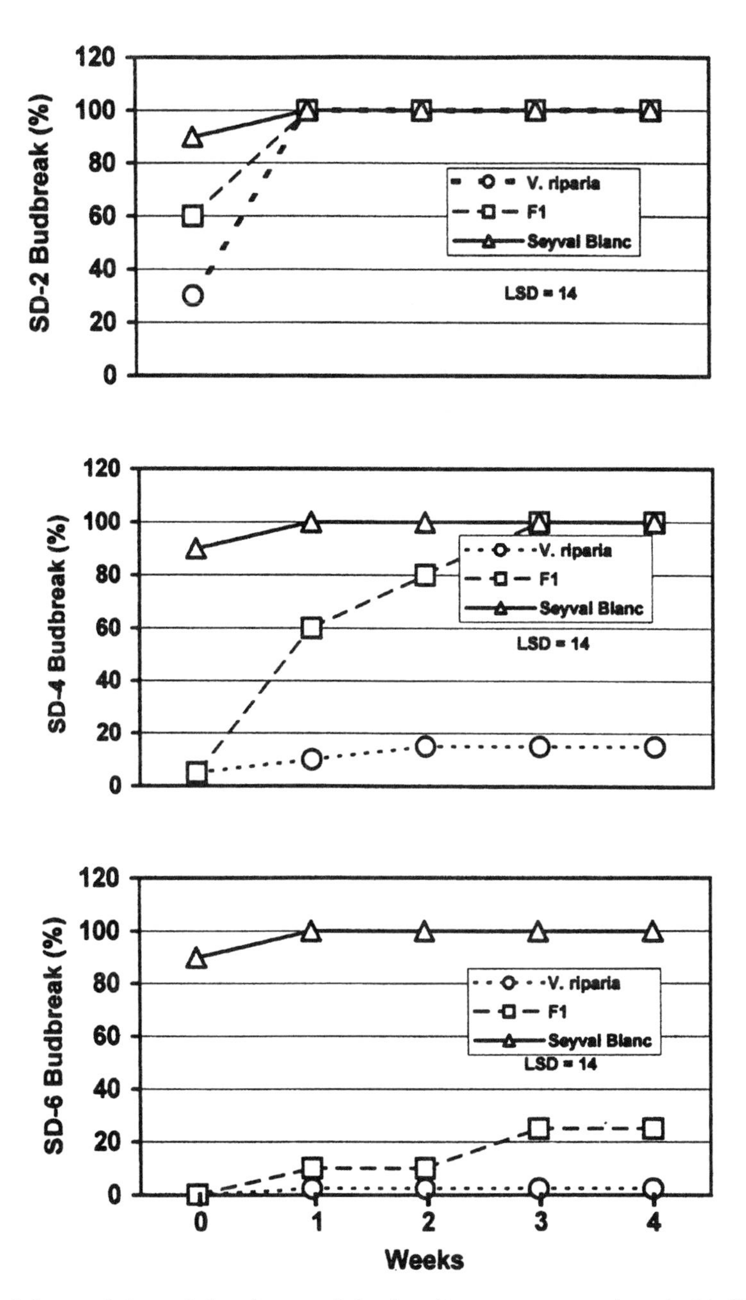

Figure 5. Influence of photoperiod on dormancy inductrion. Dormancy rate was determined in *V. riparia*, F1 and Seyval Blanc at 2, 4, and 6 weeks of LD or SD treatment. Student Newman Kuels LSD at $P \leq 0.05$, n=8. (Reproduced with permission from Physiologia Plantarum 109:203-210)

Figure 6. *V. riparia* whole plant dormancy assay. Plants were spur pruned and placed in LD (15 h photoperiod) after 2, 4, 6 weeks of LD or SD (8h, photoperiod). Photographs were taken 4 weeks after the 6 week plants had been placed in LD to monitor budbreak.

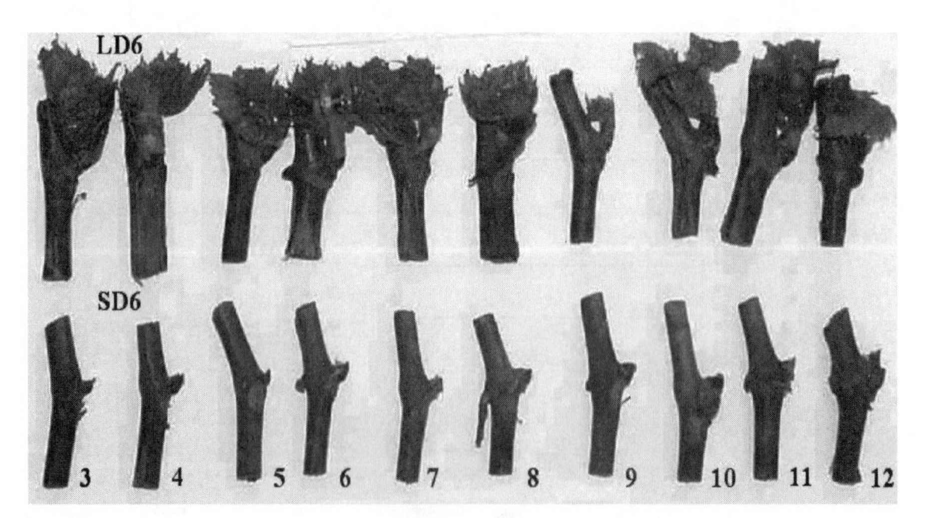

Figure 7. *V. riparia* single node dormancy assay. Nodes 3 to 12 of shoots from 6 week LD and SD treatment. Photograph was taken 4 weeks after placing node sections in LD conditions.

Table 1. Killing temperature of the primary bud in *V. riparia*, Seyval Blanc and F1 after 2, 4, and 6 weeks of LD or SD treatment. Values are means of replicated freezing trials conducted in three separate years.

	Primary Bud Killing Temperature (°C)[a]					
	2 weeks		4 weeks		6 weeks	
	LD	SD	LD	SD	LD	SD
V. riparia	-4.0 d	-4.0 d	-4.6 d	-7.5 b	-5.6 c	-11.0 a
F1	-4.0 d	-4.0 d	-4.0 d	-4.7 d	-4.0 d	-8.0 b
Seyval Blanc	-4.0 d	-4.0 d	-4.2 d	-4.0 d	-4.5 d	-6.0 c

[a]Values are expressed as the mean of 7 replications. Values with different letters are different at P=0.05 level as determined by Student Newman Kuels mean separation.

Bud dormancy did not occur in LD plants of all three genotypes. Bud dormancy was induced but reversible in *V. riparia* after 2 weeks of SD as revealed by the delay in budbreak (Fig. 5). The majority of *V. riparia* plants were dormant after 4 weeks of SD and all plants were dormant after 6 weeks of SD (Fig. 5,6). Bud dormancy was initiated in F1 after 4 weeks of SD, but still reversible as indicated by the 7 to 14 day delay in budbreak relative (Fig. 5). The majority of *V. riparia* plants were dormant after 4 weeks of SD and all plants were dormant after 6 weeks of SD (Fig. 5,6). Bud dormancy was initiated in F1 after 4 weeks of SD, but still reversible as indicated by the 7 to 14 day delay in budbreak relative to the LD plants. Seyval Blanc did not become dormant under SD treatments. Dormancy response was very uniform throughout the shoot in all genotypes and single node (3 to 12) sections exhibited the same bud break response as the whole plants (Fig. 7).

Acclimation is induced by SD treatment in *V. riparia* and correlates with the induction of dormancy (Table 1). Primary bud killing temperature of *V. riparia* begins decreasing with 4 weeks of SD and reaches -11°C during the 6 weeks of SD. There is a small change in freezing tolerance in the LD buds over the six week period, resulting in a 5.5°C greater freezing tolerance in *V. riparia* SD than LD primary buds. Freezing tolerance in the F1 increases 4°C after 6 weeks of SD, with no change in LD freezing tolerance. There is a small (1.5°C) increase in freezing tolerance in Seyval Blanc after 6 weeks of SD and no change in LD bud freezing tolerance. This SD acclimation response corresponds with the timing of dormancy induction in *V. riparia* and the F1 genotype, while the minimal change in freezing tolerance in Seyval Blanc corresponds with the lack of bud dormancy.

3.2. Biophysical Characterization

Changes in the state of water from free to bound have been correlated with dormancy

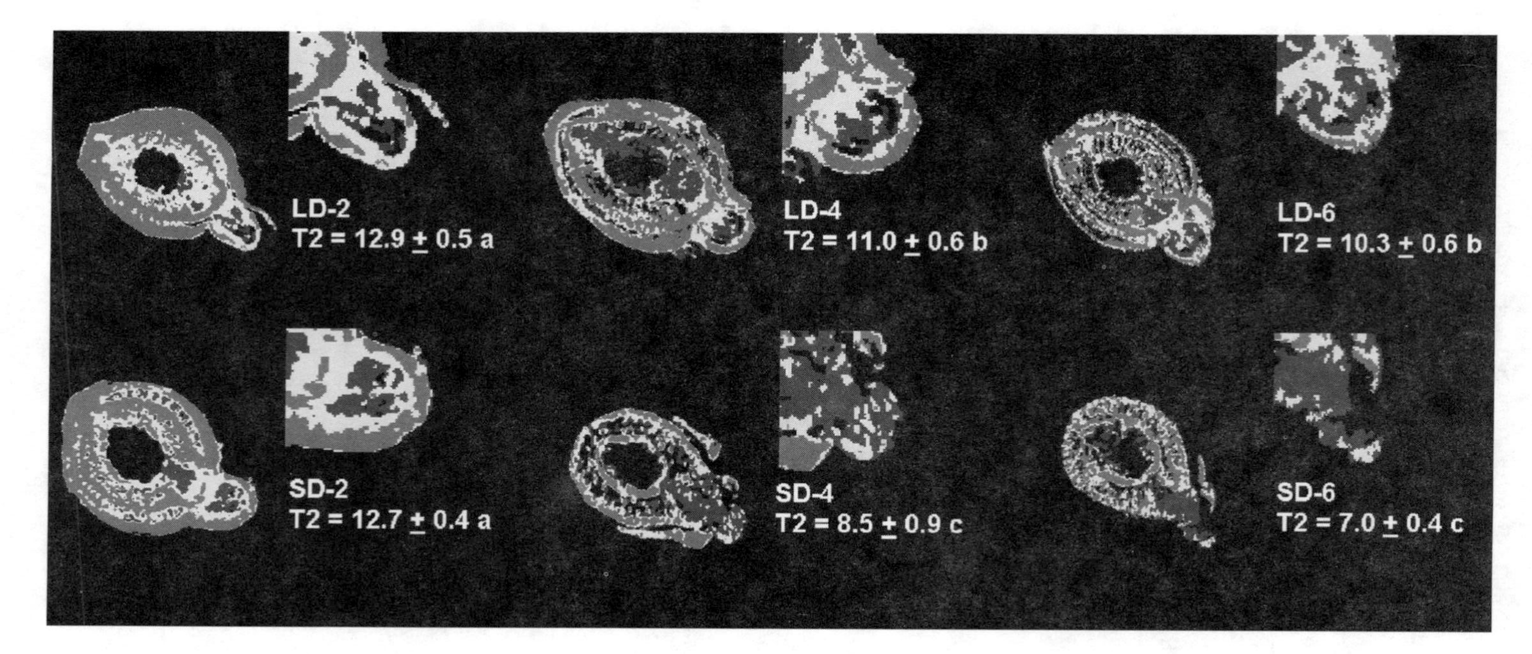

Figure 8. MRI of primary bud at 2, 4, and 6 weeks of long photoperiod (LD) or short photoperiod (SD). T_2 relaxation times are medium gray = $T_2 \geq 16$ ms, white 11 ms $\leq$ 16 ms, dark gray $T_2 < 11$ ms. Mean T_2 (n=7 for each treatment are listed below the representative node section. Values with different letters are significantly different across all daylength and treatment weeks at $P = 0.05$ as determined by SNK mean separation. (Reproduced from Fennell and Line 2001, J. Amer. Soc. Hort. Sci. 126:681-688).

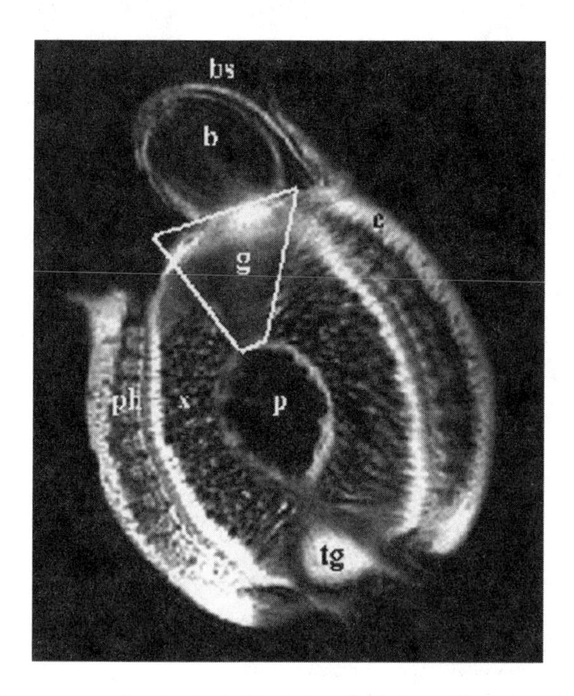

Figure 9. Magnetic resonance showing bud (b), bud scale (bs), periderm (e), phloem (ph), xylem (x), cortex/gap tissue (g) adjacent to the bud, pith (p), and tendril gap (tg). Lines delineate the gap region that was analyzed separately from the rest of the stem.

Table 2. T_2 times for bud, leaf gap and stem tissue at 2, 4 and 6 weeks of long-day (LD) or short-day (SD) photoperiod treatment. (Fennell and Line, 2001, J. Amer. Soc. Hort. Sci. 126:681-688.)

	T_2 (ms)[a]					
	2 weeks		4 weeks		6 weeks	
Tissue	LD	SD	LD	SD	LD	SD
Bud	10.61 cd	10.04 cd	10.24 cd	9.31 cde	8.49 de	5.75 f
Leaf Gap	12.78 b	10.96 c	9.26 cde	7.39 e	10.95 c	5.49 f
Stem	16.13 a	15.62 a	15.95 a	13.60 b	16.66 a	13.24 b

[a]Values are expressed as the mean of 7 replications. Values with different letters are different at P=0.05 level as determined by Student Newman Kuels mean separation.

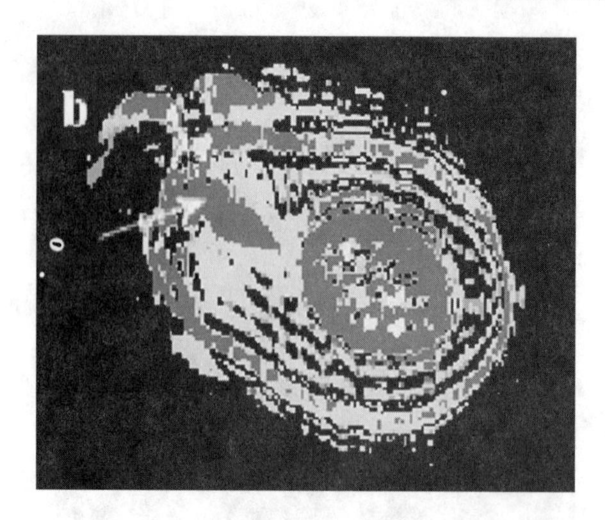

Figure 10. MRI of 4 week SD *V. riparia* bud during freezing. Light gray areas froze between -13 to -17°C and dark gray regions froze between -18 to -22°C, (b = bud, arrow indicates gap tissue subjacent to bud).

in several woody plants and has been proposed as a mechanism controlling dormancy in woody plants (Faust et al. 1991, 1997). The state of water is determined by the relaxation characteristics of protons. Water with restricted motional freedom (associated with macromolecules) has short relaxation times (T_2) and water that is free (unassociated with macromolecules) has longer relaxation times.

The anatomical and biophysical information provided by MRI allows differential tissue responses to be detected during induction of acclimation and dormancy. An increase in bound water for the entire node in response to SD is readily noted in the decreased T_2 times (Fig. 8). There was a slight decrease in T_2 in response to increasing age (LD-6 weeks); however, T_2 times in *V. riparia* were significantly different after 4 weeks of SD. When a gray scale was assigned to the range of T_2 times in the node section, it provided a visual map of differential tissue response. The bud and gap tissues subjacent to the bud appeared to have more bound water than the rest of the stem. When average T_2 times were calculated for separate portions of the node (bud, gap tissue subjacent to the bud, and the conducting portion of the stem), the differences in tissue response became more apparent (Fig. 9). Analysis of the separate areas indicated that there was a decrease in T_2 in the gap region after 2 weeks of SD and it continued to decrease throughout the study (Table 2). There was no difference in the average T_2 time for LD and SD bud or conducting stem tissue until 6 weeks of SD. However, further analysis of the distribution of the population of T_2 times in the bud indicated that there was a greater proportion of short T_2 times than longer T_2 times in the bud at 4 weeks of SD (Fennell and Line, 2001). This suggested that an increase in bound water corresponded with the induction of dormancy and acclimation found in the bud at 4 weeks of SD.

MRI analysis of LD and SD *V. riparia* node sections during decreasing freezing temperatures indicated changes in the nodes were also related to acclimation and freezing

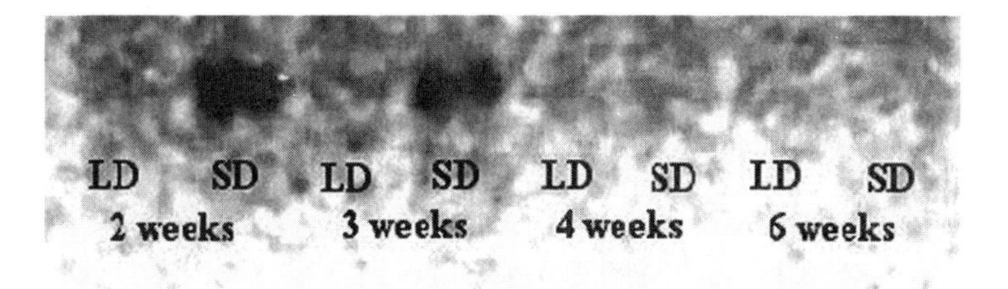

Figure 11. Northern blot analysis for SD495 in *V. riparia* at 2,3, 4, and 6 weeks of LD or SD photoperiod treatment.

tolerance. Differences in freezing temperatures of tissues were related to their acclimation state. Most of the water in SD-treated node sections froze at lower temperatures than water in the LD-treated node sections (data not shown). In the SD nodes the tissue subjacent to the bud freezes at a lower temperature (-18 to -22°C) than the other stem tissues (Fig.10). This difference was not detected in the LD nodes. Although the grape stem tissues freeze extracellularly, the grape bud supercools to avoid freezing injury. Recently Jones et al. (2000) found that there was a decrease in apoplastic permeability at the base of the grape bud during acclimation and suggested that it could provide a barrier for ice propagation into the bud. The decreased T_2 times observed in the gap tissues shown in Figure 9 corresponded to a similar area of tissue that had lower freezing temperatures than the rest of the node tissues. The combination of these changes in the tissues directly adjacent to the bud could contribute to the bud's isolation by inhibiting ice propagation from the stem thereby, enhancing the buds ability to supercool. The potential role of changes in these tissues promoting supercooling is further supported by previous studies that have shown that a small portion of the stem needs to remain with the grape bud to detect supercooling using differential thermal analysis (Andrews et al., 1984, Quamme, 1986).

3.3. Molecular Characterization

Previous studies in *V. riparia*, F1 and Seyval Blanc have shown peptides that increase in all genotypes under SD (Wake and Fennell, 2000). Using differential display we isolated differentially expressed fragments and obtained the full length cDNAs from a *V. riparia*-SD cDNA library. One fragment SD495 was differentially expressed during the induction of dormancy and acclimation (Fig. 11). In *V. riparia*, it is expressed at 2 and 3 weeks of short day and at 3 weeks of SD in F1. It is not expressed in Seyval Blanc. SD495 expression is bud specific. Sequence analysis suggests that it is a pathogenesis related protein and further analysis is being conducted.

4. SUMMARY

Freezing injury in the fall and early winter is a common problem that limits the productivity of grapes in many northern production regions, and an early acclimation could contribute to greater survival and productivity. Induction of dormancy and acclimation is controlled by day length in many woody species; however, in grapes there is considerable variation in response to photoperiod (Fennell and Hoover, 1991; Schnabel and Wample, 1987). *V. riparia* exhibits a classic two stage acclimation response with induction of dormancy and acclimation occurring in response to short photoperiod at high temperatures. Under short photoperiods, growth ceases, node tissues mature, dormancy is induced, freezing tolerance of the bud increases, and the ability to supercool is enhanced by changes in the stem tissues adjacent to the bud. Subsequent exposure to subzero temperatures induces a rapid increase in freezing tolerance in *V. riparia.* In contrast, Seyval Blanc does not acclimate in response to decreasing photoperiod. Under short photoperiods, growth continues and dormancy and acclimation are not induced in Seyval Blanc. The F1 (*V. riparia* x Seyval Blanc) shows an intermediate response to its parents with short photoperiods promoting growth cessation and acclimation more slowly than in *V. riparia.*

MRI analysis indicated differential tissue responses during short photoperiodic induction of dormancy and acclimation changes that appear to be strongly related to acclimation responses. Using this genetic model system, we are able to identify changes in gene expression that are specific to the bud and correspond to the induction of early acclimation and dormancy in grapes. An F2 population shows a 3:1 segregation ratio for photoperiod responsiveness for the induction of dormancy. A population of 140 F2 progeny have been propagated for characterization of dormancy induction and release responses under natural field conditions. This novel system and its segregating F2 population provide ideal materials to study the interrelationship between cold hardiness and dormancy.

5. ACKNOWLEDGEMENTS

This research was supported by the National Science Foundation under grant OSR-9452894, United States Department of Agriculture/National Research Initiative Competitive Grants Program under grant 99-35311-8324, the South Dakota Future Fund, and South Dakota State University Agricultural Experiment Station Funds.

REFERENCES

Andrews, P. K., Sandidge III, C. R., and Toyama, T. K., 1984, Deep supercooling of dormant and deacclimating *Vitis* buds, *Am. J. Enol. Vitic.* **35**: 175-177.

Bordelon, B. P., Ferree, D. C., and Zabadal, T .J., 1997, Grape bud survival in the midwest following the winter of 1993-1994, *Fruit Var. J.* **51**: 53-59.

Clore, W. J., Wallace, M. A., and Fay, R. D., 1974, Bud survival of grape varieties at sub-zero temperatures in Washington, *Amer. J. Enol. Vitic.* **25**: 24-29.

Faust, M., Liu, D., Millard, M. M., and Stutte, G. W., 1991, Bound versus free water in dormant apple buds - A theory for endodormancy, *HortScience* **26**: 887-890.

Faust, M., Erez, A., Rowland, L. J., Wang, S. Y., and Norman, H. A., 1997, Bud dormancy in perennial fruit trees: Physiological basis for dormancy induction maintenance and release, *HortScience* **32**: 623-629.

Fennell, A., and Hoover, E., 1991, Photoperiod Influences Growth, Bud Dormancy, and Cold Acclimation in *Vitis labruscana* and *V. riparia*, *J. Amer. Soc. Hort. Sci.* **116**: 270-273.

Fennell, A., Wake, C., and Molitor, P., 1996, Use of ^{1}H-NMR to determine grape bud water state during the photoperiodic induction of dormancy, *J. Amer. Soc. Hort. Sci.* **121**:1112-1116.

Fennell, A., and Line, M. J., 2001, Identifying tissue response in grapes during induction of endodormancy using magnetic resonance imaging, *J. Amer. Soc. Hort. Sci.* **126**: 681-688.

Jones, K. S., McKersie, B. D., and Paroschy J. , 2000, Prevention of ice propagation by permeability barriers in bud axes of *Vitis vinifera*, *Can. J. Bot.* **78**: 3-9.

Liang P., and Pardee, A. B., 1992, Differential display of eukaryotic messenger RNA by means of the polymerase chain reaction, *Science* **257**: 967-971.

Liu, D., Faust, M., Millard, M. M., Line, M. J., and Stutte, G. W., 1993, States of water in summer-dormant apple buds determined by proton magnetic resonance imaging, *J. Amer. Soc. Hort. Sci.* **118**: 632-637.

Millard, M. M., Veisz, O. B., Krizek, D. T., Line, M., 1995, Magnetic Resonance Imaging (MRI) of Water During Cold Acclimation and Freezing in Winter Wheat, *Plant Cell Environ.* **18**: 535-544.

Quamme, H. A., 1986, Use of thermal analysis to measure freezing resistance of grape buds, *Can J. Plant Sci.* **66**: 945-952.

Schnabel, B.J. and R.L. Wample, 1987, Dormancy and Cold Hardiness in *Vitis vinifera* L. cv. White Riesling as Influenced by Photoperiod and Temperature, *Amer. J. Enol. Vitic.* **38**:265-272.

Seyedbagheri, M.M., and Esmaeil, F., 1994, Physiological and environmental factors and horticultural practices influencing cold hardiness of grapevines, *J. Small Fruit Vitic.* **2**: 3-38.

Wake, C. F. M. and Fennell, A., 2000, Morphological, physiological and dormancy responses of three *Vitis* genotypes to short photoperiod, *Physiol. Plant.* **109**: 203-210.

9

SURVIVAL OF TROPICAL APICES COOLED TO –196°C BY VITRIFICATION

Development of a potential cryogenic protocol of tropical plants by vitrification

Akira Sakai, Toshikazu Matsumoto, Dai Hirai, and Rommanee Charoensub*

1. SUMMARY

Tropical apical meristems excised from *in vitro*-grown plants which were sufficiently dehydrated with a highly concentrated vitrification solution (designated PVS2, 7.8 M) survived subsequent plunging into liquid nitrogen (LN) and regenerated plants (recovery growth 80%). Excised meristems of cassava (*Manihot esculenta* Grantz) were precultured with 0.3 M sucrose for 16 hr and then enhanced for tolerance to PVS2 with a mixture of 2 M glycerol and 0.4 M sucrose (LS) for 20 min at 25 °C. These osmoprotected apices were then sufficiently dehydrated with PVS2, so that the cytosolic concentration required for vitrification was attained upon rapid cooling into LN. Vitrification refers to a phase transition from a liquid into amorphous glass, while avoiding crystallization. In the vitrification protocol, enhancing tolerance to PVS2 and the mitigation of injurious effects during dehydration were crucial for ensuring the survivals.

The osmotic dehydration by PVS2 which is enable cells and tissues to survive at -196 °C by vitrification is significantly advantaged over the freeze-induced dehydration in dehydrating cystosol more effectively, uniformly, speedy and less injuriously at non-freezing temperature, even in tropical plants.

* Akira Sakai, Hokkaido Univ.(retired), Asabucho 1-5-23, Sapporo, 001-0045, Japan. Toshikazu Matsumoto, Shimane Agr. Exp. St., Izumo, Shimane 693-0035, Japan. Dai Hirai, Hokkaido Central Pref. Agr. Exp. St., Naganuma, 069-1395, Japan. Rommanee Charoensub, Scientific Equipment Center, Kasetart Univ., Bangkok 10900, Thailand.

Plant Cold Hardiness, edited by Li and Palva
Kluwer Academic/Plenum Publishers, 2002

2. INTRODUCTION

Cryopreservation is becoming a very important tool for long-term storage of plant genetic resources for future generations with a minimum space and maintenance. Recently, cryopreservation was reported to offer a real hope for long-term conservation of endangered and rare species in the advent of catastrophe (Touchell, 2000; Turner *et al.*, 2000; 2001). Cryopreservation is based on the non-injurious reduction and subsequent interruption of metabolic functions of biological materials by reducing temperature to that of LN (-196 °C). However, availability or development of simple, reliable and cost-effective protocol and subsequent plant regeneration are basic requirements for plant germplasm conservation.

To maintain the viability of hydrated cells and tissues after cooling to -196 °C, it is essential to avoid lethal intracellular freezing, which occurs during rapid cooling into LN (Sakai and Yoshida, 1967). Thus, hydrated cells or tissues cryopreserved in LN, have to be sufficiently dehydrated prior to immersion into LN. Extracellular freezing is an effective method of dehydrating hardy cells or tissues. Sakai (1956; 1960) first demonstrated that very hardy winter willow twigs prefrozen at -30 °C were successfully cryopreserved in LN for one year with subsequent plant regeneration. This pre-freezing method was successfully applied to very hardy apple buds, which were recovered by grafting (Sakai and Nishiyama, 1978; Forsline *et al.*, 1998). In addition, cryopreservation of cultured cells and meristems was achieved by controlled slow pre-freezing to about -40 °C in the presence of suitable cryoprotectants (Sugawara and Sakai, 1974; Withers, 1985).

Almost 10 years ago, some new cryogenic procedures such as vitrification (Langis *et al.*, 1989; Sakai *et al.*, 1990) and an encapsulation-dehydration technique (Fabre and Dereuddre, 1990) were presented. These new procedures dehydrate a major part of the freezable water of specimens at non-freezing temperatures and enable them to be cryopreserved by direct plunging into LN without the need for a freeze-induced dehydration step. These alternative dehydration procedures have simplified cryogenic protocols, (e.g. by removing the need for controlled freezing), and have eliminated concerns regarding the potentially damaging effects of intra- and extracellular crystallization, providing high levels of post-thaw recovery and greater potential for broad applicability, especially non-hardy tropical plants.

Vitrification, a physical process, that can be defined as a glass transition of aqueous concentrated cryoprotective solutions from a liquid into amorphous glass, at the glass transition temperature (Tg), while avoiding crystallization (Rasmussen and Luyet, 1970). Thus, vitrification is a freeze- avoidance mechanism, which enables hydrated cells and meristems to survive into LN. As glass fills space in a tissue, it may contribute to preventing additional tissue collapse, solute concentration and pH alteration during dehydration. Operationally, a glass is expected to exhibit a lower water vapor pressure than the corresponding crystalline solid thereby prevents further dehydration. As glass is exceedingly viscous and stops all chemical reactions that require molecular diffusion, its formation leads to stability over time (Burke, 1986).

This paper describes our vitrification protocol using PVS2 for cryopreserving apical meristems excised from in vitro-grown plants of both temperate and tropical origins, and relates key factors of the protocol, including: induction of osmotolerance, mitigation of injurious effects during dehydration, and the quality of explants to be employed.

3. VITRIFICATION PROTOCOL

The vitrification protocol utilizes a highly concentrated vitrification solution, which sufficiently dehydrates cells without causing injury so that both cells and the surrounding solution turn into a stable glass when plunged into LN. We have developed a glycerol-based, low-toxicity plant vitrification solution (7.8 M) designated PVS2 (Sakai *et al.*, 1990). PVS2 contains 30%(w/v) glycerol, 15%(w/v) ethylene glycol (EG), 15%(w/v) dimethysulfoxide (DMSO) in the Murashige and Skoog (MS) basal medium containing 0.4 M sucrose (pH 5.8). Different plant vitrification solutions such as PVS2, PVS3 have been reported (Nishizawa *et al.*, 1993; Langis *et al.*, 1989; Langis and Steponkus, 1990; Tannoury *et al.*, 1991). At present, PVS2 is the most commonly used plant vitrification solution (Sakai, 1997; Thinh *et al.*,1999; 2000; Turner, 2000; 2001). PVS2 includes 15% DMSO which is considered to have a different mode of action to other cryoprotectants. However, the studies by Turner *et al.* (2001) in which PVS2 was modified by replacing 15% DMSO with 15% propylene glycol produced very similar survival, showing that DMSO acts as the same dehydrating agent as sucrose and glycerol in PVS2 solution.

Vitrification procedure involves the following main steps: enhancing tolerance to PVS2, PVS2-dehydration, rapid cooling into LN (vitrification), rapid thawing and dilution (Fig. 1). A. Preculture: excised apices were precultured for 16 h on a solidified MS medium supplemented with 0.2-0.4 M sucrose; B. Enhancing tolerance to PVS2: with a mixture of 2 M glycerol and 0.4 M sucrose for 20 min. C. PVS2-dehydration: at 25 or 0 °C. D. Vitrification: 10 apices suspended in 0.3-0.5 ml PVS2 in the 2 ml cryotube were plunged into LN (cooling rate: 300 °C /min); E. Rapid thawing: shaking 80 sec in water at 40 °C. F. Dilution: PVS2 was drained and then replaced with 1.2 M sucrose in MS medium for 10 min. G. Plating: different media depending on species.

This vitrification protocol is simple and reliable, is not time-consuming and is applicable to a wide range of plant species or cultivars totaling about 150 (Table 1). The protocol produced higher and earlier recovery growth than the encapsulation-dehydration

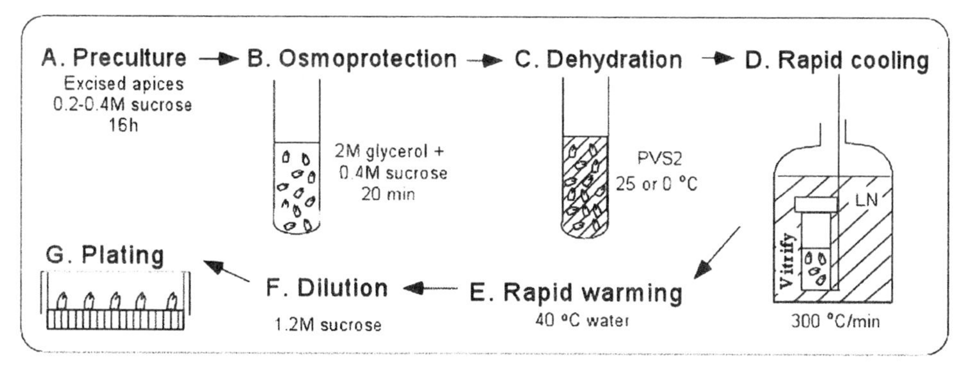

Fig. 1. General view of the vitrification protocol for excised apices from in vitro-grown plants.

Table 1. Successful cryopreservation of *in vitro*-grown apices cooled to -196 °C by vitrification using PVS2 solution (recovery growth: over 50%)

Plant	References
Temperate Plants	
# Woody Plants	
Citrus (Nucellar cells, 4 spp, cvs)	Kobayashi & Sakai 1997
Grevillea scapigera (Endangered, 7 clones)	Touchell 2000
Malus (Apple, 5 sp, cvs)	Niino *et al.* 1992a, b
Mulberry (13 sp, cvs)	Niino *et al.* 1992a
Persimmon (Winter apices, 20 sp, cvs)	Matsumoto *et al.* 2001
Poplar	Lambardi *et al.* 2000
Pyrus (Pear, 5 cvs)	Niino *et al.* 1992a
Tea plant	Kuranuki & Sakai 1995
Vitis (Grape, 10 sp, cvs)	Matsumoto *et al.* 1998, Matsumoto & Sakai 2001
# Herbaceous Plants	
Garlic (Post dormant bulbils, 12 cvs)	Niwata 1995
Lily (4 sp, cvs)	Matsumoto *et al.* 1995
Lily (19 sp, cvs, EV)	Hirai 2001
Mint (3 sp, cvs, EV)	Hirai & Sakai 1999a
Panax ginseng (Hairy roots)	Yoshimatsu *et al.* 1996, 2000
Papaver somniferum (Hairy roots)	Yoshimatsu *et al.* 2000
Solanum tuberosum (Potato, 14cvs, EV)	Hirai & Sakai 1999b
Strawberry (6 cvs, EV)	Hirai *et al.* 1998
Wasabi (*Brassicaceae*, 4 cvs)	Matsumoto *et al.* 1994
White clover (3 spp, cvs)	Yamada *et al.* 1991
Yam (Temperate, 3 spp, EV)	Hirai 2001
Tropical Plants	
Cassava (10 cvs)	Charoensub *et al.* 2001 (unpublished)
Cassava (4 cvs, EV)	Hirai & Sakai 2000
Colocasia (Taro, 6 cvs)	Thinh 1997; Takagi *et al.* 1997
Dioscorea (Yam, 6 cvs)	Kyesm & Takagi 2000
Musa (Banana, 9 cvs)	Thinh *et al.* 1999
Orchid: *Cymbidium (2cvs)*	Thinh & Takagi 2000
Orchid: *Dendrobium (Protocorm)*	Wang *et al.* 1998
Papaya (Somatic embryo)	Lu & Takagi 2000
Pineapple	Thinh 1997
Sweet potato	Pennycooke & Towill 2000
Xanthosoma (Tanier, 2 sp)	Thinh 1997

EV: encapsulation-vitrification

technique (Matsumoto *et al.*, 1994; 95; Hirai and Sakai, 1999a; b). As summarized in Table 1, the vitrification protocol with or without encapsulation for apices was successfully applied to roots and tubers, fruit trees, ornamental and medicinal plants and plantation crops.

4. MITIGATION OF INJURIOUS EFFECTS DURING PVS2-DEHYDRATION

PVS2 dehydration is an important step in the vitrification protocol. Direct exposure of less tolerant cells or meristems to a highly concentrated vitrification solution (PVS2) may lead to harmful effects due to osmotic stress or chemical toxicity. Thus, in our vitrification protocol, the injurious effects during dehydration were reduced or eliminated by dehydrating meristems by a two-step: partial dehydration by a mixture of 2 M glycerol plus 0.4 M sucrose (LS) for 20 min, followed by dehydration with PVS2 before plunging into LN. It was also necessary to optimize the exposure time to PVS2 solution. Furthermore, apices less tolerant to PVS2 were dehydrated at 0 °C to increase survival. Thinh *et al.* (2000) clearly demonstrated that the meristems such as pineapple and banana sensitive to PVS2 are difficult to dehydrate sufficiently at 25 °C prior to plunging in LN. Dehydration at 0 °C appeared to be safer and ensured their high recovery rates, although longer treatment durations were needed. Similar positive effects of dehydration at 0 °C were shown for meristems of some temperate plants (Matsumoto *et al.*, 1994; 1995).

Exposure time to PVS2 solution was essential to produce high levels of recovery growth. Properly dehydrated meristems vitrify during rapid cooling into LN, preventing lethal intracellular freezing, however, overexposure can cause toxic effects to PVS2.

Exposure to PVS2 produced time dependent shoot formation (Yamada *et al.*, 1991). In the vitrified clover meristems cooled to -196 °C, the shoot formation reached the highest rate (about 95%) at 5 min exposure at 25 °C or at 0 °C for 20 min, respectively. Rates of shoot formation of the meristems dehydrated with PVS2 for 5 min at 25 °C without cooling in LN (treated control) were similar to those of the vitrified meristems cooled to -196 °C. Thus, the rapid cooling (vitrification) into LN and subsequent rapid rewarming (devitrification) did not cause additional loss beyond that produced during dehydration process with PVS2 at non-freezing temperature. The same trend was observed in the vitrified tobacco cultured cells (Reinhold, 1996) and various meristems (Matsumoto *et al.*, 1998a; Niino *et al.*, 1992a; b). These results clearly demonstrate that tolerance to PVS2 solution is sufficient for cells and meristems to survive the vitrification procedure. The time of optimum exposure to PVS2 and the dehydration time of meristems appear to be species-specific and to increase with the size of excised meristems (Charoensub *et al.*, 1999; Niino *et al.*, 1992a).

5. ENHANCING TOLERANCE TO PVS2

Whatever cryogenic method is applied to cultured apices or cells, inducing higher levels of dehydration tolerance is required for successful cryopreservation. As summarized in Table 2, in some temperate plants, preculture with sorbitol or sucrose enriched medium for 1-2 days with or without cold hardening enhanced tolerance to PVS2 (Niino *et al.*, 1992b; Touchell, 2000; Turner *et al.*, 2000; 2001). However, in many temperate and tropical plants, preculturing meristems with sucrose or sorbitol alone did not lead to a substantial increase in the survival of vitrified meristems (Table 2), especially in tropical plants. We found that the treatment with a mixture of 2 M glycerol and 0.4 M sucrose (LS) for 20 min following preculture with 0.3 M sucrose for 16 h was very effective in enhancing tolerance to PVS2 (Matsumoto *et al.*, 1994). The significant positive effects with LS treatment were demonstrated in several tropical monocots such as banana, taro, orchid, pineapple (Thinh, 1997; Thinh and Takagi, 2000; Thinh *et al.*,

Table 2. Effective way enhancing tolerance of cultured cells and meristems to PVS2

Plant, explant, the recovery growth(%)	Effective way enhancing tolerance to PVS2	References
1. Temperate plants (herbaceous)		
Asparagus (embryogenic cultured cells, 90%*)	LS	Nishizawa *et al.* 1993
Tobacco (cultured cells: 55%*)	0.3 M mannitol (77 h)	Reinhoud 1996
Angozanthos viridis (AM: 41%)	0.4 M sorbitol (2days)	Turner *et al.* 2000
Horseradish (hairy root: 69%)	0.5 M Suc+1 M Gly (16h)	Phunchindawan *et al.* 1997
Macropidia fuliginosa (SE: 90%)	0.8 M Gly (2 days)	Turner 2000, 2001
Mint (AM**, 3cv: av. 80%)	CH (2 weeks), LS	Hirai & Sakai 1999a
Panax ginseng (hairy root: 69%)	0.3 M Suc (1-3 days), LS	Yoshimatsu *et al.* 1996
Potato (AM**, 14cv: av. 70%)	LS	Hirai & Sakai 1999b
Sugar beet (AM, 18 clones: 60-100%)	CH, 0.3 M Suc (1 day), LS	Vanderbussche *et al.* 2000
Wasabi (AM, 5 cv: av.90%)	0.3 M Suc (16 h), LS	Matsumoto *et al.* 1994
Wasabi (AM: 85%)	0.3 M Suc+0.5 M Gly (16h)	Matsumoto *et al.* 1998a
2. Temperate plants (woody plants)		
Grape (*V. vinifera*, AM, 10 cv: av. 64%)	0.3 M Suc (1 day), LS	Matsumoto & Sakai 2001
Grevillea scapigera (9 clones: av. 56.9%)	0.6 M sorbitol (2 days)	Touchell 2000
Mulberry (AM, 13 cv: av. 65%)	CH (3 weeks), 0.7 M Suc (16 h)	Niino *et al.* 1992a
Populus (AM: 64%)	CH (3 weeks), 0.09 M Suc (2 days), LS	Lambardi *et al.* 2000
3. Tropical plants		
Banana (AM, 10 cv: av. 69%)	LS	Thinh 1997, Thinh *et al.* 2000a
Cassava (AM, 10 cv: av. 70%)	0.3 M Suc (16 h), LS	Charoensub *et al.* 1999
Cymbidium (AM, 2cv: 63%)	LS	Thinh 1997, Thinh *et al.* 2000a
Cymbopogen (AM: 60%)	LS	Thinh *et al.* 2000a
Dendrobium (protocom: 60%)	ABA (3 weeks)	Wang *et al.* 1998
Doritaenopsis (culture: 64%)	0.1 M Suc (7 days), LS	Tsukazaki *et al.* 2000
Innala (*Solemostenon*, AM: 85%)	0.3 M Suc (2 days), LS	Niino *et al.* 2001
Pineapple (AM: 67%)	LS	Thinh 1997, Thinh *et al.* 2000b
Sweet potato (AM: 40% survival)	0.3 M Suc (1 day), LS	Pennycooke & Twill 2000
Taro (AM, 5cv: av. 65%)	0.3 M Suc (16 h), LS	Takagi *et al.* 1997
Yam (AM, 6cv: av.67%)	0.1 M Suc (7 days), LS	Kyesum & Takagi 2000

LS: a mixture of 2 M Glycerol + 0.4 (0.6) M Sucrose in MS medium for 20 to 30 min. at 25 °C; AM: apical meristems excised from *in vitro*-grown plants; CH: cold hardening at 5 °C; * :cell survival (%); **: encapsulation-vitrification

1999; 2000), cassava (Charoensub, 1999), innala (Niino *et al.,* 2000), yam (Kyesm and Takagi, 2000) and sweet potato (Pennycooke and Towill, 2000). Thus, the LS treatment following preculture with sucrose enriched medium is a very important step of successful cryopreservation of tropical apices by vitrification.

During 20 min treatment of meristems with LS solution, meristematic cells were osmotically dehydrated and considerably plasmolyzed. However, little or no permeation of glycerol or sucrose was observed over the 20 min in the plasmolyzed cells based on volumic change in the cytosols. The plasmolyzed cells were then subjected to PVS2 solution, in which cytosolic volume successively decreased. Thus, there did not seem to be an appreciable influx of additional cryoprotectants into cytosol due to the difference in the permeability coefficient for water and solute permeation, especially in the highly viscous solution for 20 min. As a result, the cells of apices remained osmotically dehydrated and the increase in the cytosolic concentration required for vitrification was attained by dehydration. Matsumoto *et al.* (1998a) observed that cells in longitudinal sections from wasabi were intensively convexly plasmolyzed when dehydrated with PVS2 for optimal time of exposure prior to plunging in LN.

The protective effect of brief incubation (loading) with LS for 20 min after preculture might be due to the osmotic dehydration, resulting in the concentration of cytosolic stress-responsive solutes (Reinhoud, 1996) that accumulated during preculture with sucrose or sugar-alcohol enriched medium, and in the protective effects of plasmolysis. The presence of highly concentrated cryoprotective solution in the periprotoplasmic space of plasmolyzed cells may mitigate the mechanical stress caused by successively severe dehydration (Hellergren and Li, 1981; Tao *et al.*, 1981; Jitsuyama *et al.*, 1997). These intra- and extracellular protective effects may minimize the injurious membrane changes during dehydration, though the actual protective action mechanism is poorly understood.

6. ADVANTAGE OF PVS2-DEHYDRATION

Osmotic dehydration with PVS2 has various significant advantages over the freeze-induced dehydration. In the dormant hardy axillary buds of persimmon, the apical meristems is the hardiest tissue which tolerated marginally hardy to –20 °C. The meristems prefrozen slowly to –20 or –30 °C did not survive subsequent plunging into LN regardless of thawing rates, which were followed by recovery in vitro (Matsumoto and Sakai, 2001). However, the meristems excised from axillary buds, which were dehydrated by PVS2 for 20 min at 25 °C and then directly immersed into LN, remained alive and developed shoots directly without intermediary callus formation. The recovery growth amounted to 89%. These results clearly demonstrated that the osmotic dehydration by PVS2 is significantly advantaged over the freeze-induced dehydration in dehydrating cytosol more effectively, uniformly, speedy, less injuriously at non-freezing temperature. Plasmolysis, which does not occur in natural conditions, plays very important role in enhancing osmotolerance and mitigating injurious effects during the dehydration process (Sakai and Yoshida, 1968; Jitsuyama *et al.*, 1997; Tao *et al.*, 1983).

It is very interesting to note that in vitro-grown shoot tips of cassava (Fig. 2), a tropical dicot was successfully cryopreserved using nearly the same vitrification protocol used for temperate plants such as Japanese horseradish (Matsumoto *et al.*, 1994), lily (Matsumoto *et al.*, 1995), potato (Hirai and Sakai,1999b) and some tropical monocots (Thinh *et al.*, 1999; 2000). These results strongly suggests that in vitro-grown apices of both temperate and tropical origins (Thinh, 1997; Thinh and Takagi, 2000; Thinh *et al.*, 1999; 2000), the meristematic zone, composed of a relatively homogeneous population of small, actively dividing cells, with a high nucleo-cytoplasmic ratio and little vacuolation, has a high potential to induce osmotolerance required for successful cryopreservation. These characteristics, along with its micropropagation potential and genetic stability, will make apical meristems an excellent candidate for cryopreservation of tropical plants.

Fig. 2. Shoot formation of encapsulated and vitrified shoot tip of cassava cooled to -196 °C by encapsulation-vitrification. 4 weeks after plating. White bar indicates 1 cm. Cultivar AMM22.

7. DISCUSSION

Excised apices employed for cryopreservation must be in an optimal physiological state (Dereuddre *et al.*, 1988) in order to acquire higher levels of dehydration tolerance and for the production of vigorous recovery growth. The simple micropropagation protocol can supply a large number of uniform and adequate meristems required for cryopreservation. Hirai clearly demonstrated that successful micropropagation in mint, using dense planting of about 60 nordal segments with a pair of leaves from in vitro-grown plants on a solidified culture media in a Petri dish under the light conditions. These nordal segments produced young lateral buds about 10 days later. The apical meristems excised from the lateral buds produced very high levels (80%) of recovery growth by encapsulation -vitrification (Hirai and Sakai, 1999a, b). This simple micropropagation protocol enabled production of a large number of adequate apices homogeneous in size, cellular composition, physiological state, and thus increased the chance of positive and uniform responses to subsequent cryogenic treatments. This micropropagation protocol also proved to be very effective in cassava (Hirai and Sakai, 2000).

For many vegetatively propagated species or cultivars of temperate origins, cryopreservation techniques using vitrification are sufficiently advanced to envisage their immediate utilization for large-scale experimentation in gene-banks. However, research is much less advanced for tropical species. This is because of the comparatively limited level of research activities aiming at improving the conservation of these species. However, various technical approaches can be explored to improve efficiency and to increase applicability, provided that the tissue cultures such as apical meristems and somatic embryo culture, are sufficiently operational for the species, as was clearly demonstrated by Thinh (1997), Thinh and Takagi (2000) and Thinh *et al.* (2000).

It is hoped that a critical issues such as understanding and controlling dehydration tolerance, will contribute significantly to the development of improved cryopreservation techniques for difficult tropical plant species.

REFERENCES

Burke, M. J., 1986, The glassy state and survival of anhydrous biological systems, in: *Membrane, Metabolism and Dry Organisms*, A. C. Leopold, ed, Cornell Univ. Press, Ithaca, New York. pp. 358-364.

Charoensub, R., Phansiri, S., Sakai, A.and Yongmanitchai, W., 1999, Cryopreservation of cassava in vitro-grown shoot tips cooled to -196 °C by vitrification, *Cryo-Lett.* **20**: 89-94.

Dereuddre, J., Fabre, J. and Basaglia, C.,1988, Resistance to freezing in liquid nitrogen of carnation (*Dianthus caryophyllus* L. var. Kolo) apical and axillary shoot tips excised from different aged *in vitro* plants, *Plant Cell Rep.* **7**: 170-173.

Fabre, J. and Dereuddre, J., 1990, Encapsulation-dehydration: A new approach to cryopreservation of *Solanum* shoot tips, *Cryo-Lett.* **11**: 413-426.

Forsline, P. L, Towill, L. E., Wardell, W., Stuschnoff, C., Lamboy, W. F. and Ferson, J. R., 1998, Recovery and longevity of cryopreserved dormant apple buds, *J. Amer. Soc. HortSci.* **123**: 365-370.

Hellergren, J. and Li, P.H., 1981, Survival of *Solanum tuberosum* suspension cultures to -14°C: The mode of action of proline, *Physiol. Plant.* **52**: 449-453.

Hirai, D., Shirai, K., Shirai, S. and Sakai, A., 1998, Cryopreservation of in vitro-grown meristems of strawberry (*Fragaria x ananassa* Duch.) by encapsulation-vitrification, *Euphytica* **101**: 109-115.

Hirai, D. and Sakai, A., 1999a, Cryopreservation of in vitro-grown axially shoot tip meristems of mint (*Mentha spicata* L.) by encapsulation vitrification, *Plant Cell Rep.* **19**: 150-155.

Hirai, D. and Sakai, A., 1999b, Cryopreservation of in vitro-grown meristems of potato (*Solanum tuberosum* L.) by encapsulation-vitrification, *Potato Res.* **42**: 153-160.

Hirai, D. and Sakai, A., 2000, Cryopreservation of tropical crop cassava by encapsulation vitrification, *Rep. Hokkaido Branch Jap. Soc. Breeding and Crop Sci. Soc. Jap.* **41**: 85-86.

Hirai, D., 2001, Studies on cryopreservation of vegetatively propagated crops by encapsulation vitrification method, *Rep. Hokkaido Pref. Agri. Exp. Sta.* 99:1-58.

Jitsuyama, Y., Suzuki, T., Harada, T. and Fujikawa, S., 1997, Ultrastructural study of mechanism of increased freezing tolerance to extracellular glucose in cabbage leaf cells, *Cryo-Lett.* **18**: 33-44.

Kobayashi, S. and Sakai, A., 1997, Cryopreservation of *Citrus sinensis* cultured cells, in: *Conservation of Plant Genetic Resources in vitro. Vol.1: General Aspects,* M. K. Razdan and E. C. Cocking, eds., Science Publishers, U.S.A., pp. 201-223.

Kuranuki, Y. and Sakai, A., 1995, Cryopreservation of *in vitro*-grown shoot tips of tea (*Camellia sinensis*) by vitrification, *Cryo-Lett.* **16**: 345-352.

Kyesmu, P. M. and Takagi, H., 2000, Cryopreservation of shoot apices of yams (*Dioscorea* species) by vitrification, in: *Cryopreservation of Tropical Plant Germplasm,* F. Engelmann and H. Takagi, eds., Tsukuba, JIRCAS, Japan, pp. 411-413.

Langis, R., Schnabel, B., Earle, E. D. and Steponkus, P. L., 1989, Cryopreservation of *Brassica campestris* L. cell suspensions by vitrification, *Cryo-Lett.* **10**: 421-428.

Langis, R. and Steponkus, P. L., 1990, Cryopreservation of rye protoplasts by vitrification, *Plant Physiol.* **92**: 666-671.

Lambardi, M., Fabbri, A. and Caccavale, A., 2000, Cryopreservation of white poplar (*Populus alba* L.) by vitrification of in vitro-grown shoot tips, *Plant Cell Rep.* **19**: 213-218.

Lu, T. and Takagi, H., 2000, Cryopreservation of somatic embryos of papaya (*Carica papaya* L.) by vitrification. 2000 World Congress on In Vitro Biology, Program Issue, pp. 73.

Matsumoto, T., Sakai, A. and Yamada, K., 1994, Cryopreservation of *in vitro*-grown apical meristems of wasabi (*Wasabia japonica*) by vitrification and subsequent high plant regeneration, *Plant Cell Rep.* **13**: 442-446.

Matsumoto, T., Sakai, A. and Yamada, K., 1995, Cryopreservation of *in vitro*-grown apical meristems of lily by vitrification, *Plant Cell Tissue and Organ Cult.* **41**: 237-241.

Matsumoto, T., Sakai, A. and Nako, Y., 1998a, A novel preculturing for enhancing the survival of *in vitro*-grown meristems of wasabi (*Wasabia japonica*) cooled to -196 °C by vitrification, *Cryo-Lett.* **19**: 27-36.

Matsumoto, T., Sakai, A. and Nako, Y., 1998b, Cryopreservation of *in vitro* cultured axillary shoot tips of grape (*Vitis vinifera*) by vitrification, *Suppl. J. Japan. Soc. Hort. Sci.* **67**(1): 78.

Matsumoto, T., Mochida, K., Hamura, H. and Sakai, A., 2001, Cryopreservation of persimmon (*Diopyros kaki* Thumb.) by vitrification of dormant shoot tips, *Plant Cell Rep.* **20**: 398-402.

Matsumoto, T. and Sakai, A., 2001, Cryopreservation of *in vitro*-grown axillary shoot tips of grapevine (*Vitis vinifera*), *Cryobiology* **42** (6), in press

Murashige, T. and Skoog, F., 1962, A revised medium for rapid growth and bioassays with tobacco tissue culture. *Physiol. Plant.* **15**: 473-497.

Niino, T., Sakai, A., Enomoto, S., Magoshi J. and Kato, S., 1992a, Cryopreservation of *in vitro*-grown shoot tips of mulberry by vitrification, *Cryo-Lett.* **13**: 303-312.

Niino, T., Sakai, A., Yakuwa, W. and Nojiri, K., 1992b, Cryopreservation of *in vitro*-grown shoot tips of apple and pear by vitrification, *Plant Cell Tissue and Organ Culture* **28**: 261-266.

Niino, T., Hettiarachii, A. and Takahashi, I., 2000, Cryopreservation of lateral buds of *in vitro*-grown innala plants (*Solenostemon rotundifolium*) by vitrification, *Cryo-Lett.* **21**: 349-356.

Nishizawa, S., Sakai, A., Amano, Y. and Matsuzawa, T., 1993, Cryopreservation of asparagus (*Asparagus officinalis* L.) embryogenic suspension cells and subsequent plant regeneration by the vitrification method, *Plant Sci.* **88**: 67-73.

Niwata, E., 1995, Cryopreservation of apical meristems of garlic (*Allium sativum* L.) and high subsequent plant regeneration, *Cryo-Lett.* **16**: 102-107.

Pennycooke and Towill, L. E., 2000, Cryopreservation of shoot tips in vitro plants of sweet potato [*Ipomoea batatas* (L.) Lam.] by vitrification, *Plant Cell Rep.* **19**: 733-737.

Phunchindawan, M., Hirata, K., Sakai, A. and Miyamoto, K., 1997, Cryopreservation of encapsulated shoot primordia induced in horseradish (*Armoracia rusticana*) hairy root cultures, *Plant Cell Rep.* **16**: 469-473.

Rasmussen, D. H. and Luyet, B. J., 1970, Contribution to the establishment of the temperature-concentration curves of homogeneous nucleation in solutions of some cryoprotective agents, *Biodinamica* **11**: 33-44.

Reinhoud, P. J., 1996, Cryopreservation of tobacco suspension cells by vitrification, Doctoral thesis, Leiden Univ., Institute of Molecular Plant Sci., Leiden. pp. 1-95.

Sakai, A., 1956, Survival of plant tissue at super-low temperature, Low Temp. Sci., Ser. B., 14: 17- 23.

Sakai, A., 1960, Survival of the twig of woody plants at -196 °C, Nature, 185: 393-394.

Sakai, A., 1997, Potentially valuable cryogenic procedures for cryopreservation of cultured plant meristems, in: *Conservation of Plant Genetic Resources In Vitro,* M. K. Razdan and E. C. Cocking, eds., Science Publishers, U.S.A., pp. 53-66.

Sakai, A. and Yoshida, S., 1967, Survival of plant tissue at super-low temperatures. VI. Effects of cooling and rewarming rates on survival, *Plant Physiol.* **42**: 1695-1701.

Sakai, A. and Yoshida, S., 1968, The role of sugar and related compounds in variation of freezing tolerance, *Cryobiol.* **5**: 160-174.

Sakai A. and Nishiyama, N., 1978, Cryopreservation of winter vegetative buds of hardy fruit trees in liquid nitrogen, *Hort Sci.* **13**: 225-227.

Sakai, A., Kobayashi, S. and Oiyama, I., 1990, Cryopreservation of nucellar cells of navel orange (*Citrus sinensis* Osb. var.*brasiliensis* Tanaka) by vitrification, *Plant Cell Rep.* **9**: 30-33.

Sugawara, T. and Sakai, A., 1978, Survival of suspension-cultured sycamore cells cooled to the temperature of liquid nitrogen, *Plant Physiol.* **54**: 722-724.

Takagi, H., Thinh, N. T., Islam, O. M., Senboku, T. and Sakai, A., 1997, Cryopreservation of *in vitro*-grown shoot tips of taro (*Colocasia esculenta* (L.) Schott) by vitrification. 1. Investigation of basic conditions of the vitrification procedures, *Plant Cell Rep.* **16**: 594-599.

Tannoury, M., Ralambosoa, J., Kaminsky, M. and Dereuddre, J., 1991, Cryopreservation by vitrification of alginate-coated carnation (*Dianthus cargo* Phyllus L.) shoot tips of *in vitro* plantlets, C. R. Acad. Sci., Paris, t. 310, Serie III, pp. 633-638.

Tao, D., Li, P. H. and Carter, J. V., 1983, Role of cell wall in freezing tolerance of cultured potato cells and their

protoplasts, *Physiol. Plant.* **58**: 527-532.
Thinh, N. T., 1997, Cryopreservation of germplasm of vegetatively propagated tropical monocots by vitrification, Doctoral thesis of Kobe Univ., Dep. of Agronomy, Japan, pp.1-187.
Thinh, N. T., Takagi, H. and Yashima, S., 1999, Cryopreservation of *in vitro*-grown shoot tips of banana (*Musa spp*) by vitrification method, *Cryo-Lett.* **20**: 163-174.
Thinh, N. T. and Takagi, H., 2000, Cryopreservation of in *vitro*-grown apical meristems of terrestrial orchids (*Cymbidium* spp) by vitrification, in: *Cryopreservation of Tropical Plant Germplasm*, F. Engelmann and H. Takagi, eds., Tsukuba, JIRCAS, Japan, pp. 441-443.
Thinh, N. T., Takagi, H. and Sakai, A., 2000, Cryopreservation of *in vitro*-grown apical meristems of some vegetatively propagated tropical monocots by vitrification, in: *Cryopreservation of Tropical Plant Germplasm*, F. Engelmann and H. Takagi, eds., Tsukuba, JIRCAS, Japan, pp. 227-232.
Touchell, D., 2000, Conservation of threatened flora by cryopreservation of shoot apices, in: *Cryopreservation of Tropical Plant Germplasm*, F. Engelmann and H. Takagi, eds., Tsukuba, JIRCAS, Japan, pp. 269-272.
Tsukazaki, H., Mii, M., Kokuhara, K. and Ishikawa, K., 2000, Cryopreservation of *Doritaenopsis* culture by vitrification, *Plant Cell Rep.* **19**:1160-1164.
Turner, S. R., Touchell, D. H., Dixson, K. and Tan, B., 2000, Cryopreservation of *Anigozanthos viridis* spp *viridis* and related taxa from the south west of Western Australia, *Aust. J. Bot.* **48**: 739-744.
Turner, S. R., Senaratna, T., Bunn, E., Tunn, B., Dixon, K. W. and Touchell, D. H., 2001, Cryopreservation of shoot tips from six endangered Australian species using a modified vitrification protocol, *Ann. Bot.* **87**: 739-744.
Vandenbussche, B., Weyens, G. and De Proft, M., 2000, Cryopreservation of in vitro sugar beet (*Beta vulgaris.* L) shoot tips by a vitrification technique, *Plant Cell Rep.* **19**:1064-1068.
Wang, J. H., Ge, J. G., Liu, F., Bian, H. W. and Huang, C. N., 1998, Cryopreservation of seeds and protocorms of *Dendrobium candidum, Cryo-Lett.* **19**: 123-128.
Withers, L. A., 1979, Freeze preservation of somatic embryos and clonal plantlets of carrot (*Daucus carota*). *Plant Physiol.* **63**: 460-467.
Yamada, T., Sakai, A., Matsumura, T. and Higuchi, S., 1991, Cryopreservation of apical meristems of white clover (*Trifolium repens* L.) by vitrification, *Plant Sci.* **78**: 81-87.
Yoshimatsu, K., Touno, K. and Shimomura, K., 2000, Cryopreservation of medicinal plant resources: retention of biosynthetic capabilities in transformed cultures, in: *Cryopreservation of Tropical Plant Germplasm*, F. Engelmann and H. Takagi, eds., Tsukuba, JIRCAS, Japan, pp. 77-88.

10

EXPRESSION OF COLD-REGULATED (*cor*) GENES IN BARLEY

Molecular bases and environmental interaction

Luigi Cattivelli, Cristina Crosatti, Caterina Marè, Maria Grossi, Anna M. Mastrangelo, Elisabetta Mazzucotelli, Chiara Govoni, Gabor Galiba, and A. Michele Stanca*

1. INTRODUCTION

Barley is grown either in the northern countries close to the polar circle or on the Himalayan mountains up to 4500 m on the sea level. Such a great diffusion, despite the differences in the climatic conditions, already suggests that the barley gene pool should contain characters for wide environmental adaptability and good stress resistance. The genetic adaptation to cold climate can be achieved either by evolving a powerful frost tolerance ability or by limiting the life cycle to the short summer season (escape strategy). It is a known fact that the winter barley varieties are less hardy than winter wheat, rye and triticale, nevertheless barley is grown till the Polar Circle because spring early maturity cultivars are able to run their life cycle in the short summer season. Plant growth habit and heading date can therefore be considered as the basic traits involved in barley adaptation to environments since they allow to synchronise the plant life cycle with seasonal changes. Nevertheless because winter barley has a higher yielding potential than spring ones, there is a great interest to improve its frost resistance capacity.

Freezing tolerance, a fundamental component of winter-hardiness, is based on an inducible process known as hardening or cold acclimation that occurs when plants are exposed to low non-freezing temperatures. Studies on frost resistance in cereal have been addressed either at genetics or at physiological-molecular levels. The genetic analysis discovered a number of loci controlling frost resistance or traits related to frost resistance.

* Luigi Cattivelli, Cristina Crosatti, Caterina Marè, Maria Grossi, Elisabetta Mazzucotelli, Chiara Govoni and A. Michele Stanca, Experimental Institute for Cereal Research Via S. Protaso, 302; I-29017 Fiorenzuola d'Arda (PC) Italy. Anna M. Mastrangelo, Experimental Institute for Cereal Research , Section of Foggia, S.S. 16 km 675, 71100 Foggia, Italy. Gabor Galiba, Agricultural Institute of the Hungarian Academy of Sciences, Martonvásár, Hungary.

Plant Cold Hardiness, edited by Li and Palva
Kluwer Academic/Plenum Publishers, 2002

On the other hand, physiological and molecular researches identified compounds or mRNAs accumulated in response to cold. Several genes regulated by low temperatures (Cold-Regulated *-cor-* genes) have been isolated from the cereals genomes and their expression was found associated with the development of the frost tolerance during cold hardening. Present efforts are addressed to understand the relationship between the frost resistance loci and the physiological or molecular response associated with cold acclimation.

2. GENETIC LOCI CONTROLLING FROST RESISTANCE

Frost resistance has generally been considered as a polygenic trait, although specific loci are known to play an important role. Traditional genetics approaches were used to describe characters such as the heading-date or the growth habit and their involvement in the plant adaptation to the cold stress. More recently the application of the molecular marker technology to the analysis of quantitative traits has lead to the identification of a relative small number of quantitative trait loci (QTL) having a major effect on the ability of the plants to survive under stress conditions.

The growth habit trait, even if not involved *per se* in frost resistance, has been related to the plant winter survival ability. It is a known fact that winter cultivars are generally hardier than their spring counterparts. In barley, winter habit depends on the presence of the dominant allele at locus *Sh* and of the recessive alleles at loci *sh2* and *sh3*. All the other allele combinations among these three genes are found in spring genotypes. The loci *Sh*, *Sh2* and *Sh3* are respectively located on chromosomes 4H, 5H and 1H. The homozygote genotype *shsh* is epistatic with respect to the recessive allele *sh2* and *sh3*, and evinces a facultative behaviour towards the spring habit when sown in spring. The *Sh3* allele is epistatic with respect to alleles *Sh* and *sh2*. Without vernalization and in long-day conditions, all the *Sh3Sh3* cultivars are essentially spring types. *Sh2*, which is epistatic vis à vis alleles *Sh* and *Sh3*, has a series of multiple alleles which induce several spring-to-winter variants (Cattivelli et al., 2001). Fowler et al. (1996) have reported that the loci controlling vernalization requirement in wheat and in rye are responsible for the duration of the expression of cold-regulated genes, suggesting a relationship between growth habit, frost resistance and expression of the genes involved in cold acclimation (see below). The genetic analyses of the frost resistance in barley have found that a QTL for winter survival on barley chromosome 7 (5H) is associated with the *Sh2* locus (Hayes et al., 1993) and with QTLs for heading-data and vernalization response under long day conditions (Pan et al., 1994). These results are probably due to genetic linkage rather than to pleiotropic effects, indeed recombinants between vernalization requirement and winter survival traits has been described (Doll et al., 1989).

RFLP analysis of the homeologous 5A chromosome of wheat have proved that vernalization requirement and frost resistance are controlled by two different, but tightly linked loci (*Vrn-A1* and *Fr1* respectively) (Galiba et al., 1995). In wheat the availability of chromosome substitution lines allowed the identification of chromosome carrying loci with relevant role in frost resistance. Thus, when the 5A chromosome of the frost-sensitive variety Chinese Spring was replaced by the corresponding chromosome of the frost resistant Cheyenne variety, the frost tolerance of Chinese Spring was greatly increased (Sutka 1981; Veisz and Sutka 1989). This phenomenon was also true in the opposite direction, namely when the 5A chromosome of Chinese Spring was substituted with the corresponding chromosome originated from a highly frost sensitive *Triticum spelta* accession, the frost resistance of the recipient Chinese Spring decreased (Galiba et al., 1995). Because of its large effect on frost resistance, molecular assisted selection for the *Vrn-A1-Fr1* 5A chromosome interval has been proposed as a tool to improve cold hardiness of cultivars (Storlie et al., 1998). *Vrn-A1* locus has been found to form a homeologous series with *Vrn-B1* (formerly *Vrn2*) on chromosome 5B and *Vrn-D1* (formerly *Vrn3*) and on chromosome 5D (Snape et al., 1997). Comparison of a common set of RFLP markers suggested that *Vrn-A1* locus is homologous to *Vrn-H1* (formerly *Sh2*) located in barley chromosome arm 5HL (Laurie et al. 1995) and to *Vrn-R1* (formerly *Sp1*) located on rye chromosome arm 5RL (Plaschke et al., 1993).

3. GENE EXPRESSION AND FROST RESISTANCE

The ability of barley to withstand cold stress situations is mediated by specific sets of genes which modify the cell metabolism making cells able to cope with low temperatures. This stress can damage plants, causing changing in cell volume and membrane shape, disruption of water potential gradients, physical damages to the membranes and protein degradation.

While it is becoming increasingly clear that plants can efficiently sense cold stress and mobilise the appropriate responses, it is not so clear how this information is obtained and transduced to the cell nucleus to induce freezing tolerance. The molecular mechanisms leading to the plant response involve three steps: *a)* perception of external changes; *b)* transduction of the signal to the nucleus; *c)* gene expression. Although in the recent years a number of genes whose expression is induced or enhanced by low temperatures have been cloned (Cattivelli et al., in press), little is known about the regulatory mechanism controlling the stress responses. This is mainly due to the fact that cold stress response is a multigenic trait involving many genes that may have either redundant or additive effects and may interact with each other in different and complex ways. One of the primary targets to better understand plant cold stress responses is to clarify the functions of the stress-induced genes and how they are regulated by external changes. Most of these studies have been carried out using either barley or *Arabidopsis* as model plants, although it is generally believed that most of the molecular pathways controlling the plant response to low temperatures are well conserved in all plant species.

3.1. Barley Cold-Regulated Genes

The development of the acclimation process to low temperatures leading to frost tolerant tissue require the expression of a number of *cor* genes (Cattivelli and Bartels, 1989). In barley more than 20 cDNA clones which expression is affected by low temperatures have been isolated (see Table 1). The accumulation of *cor* mRNAs depends primarily on the low temperature stimulus, but it can also be modulated by other factors such as accumulation of ABA, drought stress or light. The analysis of the expression pattern reveals that *cor* mRNAs reach their steady-state within 2-3 days after exposure to cold, while when the plants are moved from low temperatures to 20°C the level of *cor* mRNAs drops in few hours (Cattivelli and Bartels, 1990). Their expression and the accumulation of the corresponding proteins have been correlated with frost resistance (Crosatti et al., 1996; Pearce et al., 1996).

The sequence and the expression analysis of the *cor* genes allowed the identification of several gene classes. Several genes induced by cold stress belong to the Late Embryogenesis Aboundant (LEA) gene family. The dehydrins (class 2 of the *Lea* gene family) are the main group of proteins induced by drought, salt stress, cold acclimation and ABA in barley as well as in many other species. Dehydrins have highly conserved Lys-rich stretch of 15 amino acids (EKKGIMDKIKEKLPG) known as the "K segment". In addition many dehydrins contain also a stretch of Ser, the "S segment", which can be phosphorylated, and a further consensus amino acid sequence (-V/T-DEYGNP), the "Y segment" (Close, 1996). *Dhn5*, an acid dehydrin of 75kD, is the most abundant dehydrin accumulated during cold acclimation in barley (Crosatti et al., 1994; van Zee et al., 1995). Expression of low molecular weight dehydrins (*dhn1*, *dhn2* and *dhn9*) was also found associated to the developing of frost resistance under field conditions (Zhu et al., 2000).

The members of the *blt14* gene family are post-transcriptionally up-regulated (Dunn et al., 1994) in response to cold, but not to drought or ABA. Up to date five *blt14*-related genes have been isolated; they are expressed in different plant tissues and showed different threshold induction temperatures and genotype-dependent induction kinetics (Grossi et al., 1998). *In situ* hybridisation analysis of *cor* gene expression showed that *blt14* genes are strongly expressed in perivascular cell layers in the vascular-transition zone of cold-acclimated barley crown, although they are also present in other organs and tissues (Pearce et al., 1998).

A putative function can be assigned to several barley *cor* genes on the basis of their sequence similarities. The *blt4 cor* gene family encodes for a non specific Lipid Transfer Proteins (LTP). Although non specific LTPs do not bind specifically to lipids, they are known to transfer lipids between donor and acceptor membranes *in vitro*. All members of the *blt4* gene family have an extra-cellular transport consensus signal peptide in the N-terminus suggesting a possible involvement in wax synthesis or secretion (White et al., 1994), therefore a putative role in reducing stress-induced water loss can be assumed. The cDNA clone *blt801* encodes for a RNA-binding protein, the deduced amino acid sequence contains a N-terminal amino acid stretch with a consensus RNA-binding domain and a C-terminal domain with repeated glycine residues interspersed with tyrosine and arginine (Dunn et al., 1996).

Specific COR proteins have been found to be localised in the chloroplasts. The first *cor* gene isolated with such characteristics was *cor14b* (formerly *pt59* - Crosatti et al., 1995). The accumulation of *cor14b* occurs only at low temperatures, but it is enhanced after even brief exposure of the plants to light. The *cor14b* is stored in amounts only slightly greater in the resistant cultivars than in the susceptible ones, although the former have a higher induction-temperature threshold of *cor14b* than the latter (Crosatti et al., 1996). This fact may represent an evolutionary advantage enabling the resistant varieties in the field to prepare for the cold well ahead of the susceptible ones.

The analysis of *H. spontaneum* populations revealed a clear polymorphism concerning the accumulation of COR14b. While some accessions showed the same COR14b protein as cultivated barley, in a number of accessions, the COR14b antibody cross-reacted with an additional cold-regulated protein with a relative molecular weight of about 24 kDa (COR24). The accumulation of COR24 was very often associated with the absence of COR14b (Figure 1). Northern analysis showed that COR24 is encoded by a *cor* mRNA of 0.9 kb, the same size of the sequence coding for COR14b. These findings

TABLE 1. Main cold-related genes isolated and characterised in barley

Gene	Stress-related expression	Observation	Reference
blt4 gene family	Expressed in leaves during cold, drought and ABA treatment.	Codes for non specific lipid transfer protein, involved in wax synthesis or secretion. Mapped on chr. 2H.	Hughes et al., 1992; Dunn et al., 1991; White et al., 1994.
blt14 gene family	Expressed during cold treatment only.		Dunn et al., 1990; Cattivelli and Bartels, 1990; Grossi et al., 1998
Elongation factor 1α and 1Bβ, ribosomal protein S7 and L7A	Enhanced during cold treatment only.	Component of the protein synthesis machinery	Dunn et al., 1993; Baldi et al., 2001.
blt101 gene family	Expressed during cold acclimation in the perivascular layers of the transition zone of the crown.	The protein has a leader sequence for the secretory pathway.	Goddard et al., 1993: Pearce et al., 1998.
blt801	Expressed during cold and ABA treatment.	Codes for a RNA binding protein.	Dunn et al., 1996.
cor14b	Expressed during cold treatment and enhanced by light.	Expressed only in leaves, encodes for a chloroplast localised protein. Mapped on chr. 2H.	Cattivelli and Bartels, 1990; Crosatti et al., 1995; 1999.
tmc-ap3	Constiutively expressed, enhanced by cold.	Expressed only in leaves, encodes for a chloroplast localised protein. Mapped on chr. 1H.	Baldi et al., 1999.
dhn1 dhn2 dhn5	Expressed during drought, cold and ABA treatment.	LEA group 2. Mapped on chr. 6H.	van Zee et al., 1995; Zhu et al., 2000.
af93	Expressed during drought, and cold treatment.	Lea group 2. Mapped on chr. 6H.	Cattivelli and Bartels, 1990; Grossi et al., 1995.
hva1	Expressed in seedlings during cold, salt, ABA	LEA group 3.	Sutton et al., 1992.

suggest that post-translation changes may be responsible for the varying electrophoretic mobility of the *cor* protein immunologically related to COR14b (Crosatti et al., 1995). An ecophysiological evaluation of the COR14b/COR24 polymorphism showed that the COR14b allele is rare in all *H. spontaneum* populations deriving from coastal plain and desert localities, while it displays higher frequencies in population from cooler highlands. These may suggest a relationship between the COR14b/COR24 polymorphism and the adaptation of *H. spontaneum* to different environments (Crosatti et al., 1996).

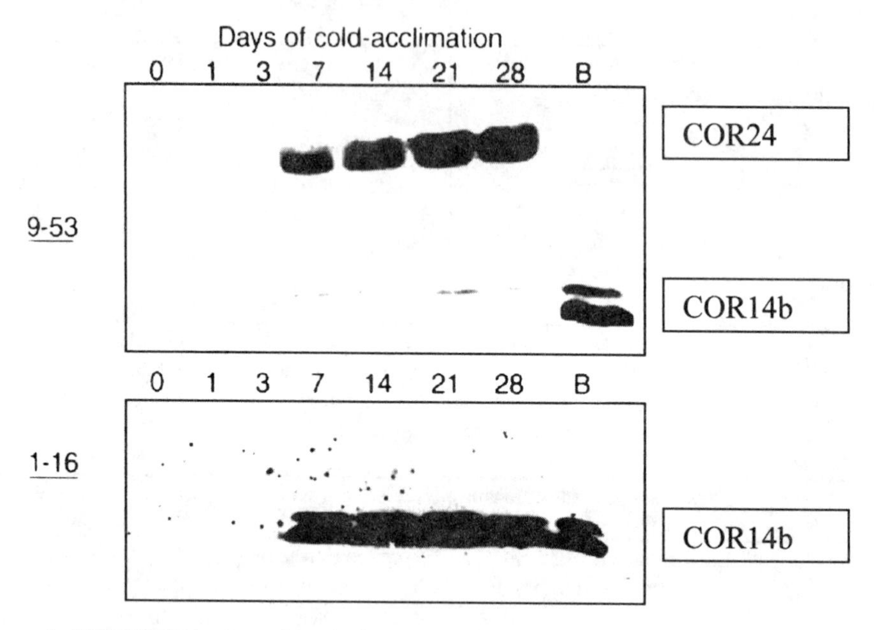

Figure 1. COR14/COR24 polymorphism in *H. spontaneum* populations. Equal amounts of proteins, extracted from *H. spontaneum* accession 9-53, carrying the COR24 allele, and from accession 1-16, carrying the COR14b allele, were separated by SDS-PAGE and hybridised with COR14b antibody. Numbers indicate the days of cold-acclimation, B = sample from cultivated barley hardened for 7d.

Besides the appearing of new mRNA species the adaptation to low temperature also involves the adjustment of constutively active metabolic processes, example are the genes involved in the amino acids transport and the genes coding for the protein synthesis machinery. The *cor* gene *tmc-ap3* encodes for a putative chloroplastic amino acid selective channel protein. *tmc-ap3* is expressed at low level under normal growing temperature, although its expression is strongly enhanced after cold treatment. A positive correlation between the expression of *tmc-ap3* and the frost tolerance was found either among barley cultivars or among cereals species. Western analysis showed that the *cor tmc-ap3* gene product is localised in the chloroplastic outer envelope membrane, supporting its putative function (Figure 2). Notably, frost-resistant cultivars accumulate COR TMC-AP3 protein more rapidly and at higher level than frost-sensitive ones. These results suggest that an increased amount of a chloroplastic amino acid selective channel

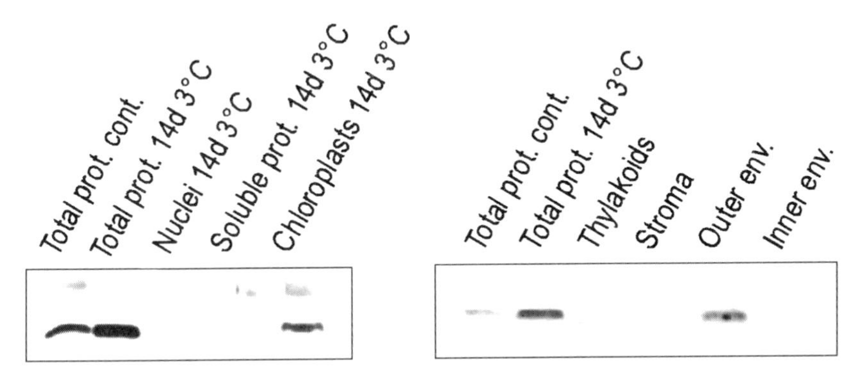

Figure 2. Sub-cellular (left panel) and sub-chloroplast (right panel) localisation of the *cor* protein TMC-AP3. Purified sub-cellular and sub-chloroplast protein fractions were separated on SDS-PAGE and tested with the TMC-AP3 specific antibody.

protein could be required for cold acclimation in cereals (Baldi et al., 1999).

Northern analysis revealed that expression of the elongation factor (EF) -1α (Dunn et al., 1993), EF1Bβ and two ribosomal protein (RP) genes S7 and L7A were enhanced following exposure to 3°C, but not during dehydration or exogenous ABA treatment (Baldi et al., 2001). The mRNA levels of EF1Bβ, RPS7 and RPL7A did not vary between cultivars with different frost tolerance, but differences in expression were found between different species.

The recent development of EST database allowed the evaluation of the relative abundance of given set of genes in the whole mRNA population. When a cDNA library constructed from cold-treated barley leaves mRNA was used as a source of cDNAs to develop an EST's database, about 8% of the clones sequenced corresponded to known *cor* genes. The most abundant *cor* mRNAs found during EST sequencing belonged to *cor413* (a sequence originally isolated in wheat), *blt4* and *blt14* gene families, these three classes of mRNAs represented about the 4% of the whole barley transcriptome at low temperature (Faccioli et al., 2001).

3.2. Field Evaluation of the Molecular Response to Cold

Many studies have been undertaken in controlled environments to demonstrate the association between gene expression and frost resistance (Pearce et al., 1996; Fowler et al., 1996) and the general conclusion is that *cor* gene expression leads to cold acclimation. Moving forward these investigations, Giorni et al. (1999) have monitored the molecular response to cold environments in barley plants grown in a field trial in order to assess the role of *cor* genes in the plant through autumn and winter. The results also show that *cor* genes are expressed all the time during winter in field grown plant, demonstrating the association between these genes and climatic conditions allowing the induction of frost resistance. Notably, barley plants in the field prepare their leaves for the cold well ahead, starting the activation of *cor* genes several weeks before frost.

Following the accumulation of *cor14b*, the highest mRNA expression level was found early in the season, while when the plants were exposed to the lowest temperatures they showed the highest amount of COR14 proteins. Fowler et al. (1996) have also demonstrated, under controlled environment, that development of frost tolerance closely follows the accumulation of COR proteins, while the expression of *cor* mRNAs peaks well before.

Interesting that winter and spring barley cultivars differ for their molecular responses when exposed to the same cold environment. The winter genotypes showed a generally higher level of COR14b accumulation than the spring ones during winter months. An example is reported in Figure 3. Winter hardiness is a very complex trait and obviously a high level of few *cor* mRNAs or proteins *per se* does not induce high frost resistance. Nevertheless, even if no attempts were made in the past to select for high expression level of COR14 protein, most winter genotypes showed all a high COR14 accumulation capacity, while reduced level of this protein was detected only in spring cultivars (Giorni et al., 1999). Because no genetic linkage exists between the genes controlling the growth habit (*Sh*, *sh2*, *sh3*, on chromosome 4H, 5H, and 1H respectively; Cattivelli et al., 2001) and the gene coding for COR14b localised onto chromosome 2H (Crosatti et al., 1996), the simple selection for winter hardiness has been effective to fix the high COR14 accumulation capacity.

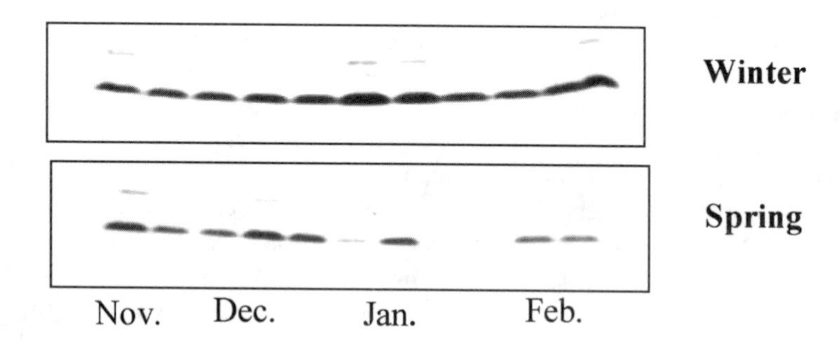

Figure 3. Field accumulation of COR14b in the winter cultivar Onice and in the spring cultivar Gitane during the winter season 1996/97. The experiment was sown in the middle of October, western analysis has been performed with total protein extracts isolated from leaves collected in the field every about 10 days from the end of November 96 till February 97.

3.3. Mechanisms Controlling Gene Expression

The precise nature of the primary sensor of cold stress is still not known, although a number mechanisms potentially involved in the perception of environmental stress (variation in membrane fluidity, Ca^{++} influx and protein kinases activation) have been identified (for details see other chapters of this book).

Works curried out with winter cereals provided increased evidence about the role of the chloroplast as primary cold/light stress sensor. Plants must constantly balance energy absorbed through the photosynthetic apparatus with energy utilised through metabolism. Low temperature in presence of light has the potential to upset this balance, leading to an overproduction of electrons which in turn can damage the reaction centre of the photosysthem II (PSII), and in particular the D1 protein, and may lead to the perturbation and inhibition of photosynthetic electron transport. This phenomenon is known as photoinhibition of photosynthesis (Andersson et al., 1992). The mechanisms leading to an inhibition of electron transport through PSII are not known, but the multiple reduction/oxidation (redox) reactions in PSII have a potential to create AOS (activated oxygen species) leading to the degradation of D1 protein (Russell et al., 1995). Some authors suggested that the redox state of PSII might be the first component in a redox sensing/signalling pathway, acting synergically to other signal transduction pathways to promote the molecular stress response. Gray et al. (1997) found that in rye the expression of the cold-induced gene *wcs19* is correlated with the reduction state of PSII. *Wcs19* can be induced either at 2°C and low light or at 20°C and high light. In agreement with the role of the chloroplast in the sensing/signalling pathway involved in cold response it has been found that mutations affecting chloroplast development modify the expression of several cold-regulated genes. In barley, the cold induced expression of *cor14b,* and *cor tmc-ap3* (two gene coding for chloroplastic localised proteins) and of the *blt14*-gene family was strongly reduced in plants carrying the a mutation at the albino locus a_n suggesting that a chloroplast deriving signal is required for cold-regulated gene expression (Grossi et al., 1998, Baldi et al., 1999, Crosatti et al., 1999) (Figure 4).

The signal transduction pathway mediated chloroplast is also interconnected with a light dependent signal. Indeed the expression of *cor14b* is also affected by light and light quality. Etiolated plants accumulate only a reduced amount of COR14b protein when exposed to low temperature in the dark. Nevertheless a flash of red or blue but not far-red or green light is able to enhance the accumulation of COR14b in etiolated plants (Figure 5), suggesting that phytochrome and blue light photoreceptors may be involved in the control of *cor14b* gene expression (Crosatti et al., 1999). Light can also acts negatively reducing the steady state transcript level of specific *cor* genes. Indeed etiolated plants accumulate *blt14*-related mRNAs at detectable level already at 22°C, when the same plants are then exposed to cold in absence of light an increased mRNA accumulation above the level present in green cold treated plants can be detected (Grossi et al., 1998) (Figure 4).

Experiments with the *albino* mutant and etiolated plants revealed that also the cold-dependent regulation of EF1Bβ, RPS7 and RPL7A transcripts is controlled by a chloroplast-related pathway impaired in the mutant. The messenger level of these genes appeared to be modulated by two distinct chloroplastic signals. The first (negative), reduced genes expression in green leaves grown at optimal temperature (22°C). In fact, *albino* mutant, lacking a normal chloroplast, showed a constitutive over-expression of RPS7, RPL7A and EF1Bβ genes which were not enhanced after exposure to cold. Etiolated plants also presented a constitutive over-expression, but when exposed to low temperature a further enhancement of gene expression was observed, indicating that in these plants and in the wild type a positive signal was present. This second signal, allowing the over-accumulation of RPS7, RPL7A and EF1Bβ mRNAs, was chloroplast-dependent. The two signals appeared to depend on chloroplast development, but while in *albino* plants both are absent, in etiolated plants only the negative one is lacking, therefore indicating distinct control mechanisms (Baldi et al., 2001).

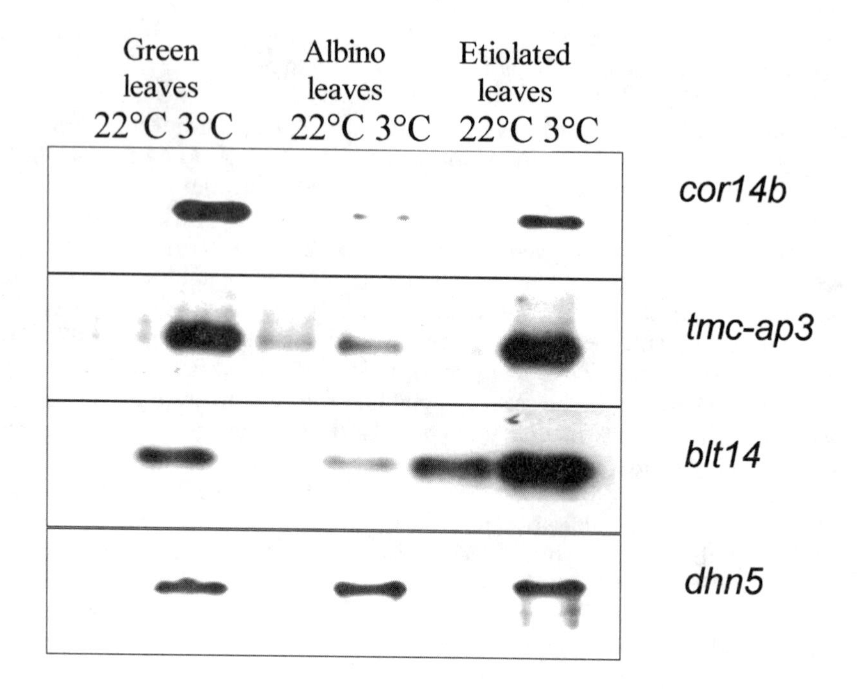

Figure 4. Expression of three chloroplast-dependent cold-regulated genes (*cor14b, cor tmc-ap3* and *blt14*) in comparison with *dhn5* a gene whose expression is not affected by chloroplast mutation. The corresponding transcripts were analysed in leaves from wild type plant (green leaves), in leaves from plants carrying the a_n mutation (albino leaves) and in leaves from etiolated wild type plants.

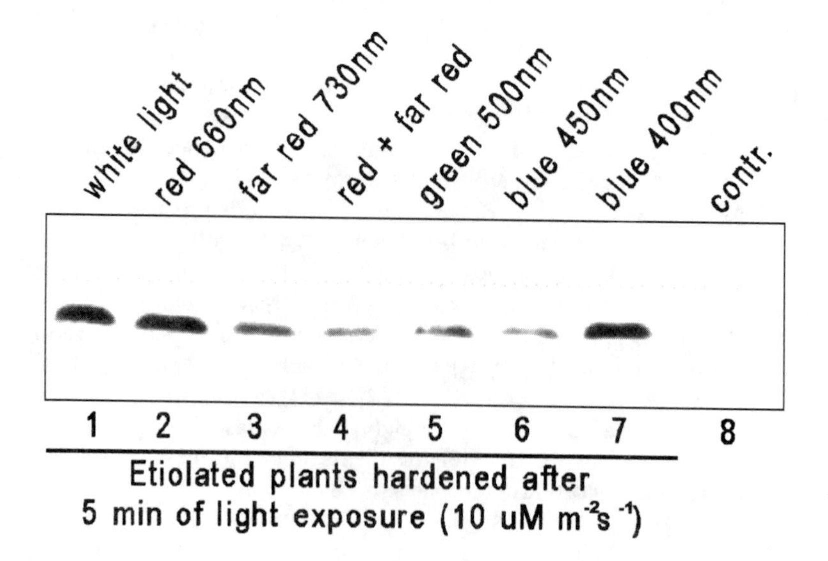

Figure 5. Effect of a flash of monocromatic light on the accumulation of COR14b in etiolated barley plants

A number of molecules have been proposed to be involved in cold stress signal transduction pathways. This is probably due to the fact that low temperatures activate, at the same time, different pathways in the cell. The interaction and reciprocal modulation among these signalling cascades, which are probably connected with each other as a net, determines the specific stress response of the cell. One of the molecules involved in stress signalling is the plant hormone abscisic acid (ABA). It is well known that low temperatures induce a temporary increase in endogenous ABA levels in plants. In barley, ABA concentration reaches a peak in 24h; then the amount of ABA returns to basal levels for the remaining time of hardening (Murelli et al., 1995). Because exogenous application of ABA under non-stressing conditions leads to a certain degree of protection against freezing stress (Chen and Gusta 1983), ABA has been proposed as a necessary mediator in triggering most of the physiological and adaptive stress responses. In barley, besides a number of ABA regulated *cor* genes (the most common are the dehydrins, Close et al., 1989) there are also many cold responsive genes which expression is not affected by ABA (Cattivelli and Bartels, 1990). A comparative study between the drought and the cold molecular responses showed that, while ABA is the key hormone in drought response (application of exogenous ABA induces the expression of most, but not all, genes induced by dehydration), its role in cold response is limited (only few cold-regulated genes are controlled by ABA) (Grossi et al., 1992). Among the stress related barley genes that equally respond to cold or drought, some show an expression profile independent from the presence of ABA. The dehydrin like gene *af93* is a typical example of a sequence activated in response to cell dehydration due to either cold or drought stress, but insensitive to ABA accumulation (Figure 6).

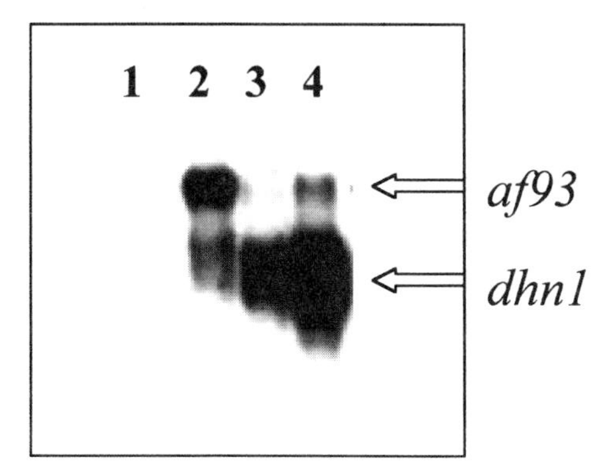

Figure 6. Different expression profile of barley stress related genes in response to different stress conditions A filter carrying barley mRNA isolated from control (lane 1), cold (lane 2), ABA (lane 3) or drought (lane 4) treated plants has been hybridised with a mixture of two probes corresponding to *af93* (about 1500 bp) and *dhn1* (about 900 bp) sequences.

The expression analysis of the barley *cor* genes under different stress conditions (cold, drought, ABA) as well as in wild type and *albino* mutants allowed the classification of the cold-responsive genes in different signal transduction pathways involving either the plant hormone ABA or the chloroplast as intermediate between signal perception and gene expression as depicted in Figure 7. Notably, all *cor* genes showing a chloroplast-dependent regulation are express in response to low temperature only. On the contrary, *cor* genes showing a chloroplast-independent induction are also expressed when plant are exposed to drought.

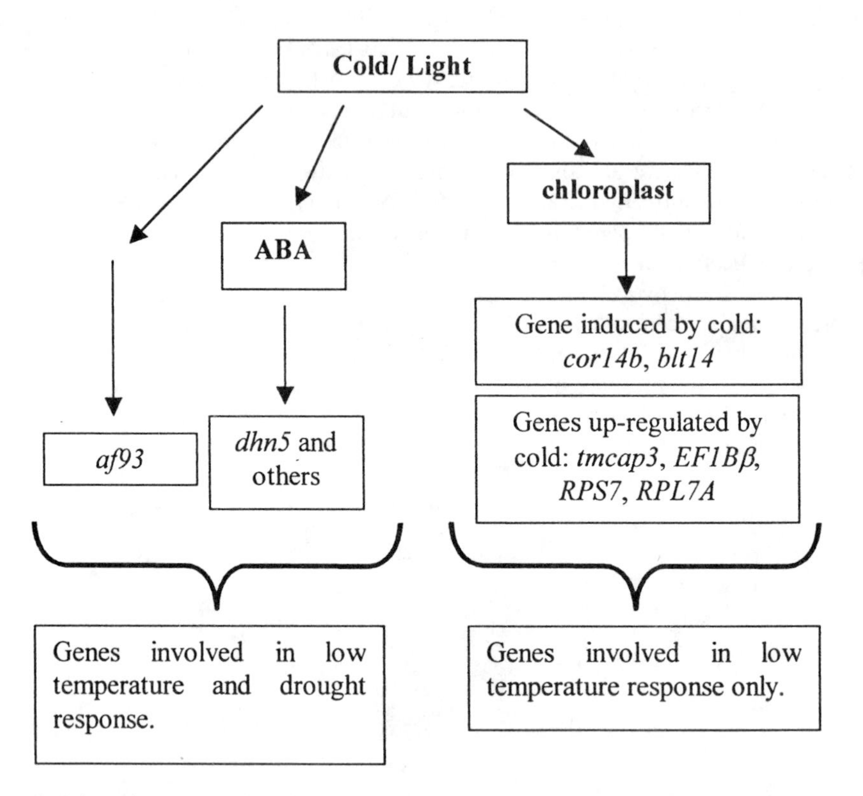

Figure 6. Schematic representation of the cold-responsive signal transduction pathways described in barley involving either the plant hormone ABA or the chloroplast as intermediate between signal perception and gene expression.

Transduction pathways activated by cold stress lead to gene activation. One of the approaches to understand how stress-activated genes are regulated is to analyse their promoters in order to characterise the sequences (*cis*-acting elements) requested for gene activation and to identify transcription factors which bind to these sequences. One of the best characterised *cis*-acting elements is the ABA-responsive element (ABRE) (Marcotte et al., 1989), which contains the palindromic motif CACGTG with the G-box ACGT core element. The ACGT-like elements have been observed in a number of *cor* genes induced by low temperature after increase of intracellular ABA concentration.

The drought response element (DRE, sequence: TACCGACAT) was found in the promoter sequence of several *Arabidopsis* cold and drought responsive genes (Yamaguchi-Shinozaki and Shinozaki, 1994). The DRE core motif, CCGAC, was also present in the promoter sequences of cold-regulated genes from wheat (Ouellet et al., 1998) and barley (Dunn et al., 1998). The promoter of the wheat cold-induced gene *wcs120* contains several putative low temperature responsive elements (LTRE) with the core motif CCGAC. This promoter sequence was active in both monocotyledon and dicotyledon species, suggesting that universal transcription factors responsive to low temperature may be present in all plants (Ouellet et al., 1998). The promoter of the barley cold-induced gene *blt4.9* contains a CCGAC element as well as the related sequence CCGAAA. Nevertheless only the CCGAAA element acted as binding site for a low-mobility nuclear protein complex in electrophoretic mobility shift assay, suggesting that CCGAAA and not CCGAC works as LTRE in barley (Dunn et al., 1998).

Post-transcriptional control of cold-responsive genes is also known. In barley, particularly, the low temperature responsive genes are regulated either at the level of gene activation or at the level of mRNA stability (Dunn et al., 1994) and low temperature-dependent protein factors are involved to modulate mRNA stability (Phillips et al., 1997). The activity of the low temperature-dependent protein factors can also be modified by the presence of green chloroplast. Indeed, etiolated barley plants accumulate the mRNAs corresponding to the cold-regulated gene *blt14* at detectable level already at 22°C. When the same plants are exposed to cold in absence of light an increased mRNA accumulation above the level present in green cold treated plants can be detected (Grossi et al., 1998). Several low temperature-induced RNA-binding proteins, which might stabilise or activate mRNA, have been isolated in *Arabidopsis* and barley (Carpenter et al., 1994; Dunn et al., 1996).

4. ASSOCIATIONS BETWEEN QTLs AND STRESS-RELATED GENES

The genetic and molecular dissection of stress tolerance have led to the identification of either genomic regions involved in stress tolerance (major loci or QTLs), or DNA sequences known to play a role in molecular stress responses (stress-related genes, *cis*-acting elements and transcription factors). Although QTL analysis and gene cloning have been used to investigate the same stress responses, the relationship between QTLs and stress-related sequences is still far from understood and will be a challenge for the near future.

Molecular analysis of the stress response identified a number of genes which are thought to have a role in stress tolerance. Therefore, they were used as "candidate genes" to search for possible co-segregation with known QTLs. Many stress-related genes have been mapped and some of them have been shown to co-segregate with stress tolerance QTLs. Two *Dhn* loci (*Dhn1/Dhn2* and *Dhn9*) are located in the same region of chromosome group 5 where *Vrn-1* and *Fr1* major loci, cold and salt tolerance QTLs and ABA accumulation QTLs have been mapped. Nevertheless, only a few studies provide genetic evidence suggesting that the stress tolerance effect explained by a given QTL can be attributed to a co-mapping stress-related gene. In cowpea the accumulation of the 35 kDa dehydrin was found involved in chilling tolerance during seedling emergence. Allelic differences in the coding region of the dehydrin structural gene map to the same position as the dehydrin protein presence/absence trait which in turn is associated with chilling tolerance/susceptibility (Ismail et al., 1999). These results also demonstrate that allelic variations in a stress-related gene can significantly alter plant stress tolerance ability.

Besides the examples described above there is a number of stress-related genes located clearly outside of any stress tolerance QTLs. Although the barley cold-regulated genes *cor14b* and *cor tmc-ap3* (chromosomes 2H and 1H, respectively) are expressed at higher levels in frost resistant than in susceptible cultivars, none of them maps on chromosome 5H where almost all cold tolerance QTLs have been located [4, 15, 82]. These results raise the possibility that the molecular basis of a QTL for stress tolerance could be explained by a regulatory gene able to control the expression of many stress-related genes. To investigate the genetic relationship between frost tolerance and the expression of cold-regulated genes, the expression and regulation of the wheat gene homologous to the barley cold-regulated *cor14b* gene was compared in frost sensitive and frost tolerant wheat genotypes at different temperatures. At 18/15°C (day/night temperatures) frost tolerant plants accumulated *cor14b* homologous mRNAs and expressed COR14b proteins whereas the sensitive plants did not. Nevertheless, when the temperature was decreased down to 2°C both tolerant and susceptible genotypes expressed *cor14b* mRNA at the same level. This result indicates that the threshold induction temperature of the wheat gene homologous to *cor14b* is higher in frost resistant plants. Parallel experiments made with tolerant and susceptible barley genotypes confirmed the results obtained in wheat, although the threshold induction temperature difference between spring and winter barley was rather small (Crosatti et al., 1996). This may reflect the smaller genetic variability for frost resistance found in barley in comparison with that described in wheat (Homo 1994; Stanca et al., in press).

Studies made with wheat chromosome substitution lines showed that the threshold induction temperature polymorphism of the *cor14b* gene in wheat is controlled by locus(i) located on the chromosome 5A, while *cor14b* gene was mapped onto the long arm of the chromosome 2A. The analysis of single chromosome recombinant lines derived from the cross between Chinese Spring/*Triticum spelta* 5A and Chinese Spring/Cheyenne 5A identified two loci (*Rcg1* and *Rcg2*) with additive effect involved in the genetic control of *cor14b* homologous mRNAs accumulation. The first locus was positioned tightly linked with marker *psr911*, while the second one was located between marker *Xpsr2021* and *Frost resistance* gene *Fr1* (Vagujfalvi et al., 2000). Because of the high degree of sinteny among the *Triticeae* homeologous group 5 in the *Vrn-A1-Fr1* region (Galiba et al., 1995) orthologous loci controlling the expression of *cor* genes are likely to be present in all *Triticeae* genomes. Similarly, Fowler et al. (1996) have reported that loci of the *Vrn-1A*/*Fr1* region explain the different expression levels of *wcs120* found in resistant and susceptible wheat and rye cultivars. The fact the frost tolerant and frost susceptible genotypes (either barley or wheat) differ more for the threshold induction temperature of the *cor* genes than for the stress related genes *per se,* suggests that different alleles at the loci controlling *cor* gene expression may explain an important part of the genetic variability for frost resistance. After the discovery of the *Arabidopsis* DRE-CBF transcription factors it has been suggested that the cereal loci controlling stress tolerance (i.e. the *Fr1* locus) could represent the DRE-CBF cereal homologous genes (Sarhan et al., 1998).

The recent advances in the molecular understanding of cold resistance have produced a number of molecular tools that will contribute to the improvement of barley adaptation to cold environments. It is generally accepted that the expression of stress related genes is an essential part of the plant adaptation processes, although for many genes is still lacking a direct evidence of their function. The use of transgenic plants, the developing of new genomic tools such as the identification of mutants with altered response to cold will contribute to further understand the molecular bases of frost resistance in cereals.

5. ACKNOWLEDGEMENTS

This work was supported by CNR "Agenzia 2000" and by the CNR-MTA co-operation program.

REFERENCES

Andersson B., Salter, A.H. Virgin, I. Vass I. and Styring S., 1992, Photodamage to photosystem II-Primary and secondary events, *J. Photochem. Photobiol. Ser B Biology* **15**: 15-31.

Baldi P., Grossi M., Pecchioni N., Valè G. and Cattivelli L., 1999, High expression level of a gene coding for a chloroplastic amino acid selective channel protein is correlated to cold acclimation in cereals, *Plant Mol. Biol.* **41**: 233-243.

Baldi P., Valè G., Mazzucotelli E., Govoni C., Faccioli P., Stanca A.M. and Cattivelli L., 2001, The transcrpts of several components of the protein synthesis machinery are cold regulated in a chloroplast-dependent manner in barley and wheat, *J. Plant Physiol.* **158**: 1541-1546.

Carpenter C.D., Kreps J.A. and Simon A.E., 1994, Genes encoding glycine-rich *Arabidopsis thaliana* proteins with RNA-binding motifs are influenced by cold treatment and an endogenous circadian rhythm, *Plant Physiol.* **104**: 1015-1025.

Cattivelli L, Baldi P., Crosatti C., Di Fonzo N., Faccioli P., Grossi M., Mastrangelo A.M., Pecchioni N. and Stanca A.M., Chromosome regions and stress-related sequences involved in resistance to abiotic stress in *Triticeae*, *Plant Mol. Biol.* in press.

Cattivelli L., Baldi P., Crosatti C., Grossi M., Valè G. and Stanca A.M., 2001, Genetic bases of barley physiological response to stressful conditions, in: *Barley Science: Recent Advances from Molecular Biology to Agronomy of Yield and Quality* G.A. Slafer, J.L. Molina-Cano, J.L. Araus, R. Savin, I. Ramagosa, eds., Food Product Press, New York, USA, pp. 269-314.

Cattivelli L. and Bartels D., 1989, Cold-induced mRNAs accumulate with different kinetics in barley coleoptiles. *Planta* **178**: 184-188.

Cattivelli L. and Bartels D., 1990, Molecular cloning and characterization of cold-regulated genes in barley, *Plant Physiol.* **93**, 1504-1510.

Chen T.H.H. and Gusta L.V., 1983, Abscisic acids-induced freezing resistance in cultured plant cells, *Pant Physiol.* **71**: 362-365.

Close T.J., 1996, Dehydrins: emerge of a biochemical role of a family of a plant dehydration proteins, *Physiol. Plant.* **97**: 795-803.

Crosatti C., Rizza F. and Cattivelli L., 1994, Accumulation and characterization of the 75kDa protein induced by low temperature in barley, *Plant Sci.* 97: 39-46.

Crosatti C., Soncini C., Stanca A.M. and Cattivelli L., 1995, The accumulation of a cold-regulated chloroplastic protein is light-dependent, *Planta* **196**: 458-463.

Crosatti C., Nevo E., Stanca A.M. and Cattivelli L., 1996, Genetic analysis of the accumulation of COR14 proteins in wild (*Hordeum spontaneum*) and cultivated (*Hordeum vulgare*) barley, *Theor. Appl. Genet.* **93**: 975-981.

Crosatti C., Polverino de Laureto P., Bassi R. and Cattivelli L., 1999, The interaction between cold and light controls the expression of the cold-regulated barley gene *cor14b* and the accumulation of the corresponding protein, *Plant Physiol.* **119**: 671-680.

Doll, H., Hahr V. and Sogaard B., 1989, Relationship between vernalization requirement and winter hardiness in double haploid of barley, *Euphytica* **42**: 209-213.

Dunn M.A., Hughes M.A., Pearce R.S. and Jack P.L., 1990, Molecular characterization of a barley gene induced by cold treatment, *J. Exp. Bot.* **41**: 1405-13.

Dunn M.A., Morris A., Jack P.L. and. Hughes M.A, 1993, A low temperature-responsive translation elongation factor 1α from barley (*Hordeum vulgare L*), *Plant Mol. Biol.* **23**: 221-225.

Dunn M.A., White A.J., Vural S. and Hughes M.A., 1998, Identification of promoter elements in a low-temperature-responsive gene (*blt4.9*) of barley (*Hordeum vulgare L*), *Plant Mol. Biol.* **38**: 551-564.

Dunn M.A., Brown K., Lightowlers R.L. and Hughes M.A., 1996, A low temperature-responsive gene from barley encodes a protein with single stranded nucleic acid binding activity which is phosphorylated *in vitro, Plant Mol. Biol.* **30**: 947-959.

Dunn M.A., Hughes M.A., Zhang L., Pearce R.S., Quigley A.S. and Jack P.L., 1991, Nucleotide sequence and molecular analysis of the low-temperature induced cereal gene, *blt4, Mol. Gen. Genet.* **229**: 389-394.

Dunn M.A., Goddard N.J., Zhang L., Pearce R.S. and Hughes M.A., 1994, Low-temperature-responsive barley genes have different control mechanisms, *Plant Mol. Biol.* **24**: 879-888.

Faccioli P., Pecchioni N., Cattivelli L., Stanca A.M. and Terzi V., 2001, Expressed sequence tags (ESTs) from cold acclimated barley identify novel plant genes, *Pant Breed.* in press.

Fowler D.B., Chauvin L.P., Limin A.E. and Sarhan F., 1996, The regulatory role of vernalization in the expression of low-temperature-induced genes in wheat and rye, *Theor. Appl. Genet.* **93**: 554-559.

Galiba G., Quarrie S.A., Sutka J., Morgounov A. and Snape J.W., 1995, RFLP mapping of the vernalization (*Vrn1*) and frost resistance (*Fr1*) genes on chromosome 5A of wheat. *Theor. Appl. Genet.* **90**: 1174-1179.

Giorni E., Crosatti C., Baldi P., Grossi M., Marè C., Stanca A.M. and Cattivelli L., 1999, Cold-regulated genes expression during winter in frost tolerant and frost susceptible barley cultivars grown under field conditions, *Euphytica*, **106**: 149-157.

Goddard N. J., Dunn M.A., Zhang L., White A.J., Jack P.L. and Hughes M.A., 1993, Molecular analysis and spatial expression pattern of a low-temperature-specific barley gene, *blt101, Plant Mol. Biol.* **23**: 871-879.

Gray G.R., Chauvin L-P., Sarhan F. and Huner N.P.A., 1997, Cold acclimation and freezing tolerance. A complex interaction of light and temperature, *Plant Physiol.* **114**: 467-474.

Grossi M., Giorni E., Rizza F., Stanca A.M. and Cattivelli L., 1998, Wild and cultivated barleys show differences in the expression pattern of a cold-regulated gene family under different light and temperature conditions, *Plant Mol. Biol.* **38**: 1061-1069.

Grossi M., Cattivelli L., Terzi V. and Stanca A.M., 1992, Modification of gene expression induced by ABA, in relation to drought and cold stress in barley shoots, *Plant Physiol. Biochem.* **30**: 97-103.

Grossi M., Gulli M., Stanca A.M. and Cattivelli L., 1995, Characterization of two barley genes that respond rapidly to dehydration stress, *Plant Sci.* **105**: 71-80.

Hayes P.M., Blake T., Chen T.H.H., Tragoonrung S., Chen F., Pan A. and Liu B., 1993, Quantitative trait loci on barley (*Hordeum vulgare* L.) chromosome-7 associated with components of winterhardiness, *Genome* **36**: 66-71.

Hommo L.M., 1994. Hadening of some winter wheat (*Triticum aestivum* L.), rye (*Secale cereale* L.), triticale (X *Triticosecale* Wittmack) and winter barley (*Hordeum vulgare* L.) cultivars during autumn and final winter survival in Finland, *Plant Breed.* **112**: 285-293.

Hughes M A., Dunn M.A., Pearce R.S., White A.J. and Zhang L., 1992, An abscisic acid responsive low temperature barley gene has homology with a maize phospholipid transfer protein, *Plant Cell Env.* **15**: 861-866.

Ismail, A.M., Hall A.E. and Close T.J., 1999, Allelic variation of a dehydrin gene cosegregates with chilling tolerance during seedling emergence, *Proc. Natl. Acad. Sci. USA* **96**: 13566-13570.

Laurie D.A., Pratchett N., Bezant J.H. and Snape J.W., 1995, RFLP mapping of five major genes and eight quantitative trait loci controlling flowering time in a winter X spring barley (*Hordeum vulgare* L.) cross, *Genome* **38**: 575-585.

Marcotte, W.R. Jr, Russel S.H., and Quatrano R.S., 1989, Abscisic acid response sequences from the *Em* gene of wheat, *Plant Cell* **1**: 969-976.

Murelli C., Rizza F., Marinone Albini F., Dulio A., Terzi V. and Cattivelli L., 1995, Metabolic changes associated with cold-acclimation in contrasting genotypes of barley, *Physiol. Plant.* **94**: 87-93.

Ouellet, F., Vazquez-Tello A.and Sarhan F., 1998, The wheat *wcs120* promoter is cold-inducible in both monocotyledonous and dicotyledonous species, *FEBS let.* **423**: 324-328.

Pan A., Hayes P.M., Chen F., Chen T.H.H., Blake T., Wright S., Karsai I. and Bedö Z., 1994, Genetic analysis of the components of winterhardiness in barley (*Hordeum vulgare* L.), *Theor. Appl. Genet.* **89**: 900-910.

Pearce R. S., Houlston C.E., Atherton K.M., Rixon J.E., Harrison P., Hughes M.A. and Dunn M.A., 1998, Localization of expression of three cold-induced genes, *blt101*, *blt4.9* and *blt14* in different tissues of the crown and developing leaves of cold-acclimated cultivated barley, *Plant Physiol.* **117**: 787-795.

Pearce R.S., Dunn M.A., Rixon J., Harrison P. and Hughes M.A., 1996, Expression of cold-inducible genes and frost hardiness in the crown meristem of young barley (*Hordeum vulgare* L. cv. Igri) plants grown in different environments, *Plant Cell Env.* **19**: 275-290.

Phillips J.R., Dunn M.A. and Hughes M.A., 1997, mRNA stability and localisation of the low temperature-responsive gene family *blt14, Plant Mol. Biol.* **33**: 1013-1023.

Plaschke J., Borner A., Xie D.X., Koebner R.D.M., Schlegel R. and Gale M.D., 1993, RFLP mapping of genes affecting plant height and growth habit in rye, *Theor. Appl. Genet.* **85**: 1049-1054

Russel A.W., Critchley C., Robinson S.A., Franklin L.A., Seaton G.G.R., Chow W-S., Anderson J. and Osmond C.B., 1995, Photosystem II regulation and dynamics of the chloroplast D1 protein in *Arabidopsis* leaves during photosynthesis and photoinhibition, *Plant Physiol.* **107**: 943-952.

Sarhan, F. and Danyluk J., 1998, Engineering cold-tolerant crops throwing the master switch, *Trends Plant Sci.* **3**: 289-290.

Snape J.W., Semikhodskii A., Fish L., Sarma R.N., Quarrie S.A., Galiba G. and Sutka J., 1997, Mapping frost tolerance loci in wheat and comparative mapping with other cereals, *Acta Agr. Hung.* **45**: 265-270

Stanca A.M., Romagosa I., Takeda K., Lundborg T., Terzi V., Cattivelli L., Diversity in abiotic stresses, in: *Diversity in barley (Hordeum vulgare L.),* R. von Bothmer, H. Knüpffer, T. van Hintum, K. Sato, eds., Elsievier Science, in press.

Storlie E.W., Allan R.E. and WalkerSimmons M.K., 1998, Effect of the *Vrn1-Fr1* interval on cold hardiness levels in near-isogenic wheat lines, *Crop Sci.* **38**: 483-488.

Sutka J., 1981, Genetic studies of frost resistance in wheat, *Theor. Appl. Genet.* **59**: 145-152.

Sutton F., Ding X. and Kenefrik D.G., 1992, Group 3 Lea genes *HVA1* regulation by cold acclimation and deacclimation in two barley cultivars with varying freeze resistance, *Plant Physiol.* **99**: 338-340.

Vágújfalvi A., Crosatti C., Galiba G., Dubcovsky J. and Cattivelli L., 2000, Two loci on wheat chromosome 5A regulate the differential cold-dependent expression of the *cor14b* gene in frost-tolerant and frost-sensitive genotypes, *Mol. Gen. Genet.* **263**: 194-200.

Van Zee K., Chen F.Q., Hayes P.M., Close T.J. and Chen T.H.H., 1995, Cold-specific induction of a dehydrin gene family member in barley, *Plant Physiol.* **108**, 1233-1239.

Veisz O. and Sutka J., 1989, The relationships of hardening period and the expression of frost resistance in chromosome substitution lines of wheat. *Euphytica* 43: 41-45.

White A. J., Dunn, M.A. Brown K. and Hughes M.A., 1994, Comparative analysis of genomic sequence and expression of a lipid transfer protein gene family in winter barley, *J. Exp. Bot.* **45**: 1885-1892.

Yamaguchi-Shinozaki K. and Shinozaki K., 1994, A novel *cis*-acting element in an *Arabidopsis* gene is involved in responsiveness to drought, low temperature or high-salt stress, *Plant Cell* **6**:251-264.

Zhu B., Choi D.W., Fenton R. and Close T.J., 2000, Expression of the barley dehydrin multigene family and the development of freezing tolerance, *Mol. Gen. Genet.* **264**: 145-153.

INVOLVEMENT OF GLUTATHIONE AND CARBOHYD-RATE BIOSYNTHESIS MOREOVER *COR14B* GENE EX-PRESSION IN WHEAT COLD ACCLIMATION

Gábor Galiba, Gábor Kocsy, Ildikó Kerepesi, Attila Vágujfalvi, Luigi Cattivelli, and József Sutka*

1. INTRODUCTION

The winter-hardiness, including frost tolerance has long been regarded as a trait being under complex multigenic control. However, it would still be possible that major differences in cold-adaptation between species or cultivars depend on allelic differences in a small number of genes as suggested in cold-acclimation in pea and *Solanum* (Liebefeld et al., 1986; Stone et al., 1993). Since common wheat (*Triticum aestivum* L.) is a hexaploid its vital genes are replicated. This permitted Sears (1953) to develop series of chromosome substitution lines. By comparing chromosome substitution lines with the parental lines it was possible to determine which chromosomes carry gene locus for freezing tolerance. The analysis of substitution lines showed that at least ten of the 21 pairs of the chromosomes are involved in the control of frost tolerance (Sutka 1981). However, major genes influencing frost tolerance (*Fr*) and vernalisation requirement (*Vrn*) were localized on the long arm of 5A and 5D chromosomes (Galiba et al., 1995; Snape et al. 1997). Of particular importance for adaptation to autumn sowing are the genes for vernalisation requirement. *Vrn* genes determine the needs for cold temperature required for flower development. Recent studies indicated that *Vrn1-Fr1* interval on 5A chromosome of wheat has a major effect on freezing tolerance (Storlie et al., 1998). Conservation of gene order (synteny) in *Triticeae* is well known and this is true for

*Gábor Galiba, Gábor Kocsy, Attila Vágújfalvi, and József Sutka, Agricultural Research Institute of the Hungarian Academy of Sciences, H-2462 Martonvásár, Hungary. Ildikó Kerepesi, Department of Genetics and Molecular Biology, University of Pécs, H-7601 Pécs, Hungary. Luigi Cattivelli, 3) Experimental Institute for Cereal Research Via S. Protaso, 302; I-29017 Fiorenzuola d'Arda, Italy.

Plant Cold Hardiness, edited by Li and Palva
Kluwer Academic/Plenum Publishers, 2002

the *Vrn1-Fr1* interval studied in barley, rye and *Triticum monococcum*, as well (reviewed in Galiba et al., 1997a). For example mapping of a quantitative trait loci (QTL) in barley has resulted in the identification of a 21-cM region on chromosome 7 that has a major role in frost tolerance. This region accounted for 32% of the variance in LT_{50} values and 39-79% of the variance in winter field survival (Hayes et al., 1993).

The mechanism whereby the *Vrn1-Fr1* interval affects freezing tolerance remains to be determined. One possible approach to solve this puzzle is the localization of genes affecting metabolic processes during cold hardening (acclimation). To achieve the full genetic potential of frost tolerance the plant must be hardened. Under natural conditions the cold hardening (acclimation) takes place in autumn when the temperature gradually decreases to 0°C over several weeks. Temperatures of 2-5°C and photoperiods of about 12 h are considered to be optimal for cold hardening under controlled environmental conditions. During cold acclimation a complex of responses took place in plants at the cellular, physiological and developmental levels resulting the enhancement of the frost tolerance. Among these we investigated the genetic regulation of the glutathione (GSH) biosynthesis, the regulation of carbohydrate accumulation, and the expression and regulation of the wheat homologue of the barley cold-regulated gene *cor14b* (Fig. 1.). To verify the genetic relationship between the frost tolerance and the cold-induced changes in metabolism, chromosome substitution, recombinant and deletion lines were studied.

2. MATERIALS AND METHODS

2.1. Plant Material and Growth Conditions

Frost-resistant Cheyenne (Ch) and -sensitive Chinese Spring (CS) bread wheat (*Triticum aestivum*) genotypes and a sensitive *Triticum spelta* (*Tsp.*) accession, Chinese Spring/Cheyenne (CS/Ch) 5A, and CS/*Tsp.* 5A chromosome substitution lines as well as single chromosome recombinant lines, developed from the cross between CS/*Tsp.* 5A and CS/Ch 5A (Galiba et al., 1995) were used. The CS 5A chromosome deletion lines were kindly presented by B.S. Gill (Kansas State University, USA). The *cor14b* gene was mapped using an F_2 population derived from a cross between a cultivated *T. monococcum* (DV92) and a wild *T. monococcum* ssp. *aegilopoides* (G3116) (Dubcovsky et al., 1996).

For studying the time dependent changes of glutathione (GSH) and carbohydrate biosynthesis the seedlings were raised in garden soil in wooden boxes. The plants were grown in growth chamber (Conviron, Canada) at 15/10°C day/night temperature for 2 weeks with 16 h illumination at 260 $\mu mol\ m^{-2}\ s^{-1}$, then at 10/5°C for a week (prehardening) and after that at 4/2°C (cold hardening) for 51 days. The plant material for the carbohydrate and thiol determination was harvested at the start and end of prehardening, after 3 d hardening and subsequently every 8th day. The frost tolerance of the seedlings was tested in the same time.

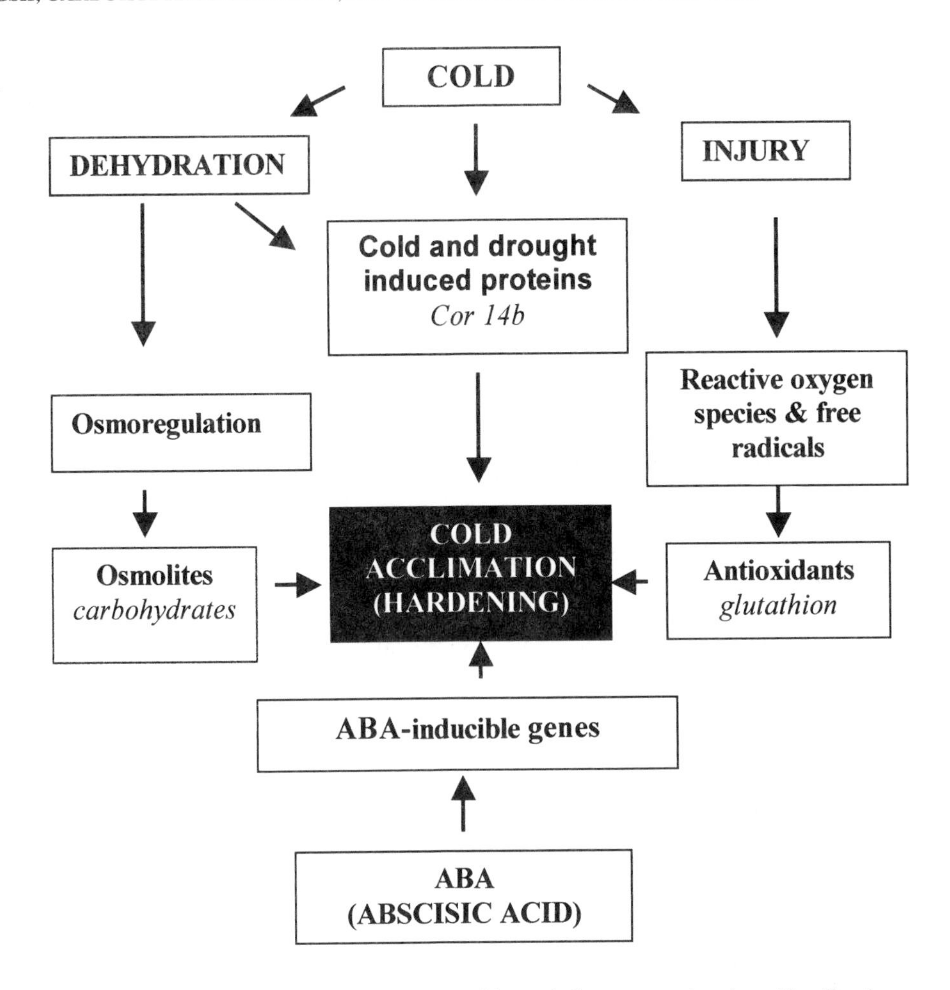

Figure 1. Schematic representation of the cold inducible metabolic processes triggering cold acclimation.

The boxes were transferred into the frost testing chamber, where the temperature was lowered by 2°C/h to -10°C, and the plants were kept at this temperature for 2 days without illumination. After freezing, the temperature was raised by 2°C/h to + 1°C for 15 h. Then, the leaves were cut off a few centimetres above the soil and the plants were let to recover (17°C/16°C day/night temperatures, Q = μmol m^{-2} s^{-1}) for 18 days. Frost tolerance was assessed in term of re-growth on a 0 (death) to 5 (undamaged) scale, and also as the percentage of survival .

For the investigation of the in vivo GSH synthesis 10-day old seedlings grown in hydroponics were transferred onto a nutrient solution containing [^{35}S]sulfate and further cultivated either at 15/10 °C or at 4/2 °C with a photoperiod of 16 h at 260 μmol m^{-2} s^{-1} for 3 days.

Hydroponically grown plants were used to compare the cold (+2°C) and the 15% PEG-4000 (Sigma) induced carbohydrate accumulation in CS and in CS 5A chromosome deletion lines for 39 days. The plants were illuminated using 260 μmol m^{-2} s^{-1} for 18 h. The PEG treated plants were maintained at 18/13°C (day/night) temperatures.

To study cold induced *cor14b* gene expression seedlings were raised hydroponically on modified Hoagland solution (Nagy and Galiba, 1995) for two weeks at 18°C 16 h light (260 μmol m^{-2} s^{-1})/13°C 8 h dark and subsequently cold hardened at 2°C. Alternatively, plants were also grown at 25°C for 6 days.

2.2. Biochemical and Molecular Analysis

Total water-soluble carbohydrate, glucose, fructose, sucrose, fructan and glucan contents were determined from fresh leaves of the plants according to Kerepesi et al. (1996). Determination of thiols GR, protein and ^{35}S-radiolabelled compounds written in details by Kocsy et al. (2000a). The details of the western analysis of COR14b protein and the northern analysis of *cor14b* mRNA are described by Crosatti et al. (1999).

3. ROLE OF GLUTATHIONE BIOSYNTHESIS IN COLD ACCLIMATION

During the cold hardening the plants are subjected to a moderate low temperature stress, which induces oxidative stress because of an imbalance between the production and removal of the reactive oxygen species (Prasad et al., 1994). The activation of oxygen by the photosystems in the presence of excessive light is probably the major site of free radicals production in leaves but other electron transport systems, including those on the mitochondria or plasmalemma, may also contribute especially in the non-photosynthetic tissues. If these reactive oxygen species like superoxide radical, hydrogen peroxide and hydroxyl radical are present in excess, they damage the nucleic acids, proteins, membrane lipids and pigments. Besides their disadvantageous effects, the reactive oxygen species are involved in a signal transduction chain, which activates different stress-responsive genes (Foyer et al., 1997; Kocsy et al., 2001a). The level of reactive oxygen species is controlled by the antioxidant system whose important component is the ascorbate-glutathione cycle. In this cycle glutathione takes part in the degradation of hydrogen peroxide. Glutathione (GSH) is a tripeptide which is synthesized in two ATP-dependent steps: first cysteine and glutamate are bound to γ-glutamylcysteine (γEC) by γEC synthetase then a glycine is added to the dipeptide by GSH synthetase (Rennenberg and Brunold, 1994). In wheat a homologue of GSH, hydroxymethylglutathione (hmGSH) is also present in which the glycine is replaced by a serine (Klapheck et al., 1991). During the removal of excess H_2O_2 GSH will be oxidised, and its oxidised form (GSSG) will be reduced by glutathione reductase (GR). This enzyme can also regenerate hmGSH from its oxidised form (hmGSSG).

The role of GSH in the defence against injuries caused by low temperature has been confirmed in field experiments in which the low-temperature-induced increase of total glutathione (TG) content has been observed in white pine during the winter (Anderson et al., 1992) and in alpine plants with increasing altitude (Wildi and Lütz, 1996). Further evidence for the protective role of GSH was found under controlled conditions, in growth chambers, where low temperature treatment resulted in a greater increase of TG in tolerant genotypes of tomato (Walker and McKersie, 1993) and maize (Kocsy et al., 1996) compared to the sensitive one. This difference between the genotypes could be explained by a faster induction of biosynthesis of cysteine, a precursor of GSH in the tolerant maize genotypes (Kocsy et al., 1997). In the leaves of *Triticum durum* at 10 °C an increase in TG was accompanied by a decrease in the GSH/GSSG ratio (Badiani et al., 1993). For the maintenance of a sufficient GSH/GSSG ratio at low temperatures an increased GR activity is necessary which was observed during the winter in spruce (Esterbauer and Grill, 1978). Besides the increase in the total GR activity, the appearance of cold-hardiness-specific GR isozymes could be important as observed in red spruce (Hausladen and Alscher, 1994). Chemical manipulation of GSH level and GR activity showed a positive quantitative relationship between these parameters and cold tolerance of maize (Kocsy et al., 2000b, 2001b). In a good agreement with these results, overexpression of GR in poplar resulted in lower sensitivity to high light and cold treatment at 5 °C (Foyer et. al., 1995).

4. CHROMOSOME 5A AFFECTS THE REGULATION OF GLUTAHIONE BIOSYNTHESIS DURING COLD ACCLIMATION

The aim of the present studies was to investigate the role of GSH and GR in cold acclimation of wheat with a genetic approach. We wanted to find out whether the 5A chromosome of wheat has a role in the regulation of GSH accumulation and GR activity during cold hardening. The total thiol contents increased during the first few days of the long-term cold hardening and later slowly decreased (not shown). Since a great initial increase of the thiol levels was observed during the first days of the hardening, a [^{35}S]sulfate-labelling experiment was done to find out whether the cysteine for GSH formation derives from a new synthesis or from protein breakdown. The amount of the labelled hmGSH and GSH increased to the highest levels as a result of the hardening in the shoots of the frost-tolerant Ch. The Ch5A chromosome significantly increased the GSH level at the CS background (Table 1). The least hmGSH and GSH levels were detected in the *T. spelta* plants. In the same time *T. spelta* proved to be the most sensitive in the freezing test, as well (Fig.2). Conversely to the Ch5A chromosome the *T. spelta* 5A chromosome decreased not only the frost tolerance (Fig 2) but both the hmGSH and GSH concentrations at the CS background (Table 1). After 35 d hardening when the frost tolerance reached its maximum (Fig. 2), the amount of hmGSH and GSH (reduced forms) was greater in the shoots of the frost-tolerant genotypes compared to the sensitive ones. No relationship was found between the amount of the oxidised forms of

Table 1. ^{35}S-Radiolabelled hydroxymethylglutathione (hmGSH) and glutathione (GSH) in shoots of five wheat genotypes after 35 days cold hardening with [^{35}S]sulfate. Mean values ± SD of three independent experiments are presented. Values carrying different letters are significantly different at the P<5 % level (according to Kocsy et al. 2000a).

	hmGSH		GSH	
	reduced	oxidised	reduced	oxidised
Ch	145.6±15.6^{a}	21.2±2.4^{b}	633.4±59.3^{a}	60.8±5.8^{b}
CS	102.5±12.3^{c}	52.3±4.5^{d}	389.8±45.6^{c}	67.1±7.2^{b}
Tsp	52.7±6.6^{d}	22.5±3.3^{b}	251.3±19.3^{d}	59.8±4.5^{b}
CS (Ch 5A)	158.2±17.4^{a}	30.6±6.2^{b}	567.4±67.4^{a}	77.4±10.1^{b}
CS (Tsp 5A)	68.1±7.4^{e}	24.8±1.8^{b}	284.7±35.9^{d}	53.3±5.1^{b}

these thiols and the frost tolerance (Table 2). Similarly to the GSH and hmGSH contents, the highest GSH/GSSG and hmGSH/hmGSSG ratios were measured in the two tolerant genotypes (Ch and CS(Ch5A), in which the GR activity was also greater compared to the other genotypes (Fig. 3). Since the decrease in the amount of the labelled proteins in the shoot (data not shown, Kocsy et al. 2000a) was two-fold greater compared to the increase of the labelled GSH and hmGSH together (Table 1), the thiol accumulation is a result of a new synthesis rather than of a protein breakdown.

Using 5A chromosome substitution lines in which the 5A chromosome derived from wheat genotypes with different frost tolerance it was shown that this chromosome also affect the changes in GSH and hmGSH levels and GR activity during hardening. Similarly to tomato and maize (Walker and McKersie, 1993, Kocsy et al. 1996), a greater accumulation

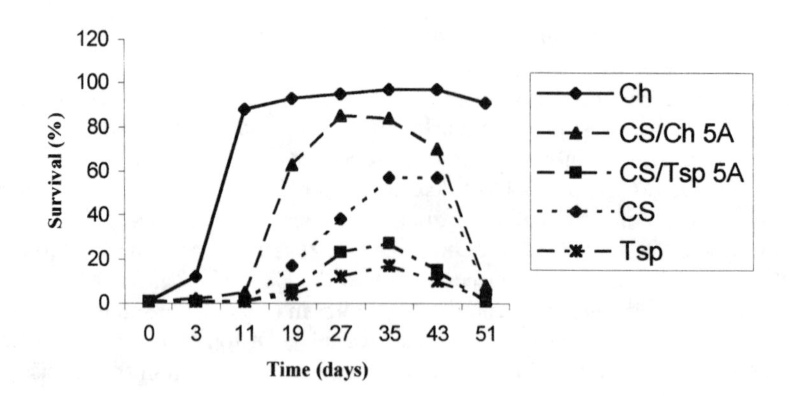

Figure 2. Time-dependent changes in frost tolerance during cold hardening at + 4/2°C for 51 days.

Table 2. Reduced and oxidised hydroxymethylglutathione (hmGSH) and glutathione (GSH) contents in shoots of five wheat genotypes after 35 days cold hardening. Mean values ± SD of six measurements from three independent experiments are presented. Values carrying different letters are significantly different at the P<5 % level (according to Kocsy et al. 2000a).

	hmGSH		GSH	
	control	hardened	control	hardened
Ch	81.2±14.2[a]	123.4±15.4[b]	417.5±38.6[a]	546.2±48.2[b]
CS	71.3±12.5[a]	102.5±16.3[ab]	346.2±41.2[ac]	449.2±37.3[a]
Tsp	65.8±13.4[a]	98.5±13.4[ab]	345.2±29.6[ac]	385.2±41.5[a]
CS (Ch 5A)	84.6±14.5[a]	121.7±16.2[b]	429.6±48.7[a]	557.4±45.1[b]
CS (Tsp 5A)	74.8±11.8[a]	87.3±14.5[a]	284.1±35.6[c]	367.1±39.5[a]

of GSH was observed in the tolerant genotypes of wheat compared to the sensitive one after cold treatment. The increase in GSH level is a result of the increased GSH synthesis also in wheat as it was observed in the [^{35}S]sulfate labelling experiments for maize (Kocsy et al. 1996). For an increased GSH synthesis a greater availability of cysteine is also necessary which was ensured by the cold-induced increase of the activity of adenosine 5'-phosphosulfate reductase, the key enzyme of assimilatory sulfate reduction in maize (Kocsy et al. 1997). The increase in the level of glycine, glutamate and serine, the other precursors of GSH and hmGSH, respectively, was also observed in cold-treated maize (Szalai et al., 1997). In a good accordance with the present results, in time course experiments with other plant

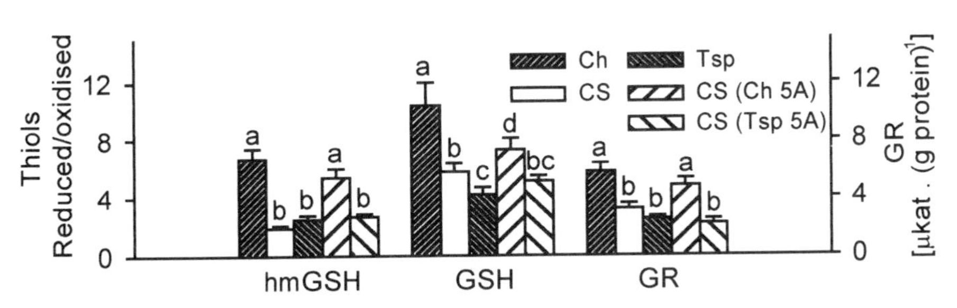

Figure 3. Activity of GR and ratio of reduced and oxidised cysteine, γEC, hmGSH and GSH contents in shoots of five wheat genotypes after 35 days cold hardening. Mean values ± SD of six measurements from three independent experiments are presented. Values carrying different letters are significantly different at the P<5 % level(according to Kocsy et al. 2000a).

species the TG level first also increased then decreased during cold treatment. The maximum TG content was detected after 3 d at 4/2 °C in wheat (present study), after 2 weeks at 5 °C in roots of jack pine (Zhao and Blumwald, 1998) and after one week at 5 °C in soybean leaves (Vierheller and Smith, 1990). Besides the changes in the TG content, the cold-induced changes in the GSH/GSSG ratio are very important in the adaptation to low temperature stress as it was shown by its higher level in the tolerant genotypes of wheat. To maintain the high GSH/GSSG ratio a sufficiently high GR activity is necessary as it was observed in wheat. Further evidence for the role of an appropriate GSH/GSSG ratio and GR activity in cold tolerance was found in tomato since the GSH/GSSG ratio decreased during low temperature stress only in the sensitive genotype, and the GR activity was also less in this genotype (Walker and McKersie, 1993). The changes in the GSH/GSSG and hmGSH/hmGSSG ratio may activate the antioxidative system and other defence reactions through a redox signalling pathway. Since the maximum thiol levels were detected after 3 d hardening, and the maximum frost tolerance was achieved only after 35 d in the most genotypes, the activation of the antioxidative system belongs to the first adaptive mechanisms during cold hardening.

5. REGULATION OF THE SOLUBLE CARBOHYDRATE CONTENT DURING COLD ACCLIMATION

The relationship between osmotic adjustment and the freezing tolerance can be easily elucidated if we consider the process of ice formation. As temperatures drop below 0 °C, ice formation is generally initiated in the cellular spaces due, in part, to the extracellular fluid having a higher freezing point (lower solute concentration) than the intracellular fluid. Because the water potential of the extracellular ice is less than the water potential of liquid water within the cells, the ice formation results in drop in water potential outside the cell. Consequently, there is movement of unfrozen water down the water potential gradient from inside the cell to the intercellular spaces. At equilibrium, the extent of dehydration is a direct consequence of temperature: with progressively greater fall in temperature below freezing, the less intracellular liquid water remains, the more strongly collapsed are the cells, and the lower is the intracellular water potential. At −10 °C, more than 90% of the osmotically active water typically moves out of the cells and the osmotic potential of the remaining fluid is greater than 5 osmolar (Pearce 1999). Consequently, the increased osmolarity of the cellular solute is an effective mechanism for avoiding cellular dehydration. In most, if not all hardy plants, cell solutes concentrate during hardening. Numerous reports point out the accumulation of various osmotically active solutes such as sucrose, fructan, quaternary ammonium compounds and proline during hardening (McKersie and Leshem, 1994). These may function as cryo-protectants helping to avoid cellular dehydration during extracellular ice formation, and many of these compounds may also stabilize membranes. Among them we have summarized here the results of our recent attempts to localize the genes responsible for the stress induced water soluble carbohydrate accumulation.

The importance of carbohydrates in the cold acclimation process and some of our attempts to localise the gene(s) regulating the cold-induced carbohydrate accumulation were reviewed in the previous volume of these cold hardiness seminars (Galiba et al. 1997a). Studies of photosynthetic acclimation of spring and overwintering crops have shown that freezing tolerance is strongly correlated with the capacity to increase photosynthesis and with the capacity to increase soluble carbohydrate pools during cold hardening (Tognetti et al., 1990; Öquist et al., 1993; Hurry et al. 1995). Furthermore, field studies have shown that plants become vulnerable to freezing injury when the fructan pool becomes depleted and simple sugars can no longer be released into the cytosol and intracellular liquid (Olien and Clark, 1993). Without unnecessary redundancy it seems valuable to mention that sucrose and fructan is considered to play key role in stress-induced metabolic processes in wheat. The role of reducing sugars (glucose and fructose) in the adaptive mechanism is more controversial, and even their accumulation can be detrimental. Glucose participates in crosslinking with protein by a complex glycolisation reaction between amino and carbonyl groups known as the Maillard reaction (Koster and Leopold, 1988). As respiratory substrates, monosaccharides promote respiration and mithochondrial electron transport which would seem to oppose the onset of quiescense and favor metabolism, energy production, and the formation of oxygen radicals (Leprince et al., 1993).

The importance of the genes located on 5A chromosome in osmoregulation became obvious when their impact on osmotic stress-induced accumulation of free amino acids and polyamines was discovered (Galiba et al., 1992, Galiba et al., 1993). The central role of this chromosome was further substantiated upon mapping the gene controls the stress induced abscisic acid (ABA) accumulation (Galiba et al. 1997a). ABA is considered as 'stress hormon' emphasizing its role in the adaptation of plants to different abiotic stress conditions like drought, salinity, heat and cold. We studied the CS(Ch) chromosome substitution lines to elucidate the possible chromosomal location of gene(s) responsible for elevated carbohydrate levels during cold stress conditions. Our preliminary experiment showed that the 5A Ch chromosome raised total sugar, fructan and invertase activity together with frost tolerance in the CS background (Housley et al., 1993). These promising results promted us to investigate again the CS, Ch cultivars and the 5A single chromosome recombinant lines to localize the gene regulating cold induced carbohydrate accumulation (Galiba et al 1997b). The carbohydrate concentration in the control plants of CS and Ch was similar. Cold treatment increased the carbohydrate concentration considerable in both cultivars. The fructan and the sucrose content increased significantly higher in the frost tolerant Ch (43 and 165 mg/g fresh weight, respectively) than in the sensitive CS (15 and 83 mg/g fresh weight, respectively) seedlings. Fructan and sucrose content was further evaluated in the 5A chromosome recombinant lines. The *T. spelta* 5A chromosome carrying the *Fr1* (frost sensitive) allele for frost tolerance and *Vrn1* (spring habit) allele for vernalization requirement did not change considerable the sucrose and fructan content at the CS background. On the other hand, the presence of Ch allele for vernalization requirement *vrn-A1* (needs vernalization) increased significantly their concentrations. Considering the line which exhibits recombination between *Vrn-A1* and *Fr1* loci gene regulating sucrose accumulation seemed to be closely associated with the V*rn1* locus but separable from the *Fr1* locus (Galiba et al. 1997b).

In this stage there were two open questions still to be answered regarding sugar accumulation. Firstly, the above cited results were based on samples taken only once, at the end of cold hardening, but no information was available on time-dependent changes in the sugar components versus freezing tolerance during the execution of cold hardening. Secondly, further research seemed to be important for the verification of the position of the gene which regulates the cold-induced sugar accumulation.

5.1. Correlation between Frost Tolerance and Sugar Accumulation During Cold Hardening

There was a great difference between the frost tolerant Ch and the frost-sensitive CS cultivars in both the level and the dynamics of frost tolerance (Fig. 2). The frost tolerance of Ch started to rise after 3 days of hardening and reached its maximum on the 19th day. The maximum tolerance was retained for a long time, and only a slight decrease was observed around the 50th day. In the case of CS the frost tolerance increased after 11th day, as did the substitution lines, and reach maximum around the 40th day. Afterwards, the tolerance decreased rapidly. As it was expected the Ch 5A increased both the duration and the maximum of the frost tolerance at the CS genetic background (Fig. 2). In the same time the substituted *T. spelta* 5A chromosome lowered the level of frost tolerance of the recipient CS from the maximum 59% survival percentage to 22%.

The sensitive CS and CS(*T. spelta* 5A) accumulated the maximum amount of sucrose on the 11th and 19th days of cold hardening, respectively, surpassing that of the tolerant one (Fig. 4). This tendency started to change around the 22nd day of cold hardening and changed completely by the 43th day, when the order corresponded to the order of frost tolerance.

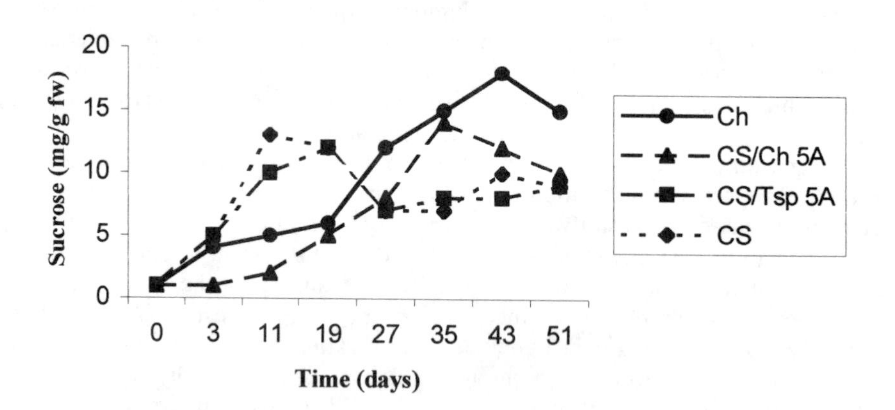

Figure 4. Changes in sucrose content during cold hardening.

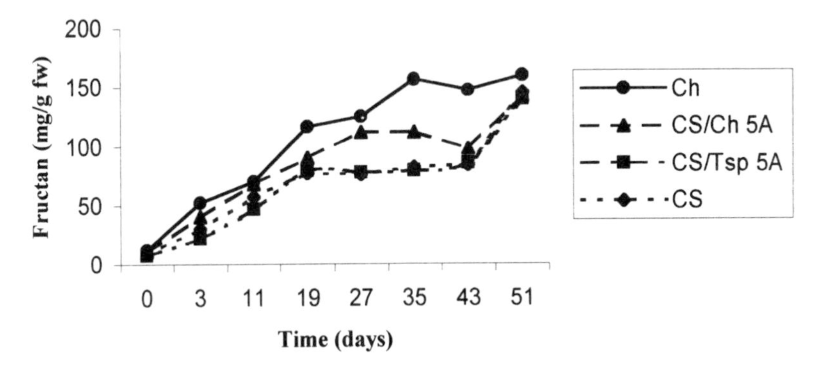

Figure 5. Changes in fructan content during cold hardening.

The fructan content showed continuos increase during the cold hardening. This study showed genotype-related cold-induced fructan accumulation during hardening, resulting in a higher fructan content in Ch and CS(Ch 5A) than in CS and CS(*T. spelta* 5A) (Fig. 5). These results show that the substituted Ch 5A chromosome increased not only the frost tolerance in a time dependent manner, but the carbohydrate accumulation as well, in the CS background. This phenomenon supports the hypothesis that the soluble sugar accumulation is genetically associated with freezing tolerance. Correlation analysis showed that fructan content and survival rate significantly correlated from the 19th day of hardening until the 43rd while in the case of sucrose at the full hardened stage (critical r values were higher than 0.81; see the details in Vágújfalvi et al. 1999).

5.2. Physical Mapping of the *Vrn-A1* and *Fr1* Genes

Recently, 436 wheat chromosome deletion lines were generated in the cultivar CS using the gametocidal chromosomes of *Aegilops cylindrica* (Endo and Gill 1996). These deletions were used for physical mapping. The removal of the 5A chromosome from CS under unvernalized conditions, as in the monosomic and nullisomic, leads to a delay in flowering (Flood and Halloran 1986). Therefore, any deletion line lacking the *Vrn-A1* locus would be expected a similar late flowering. Analysis of variance for flowering time revealed significant differences between the deletion lines grown at 20°C (Fig. 6). They could be divided into two groups. The difference in flowering between group I and II was 14 days. The division between the group of deletion lines for flowering time occurred between breakpoints 0.68 and 0.78 indicating, that the *Vrn-A1* locus situated between them. The frost resistance of the deletion lines could also be divided into two groups. Group II had an average survival of 40%, while for group I the average survival was 27% (Fig. 7). This 13% difference in survival between groups was statistically significant (Sutka et al., 1999). The division of the two groups happened between breakpoints 0.67 and 0.68, so *Fr1* located in this interval. Thus, the *Fr1* is situated closer to the centromere than *Vrn-A1*, confirming the results published by Snape et al. (1997).

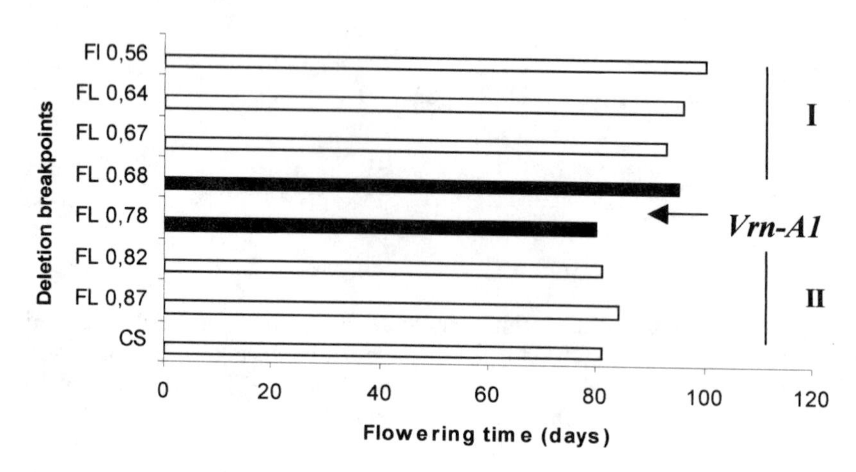

Figure 6. Flowering times of Chinese Spring (CS) and deletion lines for the 5AL chromosome arm.

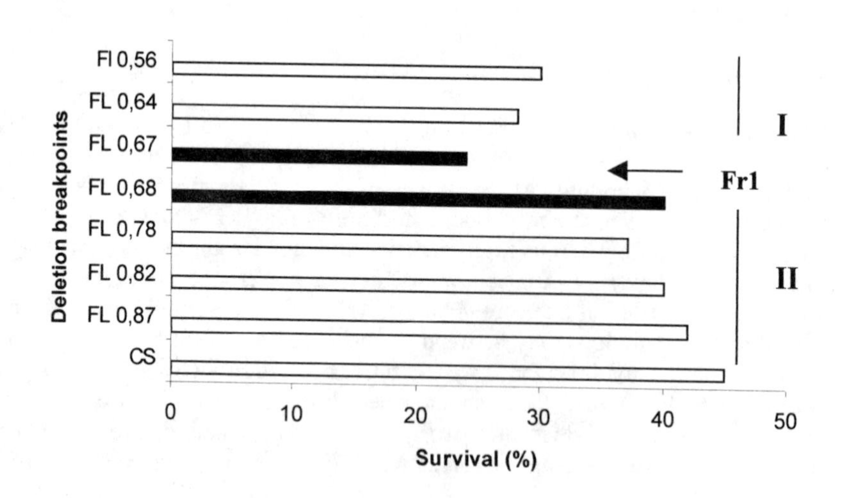

Figure 7. Freezing tolerance of Chinese Spring (CS) and deletion lines for the 5AL chromosome arm.

5.3. Physical Mapping of the Gene Regulates Carbohydrate Accumulation

For the verification of the position of the gene which regulates the stress-induced sugar accumulation four deletion lines were studied, which fraction lengths (FL) varied between 0.68 and 0.56. Non of them contained the *Vrn-A1* locus but the deletion line which FL was 0.68 still possessed the *Fr1* gene (Fig. 10). Our working hypothesis based on the following assumption: if *Vrn-A1* co-segregates with the gene regulates sugar accumulation, the sugar content in these deletion lines should be similar to each other but must differ from CS during cold treatment. To separate the effect of cold from the water deprivation the seedlings were raised hydroponically. Cold treatment was applied as +2C for 39 days. Meantime, to study the effect of osmotic stress, plants were grown in hydroponics supplemented with 15% PEG-4000 at 18/13°C (day/night).

Total water-soluble carbohydrate (WSC), glucose, fructose, sucrose, and fructan content of leaves were determined. For demonstration the results of WSC and sucrose analysis are shown here. The fructans are the main components of WSC so, the WSC curves more or less represent the course of fructan accumulation, as well (Vágújfalvi et al. 1999). The WSC content increased rapidly in all studied genotypes and reached maximum values on the 11th day of the cold treatment. The accumulation measured in the deletion lines surpassed considerably the values detected in CS both on the 3rd and 11th days (Fig 8). This phenomenon corresponded well with the behaviour of fructan e.g.: on the 11th day the fructan concentration was twice as high in the deletion lines than in CS, 40 and 80 mg/g on fresh weight (fw) bases, respectively. Considering the PEG treatment slow, gradual increase was detected in each studied genotype until the 35th day (not shown). On the last sampling date (39th) the WSC content almost doubled in the deletion lines reaching average value of 110 mg/g fw while, its concentration remained steady in CS (roughly 68 mg/g fw). Similarly, the

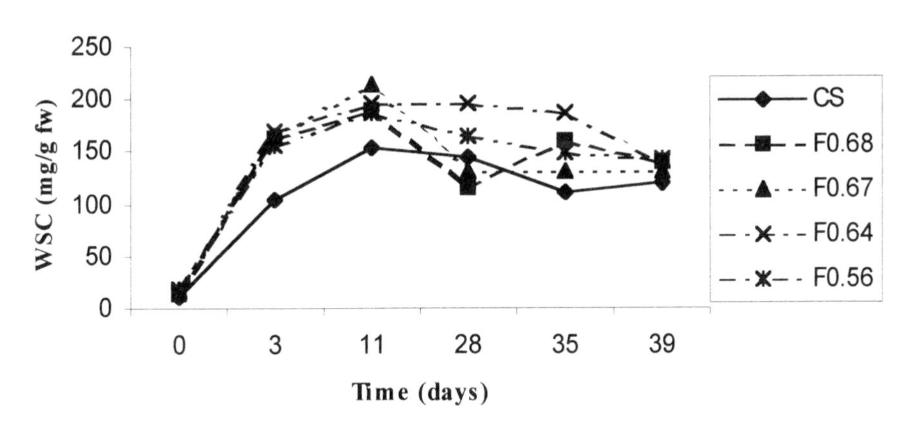

Figure 8. Changes of total water soluble carbohydrate (WSC) content in leaves of Chinese Spring (CS) and CS 5A deletion lines grown in hydroponics at +2°C.

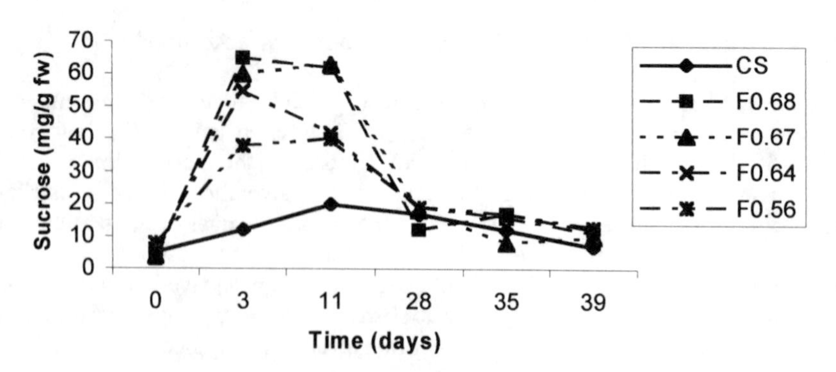

Figure 9. Changes of sucrose content in leaves of Chinese Spring (CS) and CS 5A deletion lines grown in hydroponics at +2°C.

fructan content also doubled this time in the deletion lines. The corresponding values: 22 mg/g fw in CS and around 40 mg/l fw in the deletions (not shown).

The sucrose content increased considerably higher in the deletion lines than in CS at the beginning (on the 3rd and the 11th day) of the cold treatment (Fig. 9). Later, the concentrations levelled off in all genotypes. Changes of sucrose content in the PEG-treated CS and the deletion lines showed different pattern over time. A gradual increase was detected in the deletion lines during the 39-day treatment. The sucrose concentrations were higher in CS leaves on the 28th and 35th days than in the deletion lines afterwards, its concentration has dropped to the control level (not shown).

Following conclusions can be drawn from these data: The cold and the PEG-induced osmotic stress caused different patterns in sugar content over time. The behaviour of the CS 5A deletion lines was similar to each other but distinct from the CS irrespectively from the type of stresses, applied. These results support the hypothesis that the gene regulating the stress-induced carbohydrate accumulation co-segregates with *Vrn-A1* and located distally from *Fr1* on the long arm of 5A chromosome (Fig. 10). It is necessary to emphasise that alleles of this gene, possessing the same or similar function, might be located on 5D and 7A chromosomes, as well (Vágújfalvi et al., 1999).

6. MAPPING THE *COR14B* REGULATORY LOCI ON WHEAT CHROMOSOME 5A

The function and the possible role of COR14b protein in the cold acclimation processes have already been described in the previous chapter written by Cattivelli et al. The mechanisms controlling *cor14b* gene expression were also explained in details. We introduce here our genetic system and the experimental conditions which allowed us the mapping of the gene loci regulating the temperature-dependent expression of the wheat gene homologous to the barley cold-regulated *cor14b* gene.

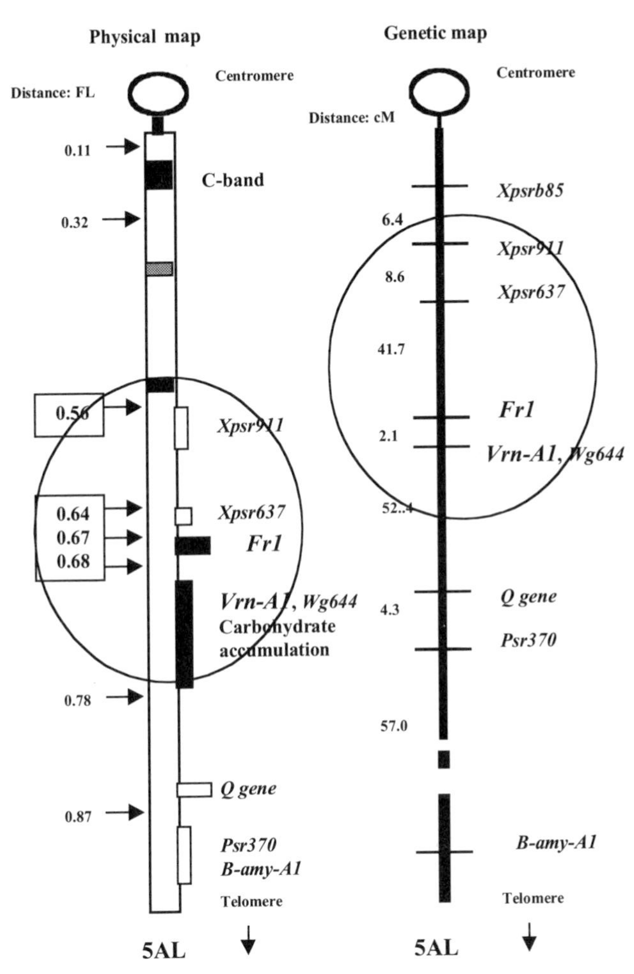

Figure 10. Comparison between the physical and genetic maps of 5AL. Fraction lengths (FL) are indicated on the left of the physical map. Marker names are indicated on the right of the chromosomes.The maps are based on Sutka et al. (1999).

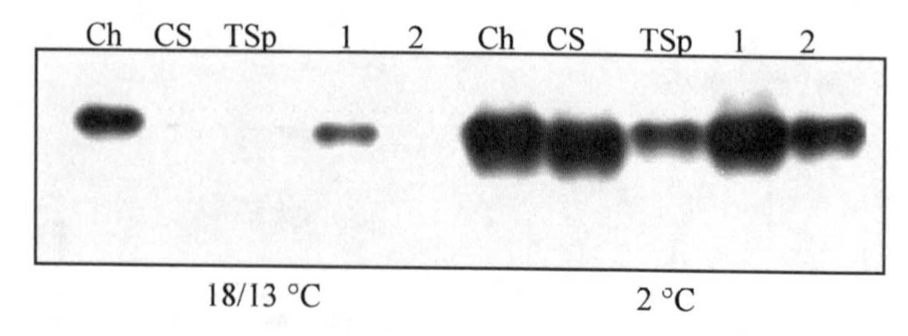

Figure 11. Temperature-dependent expression of *cor14b* gene in the parental and 5A substitution lines. Symbols: 1, CS(Ch 5A); 2, CS(*T. spelta* 5A).

Distance (cM)	Marker	CS/TSP 5A	CS/CNN 5A	7-3	32-5	34-3	4-4	56-1	12-4	58-3	46-1	38-6
				recombinant lines								
centromere												
	psrB85	T	C	T	T	T	C	C	C	T	T	C
6	psrB89	T	C	T	T	T	C	C	C	T	T	C
	psr911 (Rcg1)	T(r_1)	C(R_1)	T(r_1)	T(r_1)	T(r_1)	C(R_1)	C(R_1)	C(R_1)	T(r_1)	C(R_1)	T(r_1)
9	psr637	T	C	T	T	T	C	C	T	C	C	T
42	Xpsr2021 (Rcg2)	T (r_2)	C (R_2)	C (R_2)	T (R_2)	T (r_2)	T (r_2)	C (R_2)	nd (r_2)	nd (R_2)	nd (R_2)	nd (R_2)
9	Fr1	T	C	T	C	T	T	C	T	nd	T	nd
2	Vrn1, Xspr426	T	C	C	C	T	T	C	T	T	T	C
46	Xpsr805	T	C	T	T	C	C	T	nd	C	T	C
8	Xspr370	T	C	T	T	C	C	T	T	C	T	C
6	Xspr164	T	C	T	T	C	C	T	T	C	T	C
50	B-amy-A1	T	C	T	T	T	C	T	T	T	T	C

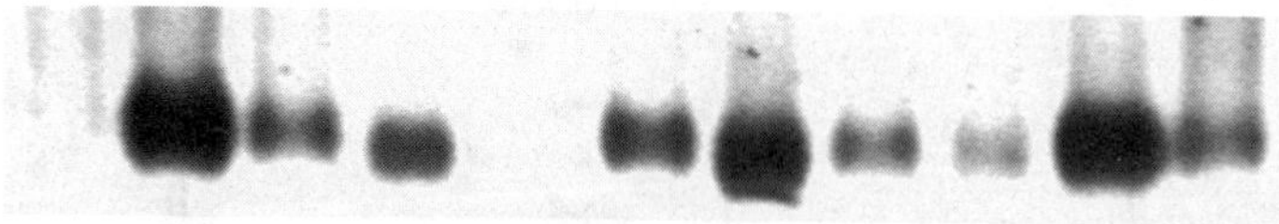

Figure 12. *Cor14b* mRNA accumulation in the parental CS(*Triticum sp.* 5A) (CS/TSP 5A) and CS (Ch 5A) (CS/CNN 5A) and in the single chromosome recombinant lines grown at 18/13°C.

Among the wheat plants raised under control (18/13°C day/night) conditions the accumulation of *cor14b* mRNA was observed in the leaves of frost-resistant genotypes, but not in those of frost-sensitive varieties and lines (Vágújfalvi et al., 2000). At higher temperature (25/18°C) there was no detectable quantity of *cor14b* mRNA in any of the genotypes. At low (2°C) temperature all the varieties accumulated mRNA. As experienced previously in barley, gene expression was temperature- and tolerance-dependent, though the temperature threshold at which the gene became expressed was higher in wheat than in barley. This was followed by the mapping of the gene regulating the expression of the *cor14b* gene. The results of Northern analyses showed that the 5A chromosome carries genes responsible for the sensing of the threshold temperature. In the CS genetic background the Ch 5A chromosome increased the quantity of *cor14b* mRNA at 2°C. The COR 14b protein was only present in demonstrable quantities in the Ch variety and in the CS(Ch 5A) line, but not in the CS parent or in the CS(*T. sp* 5A) line at 18/13°C (Fig. 11.)

The analysis of single chromosome recombinant lines derived from the cross between Chinese Spring/*Triticum spelta* 5A and Chinese Spring/Cheyenne 5A identified two loci with additive effect which control the *cor14b* homologous mRNAs accumulation. The first locus was positioned tightly linked with marker *Xpsr911*, while the second one was located between marker *Xpsr2021* and *Frost resistance* gene *Fr1* (Fig. 12). It was known that the structural gene was located on chromosome 2 in barley, but its position on the chromosome was unknown (Crosatti et al., 1996). The RFLP mapping of the structural gene was carried out using a *cor14b* gene cDNA probe on *Triticum monococcum* mapping population. The locus of the *cor14b* allele examined was mapped on the long arm of the $2A^m$ chromosome (Vágújfalvi et al., 2000).

In conclusion, the adaptation processes related to frost resistance are regulated by polygenic manner. Dubcovsky et al. (1995) mapped dehydrin and *Esi* genes involved in osmoregulation and the loci of heat shock proteins to the region of the *Vrn* gene locus on the long arm of the fifth chromosome of *T. monococcum* ($5A^mL$). These results indicate that the 5A chromosome carries an "adaptation gene complex" in the region of the *Vrn1* and *Fr1* genes (Fig. 13). The development of this gene family regulating adaptation may have represented an advantage for the evolution. Recombination frequency is low for genes located close to each other, so the genes are inherited together and form a relatively large functional unit. As the result of continual selection pressure due to the winter climate, the grouping of the genes ensured a selective advantage for progeny generations.

7. ACKNOWLEDGEMENTS

This work was supported by the grants : OTKA F025190, OTKA T34277, OMFB-02579/2000, by the CNR-MTA Hungarian – Italian co-operation program (Project number 15/MTA 401), and by the MTA-OTKA-NSF co-operation program (Project numbers: MTA 1/2001/A and OTKA N35539).

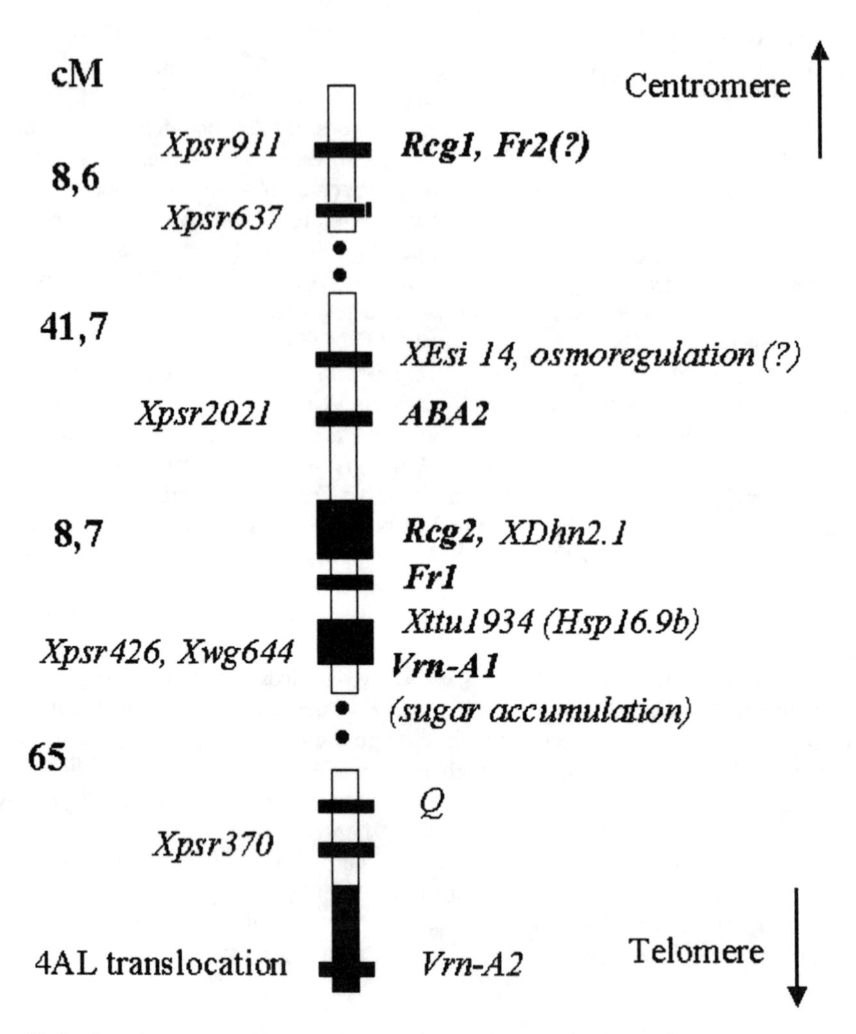

Figure 13. Position of stress inducible genes on the long arm of 5A chromosomes. This map based on: Dubcovsky et al., 1995;Galiba et al., 1995; Galiba et al., 1997; Dubcovsky etal., 1998; Vagujfalvi et al., 2000.

REFERENCES

Anderson, J. V., Chevone, B. I., and Hess, J. L., 1992, Seasonal variation in the antioxidant system of eastern white pine needles. *Plant Physiol.* **98**:501-508.

Badiani, M., Paolacci, A. R., D`Annibale, A., and Sermanni, G. G., 1993, Antioxidants and photosynthesis in the leaves of *Triticum durum* L. seedlings acclimated to low, non-chilling temperature. *J. Plant Physiol.* **142**:18-24.

Crosatti, C., Polverino de Laureto, P., Bassi, R., and Cattivelli, L., 1999, The interaction between cold and light controls of the expression of the cold-regulated barley gene *cor14b* and the accumulation of the corresponding protein. *Plant Physiol.* **119**:671-680.

Crosatti, C., Nevo, E., Stanca, A. M., and Cattivelli, L., 1996, Genetic analysis of the accumulation of COR 14 proteins in wild (*Hordeum spontaneum*) and cultivated (*Hordeum vulgare*) barley. *Theor. Appl. Genet.* **93**:975-981.

Dubcovsky, J., Lijavetzky, D., Appendino, L., and Tranquili, G., 1998, Comparative RFLP mapping of Triticum monococcum genes controlling vernalization requirement. *Theor. Appl. Genet.* **97**: 968-975.

Dubcovsky, J., Luo, M. C., and Dvorak, J., 1995, Linkage relationships among stress-induced genes in wheat. *Theor. Appl. Genet.,* **91**:795-801.

Dubcovsky, J., Lou, M. C., Zhong, G. Y., Bransteitter, R., Desai, A., Kilian, A., Kleinhofs, A., and Dvorák, J., 1996, Genetic map of diploid wheat, *Triticum monococcum* L., and its comparison with maps of *Hordeum vulgare* L. *Genetics* **143**:983-999.

Endo, T. R., and Gill, B. S., 1996, The deletion stocks of common wheat. *J. Hered.* **87**:295-307.

Esterbauer, H., and Grill, D., 1978, Seasonal variation of glutathione and glutathione reductase in needles of *Picea abies*. *Plant Physiol.* **61**:119-121.

Flood, R. G., and Halloran, G. M., 1986, Genetics and physiology of vernalization responses in wheat, in: Advances in Agronomy, N. C. Brady, ed., Academic Press, New York, vol. 39., pp. 87-125.

Foyer, C. H., Souriau, N., Perret, S., Lelandais, M., Kunert, K. J., Pruvost, C., and Jouanin, L., 1995, Overexpression of glutathione reductase but not glutathione synthetase leads to increases in antioxidant capacity and resistance to photoinhibition in poplar trees. *Plant Physiol.* **109**:1047-1057.

Foyer, C. H., Lopez-Delgado, H., Dat, J. F., and Scott, I. M., 1997, Hydrogen peroxide- and glutathione-associated mechanisms of acclimatory stress tolerance and signalling. *Physiol. Plant.* **100**:241-254.

Galiba, G., Kerepesi, I., Snape, J. W., and Sutka, J., 1997a, Mapping of genes controlling cold hardiness on wheat 5A and its homologous chromosomes of cereals, in: Plant Cold Hardiness Molecular Biology, Biochemistry, and Physiology, P. H. Li,, and T. H. H. Chen, eds., Plenum Press, New York and London, pp. 89-98.

Galiba, G., Kerepesi, I., Snape, J. W., and Sutka, J., 1997b, Location of a gene regulating cold-induced carbohydrate production on chromosome 5A of wheat. *Theor. Appl. Genet.* **95**:265-270.

Galiba, G., Kocsy, G., Kaur-Sawhney, R., Sutka, J., and Galston, A. W., 1993, Chromosomal localisation of osmotic and salt stress-induced differential alterations in polyamine content in wheat. *Plant Sci.* **92**:203-211.

Galiba, G., Quarrie, S. A., Sutka, J., Morguonov, A., and Snape, J. W., 1995, RFLP mapping of the vernalization (*Vrn1*) and frost resistance (*Fr1*) genes on chromosome 5A of wheat. *Theor. Appl. Genet.* **90**:174-1179.

Galiba, G., Simon-Sarkadi, L., Kocsy, G., Salgó, A., and Sutka, J. 1992, Possible chromosomal location of genes determining the osmoregulation of wheat. *Theor. Appl. Genet.* **85**:415-418.

Hausladen, A., and Alscher, R. G., 1994, Cold-hardiness-specific glutathione reductase isozymes in red spruce. *Plant Physiol.* **105**:215-223.

Hayes, P. M., Blaket, T., Chen, T. H. H., Tragoonrung, S., Chen, S., Pan, A., and Liu, B., 1993, Quantitative trait loci on barley (*Hordeum vulgare* L.) chromosome 7 associated with components of winter hardiness. *Genome* **36**:66-71.

Housley, L., Pollock, and C. J., 1993, The metabolism of fructan in higher plants, in: M. Suzuki, and N. J. Chatterton, eds., *Science and Technology of Fructans*. CRC Press, London, pp. 191-225.

Hurry, V. M., Strand, A., Tobiaeson, M., Gardeström, P., and Öquist, G., 1995, Cold hardening of spring and winter wheat and rape result in differential effects on growth, carbon metabolism, and carbohydrate content. *Plant Physiol.* **109**:697-706.

Kerepesi, I., Tóth, I., and Boross, L., 1996, Water-soluble carbohydrates in dried plants. *J. Agric. Food Chem.* **44**:3235-3239.

Klapheck, S., Chrost, B., Starke, J., and Zimmermann, H., 1991, γ-Glutamylcysteinylserine – a new homologue of glutathione in plants of the family *Poaceae*. *Bot. Acta* **105**:174-179.

Kocsy, G., Brunner, M., Rüegsegger, A., Stamp, P., and Brunold, C., 1996, Glutathione synthesis in maize genotypes with different sensitivity to chilling. *Planta* **198**:365-370.

Kocsy, G., Galiba, G., and Brunold, C., 2001a, Role of glutathione in adaptation and signalling during chilling and cold acclimation in plants. *Physiol. Plant.***113**:158-164.

Kocsy, G., Owttrim, G., Brander, K., and Brunold, C., 1997, Effect of chilling on the diurnal rhythm of enzymes involved in protection against oxidative stress in a chilling tolerant and a chilling sensitive maize genotype. *Physiol. Plant.* **99**:249-254.

Kocsy, G., Szalai, G., Vágújfalvi, A., Stéhli, L., Orosz, G., and Galiba, G., 2000a, Genetic study of glutathione accumulation during cold hardening in wheat. *Planta* **210**:295-301.

Kocsy, G., von Ballmoos, P., Rüegsegger, A., Szalai, G., Galiba, G., and Brunold, C., 2001b, Increasing the glutathione content in a chilling-sensitive maize genotype using safeners increased protection against chilling-induced injury. *Plant Physiol.* **127**:147-1156.

Kocsy, G., von Ballmoos, P., Suter, M., Rüegsegger, A., Galli, U., Szalai, G., Galiba, G., and Brunold, C., 2000b, Inhibition of glutathione synthesis reduces chilling tolerance in maize. *Planta* **211**:528-536.

Koster, K. L., and Leopold, A. C., 1988, Sugars and desiccation tolerance in seeds. *Plant Physiol.* **96**:302-304.

Leprince, O., Hendry, G. A. F., and McKersie, B. M., 1993, The mechanisms of desiccation tolerance in developing seeds. *Seed Sci. Res.* **3**:231-246.

Liesenfeld, D. R., Auld, D. L., Murray, G. A., and Swensen, J. B., 1986, Transmittance of winter hardiness in segregated populations of peas. *Crop Sci.* **26**:49-54.

McKersie, B. D., and Leshem, Y. Y., 1994, Stress and Stress Coping in Cultivated Plants, Kluwer Academic Publishers, Dordecht.

Nagy, Z., and Galiba, G., 1995, Drought and salt tolerance are not necessarily linked: a study on wheat varieties differing in drought tolerance under consecutive water and salinity stresses. *J. Plant Physiol.* **145**:68-174.

Olien, C. R., and Clark, J. L., 1993, changes in soluble carbohydrate composition of barley, wheat, and rye during winter. *Agron. J.* **85**:21-29.

Öquist, G., Hurry, V. M., and Huner, N. P. A., 1993, Low temperature effects on photosynthesis and correlation with freezing tolerance in spring and winter cultivars of wheat and rye. *Plant Physiol.* **101**:245-250.

Pearce, R. M., 1999, Molecular analysis of acclimation to cold. *Plant Growth Regulation,* **29**:47-76.

Prasad, T. K., Anderson, M. D., Martin, B. A., and Stewart., C. R., 1994, Evidence for chilling-induced oxidative stress in maize seedlings and a regulatory role for hydrogen peroxide. *Plant Cell* **6**:65-74.

Rennenberg, H., and Brunold, C., 1994, Significance of glutathione metabolism in plants under stress. *Prog. Bot.* **55**:144-156.

Sears , E. R , 1953, Nullisomic analysis in common wheat. *Am. Nat.* **87**:245-252.

Snape, J. W., Semikhodskij, A., Fish, L., Sharma, R. N., Quarrie, S. A., Galiba, G., and Sutka, J., 1997, Mapping frost tolerance loci in wheat and comparative mapping with other cereals. *Acta Agron. Hung.* **45**:265-270.

Stone, J. M., Palta, J. P., Bamberg, J. B., Weiss, L. S., and Harbage, J. F. 1993, Inheritance of freezing resistance in tuber bearing *Solanum* species: Evidence for independent genetic control of nonacclimated freezing tolerance and cold acclimation capacity. *Proc. Natl. Acad. Sci. USA* **90**:7869-7873.

Sutka, J., 1981, Genetic studies of frost resistance in wheat. *Theor. Appl. Genet.* **59**:45-152.

Sutka, J., Galiba, G., Vágújfalvi, A., Gill, B. S., and Snape, J. W., 1999, Physical mapping of the Vrn-A1 and Fr1 genes on chromosome 5A of wheat using deletion lines. *Theor. Appl. Genet.* **99**:199-202.

Storlie, E. W., Allan, R. E., and Walker-Simmons, M. K., 1998, Effect of the Vrn1-Fr1 interval on cold hardiness levels in near-isogenic wheat lines. *Crop Sci.* **38**:483-488.

Szalai, G., Janda, T., Bartók, T., and Páldi, E., 1997, Role of light in changes in free amino acid and polyamine contents at chilling temperature in maize (*Zea mays*). *Physiol. Plant.* **101**:434-438.

Tognetti, J.A., Salerno, G. L., Crespi, M. D., and Pontis, H. G., 1990, Sucrose and fructan metabolism of different wheat cultivars at chilling temperatures. *Physiol. Plant.* **78**:554-559.

Vágújfalvi, A., Crosatti, C., Galiba, G., Dubcovsky, J. and Cattivelli, L., 2000. Mapping of regulatory loci controlling the accumulation of the cold regulated *cor14b* mRNA in wheat. *Mol. Gen. Genet.* **263**:194-200.

Vágújfalvi, A., Kerepesi, I., Galiba, G., Tischner, T., and Sutka, J., 1999, Frost hardiness depending on carbohydrate changes during cold acclimation in wheat. *Plant Sci.* **144**:85-92.

Vierheller, T. L., and Smith, I. K., 1990, Effect of chilling on glutathione reductase and total glutathione in soybean leaves (*Glycine max* (L) Merr), in: Sulfur Nutrition and Sulfur Assimilation in Higher Plants, H. Rennenberg, C. Brunold, L. J. de Kok, and I. Stulen, eds., SPB Academic Publishing bv, The Hague, pp. 261-265.

Walker, M. A., and McKersie, B. D., 1993, Role of the ascorbate-glutathione antioxidant system in chilling resistance of tomato. *J. Plant Physiol.* **141**:234-239.

Wildi, B., and Lütz, C., 1996, Antioxidant composition of selected high alpine plant species from different altitudes. *Plant Cell Environm.* **19**:138-146.

Zhao, S., and Blumwald, E., 1998, Changes in oxidation-reduction state and antioxidant enzymes in the roots of jack pine seedlings during cold acclimation. *Physiol. Plant.* **104**:134-142.

PHOTOSYNTHESIS AT LOW TEMPERATURES

A case study with *Arabidopsis*

Vaughan Hurry, Nathalie Druart, Ana Cavaco, Per Gardeström, and Åsa Strand*

1. INTRODUCTION

One of the most variable conditions in the field is temperature and relatively severe frost, caused by temperatures below -20°C, can be expected to occur over 42% of the earth's surface (Larcher 1995). Low temperature is therefore a major determinant of the geographical distribution and productivity of plant species. Exacerbating this problem, plants from high latitudes and high altitudes are faced with short growing seasons and the need to grow at low temperatures for prolonged periods to extend the growing season. Thus, the capacity for active photosynthesis during prolonged exposure to low growth temperatures is essential in determining their successful site occupancy and subsequent productivity. Despite the importance of low temperatures in determining agricultural productivity and ecological diversity at higher latitudes and altitudes, relatively little is known about either the short-term or long-term effects of cold on the underlying biochemical responses of plant energy metabolism, processes that contribute to plant growth.

In addition to growing at low temperatures these plants must also acclimatize to survive freezing winter temperatures. In 1993 Öquist and co-workers showed a strong correlation between an increased photosynthetic capacity when measured at normal warm temperatures and freezing tolerance in different cold grown cultivars of rye and wheat (Öquist et al. 1993). This correlation raised the question of what metabolic adaptation occurred in the leaves of cold tolerant herbaceous species during this recovery of photosynthetic capacity. Subsequent studies using winter rye showed that concomitant with the increase in photosynthetic capacity there was an increase in the activities of the stromal enzyme Rubisco as well as the cytosolic enzyme sucrose-phosphate synthase

* Vaughan Hurry, Nathalie Druart, Ana Cavaco, Per Gardeström, Umeå Plant Science Centre, Department of Plant Physiology, Umeå University, S-901 87 Umeå, Sweden. Åsa Strand, Plant Biology Laboratory, The Salk Institute for Biological Studies, 10010 North Torrey Pines Road, La Jolla, CA 92037.

Plant Cold Hardiness, edited by Li and Palva
Kluwer Academic/Plenum Publishers, 2002

(SPS) (Hurry et al. 1994) and an increase in ribulose-1,5-bisphosphate (RuBP) regeneration (Hurry et al. 1996) following cold acclimation. However, when spring and winter cultivars of wheat (*Triticum aestivum*) and oilseed rape (*Brassica napus*) were compared it was shown that the spring cultivars did not increase enzyme activity per unit protein following cold acclimation and the metabolite data suggested a limited ability to regenerate RuBP in cold grown spring oilseed rape (Hurry et al. 1995b). The limited capacity of the spring cultivars to increase the activity of these enzymes was suggested to be the explanation for their failure to recover photosynthetic capacity and consequently to cold acclimate. Thus, it seems likely that the restoration of photosynthesis and carbon metabolism following cold acclimation is a mechanism facilitating the accumulation of sugars both with possible cryoprotective functions and that are essential substrates for basal metabolism during winter. We show in the following case study with *Arabidopsis* that the restoration of photosynthesis and carbon metabolism and the development of freezing tolerance represent a continuum in the low temperature acclimation response.

2. LEAF MORPHOLOGY AND FREEZING TOLERANCE

The work cited above compares leaves from plants grown under warm, non-hardening temperatures with leaves that developed weeks after the plants were shifted to cold-hardening conditions. Because much of the *Arabidopsis* literature involves comparing non-hardened leaves only with leaves shifted to cold hardening conditions for short periods, usually 2 to 10 days, we have compared leaves that have developed at warm temperatures (23°C) with leaves shifted to 5°C for 10 days and with new leaves that have fully developed after the plants were shifted to cold hardening condition. Morphologically these plants and their leaves look quite different. Cold developed *Arabidopsis* leaves are smaller, more compact and robust, with shorter and thicker petioles compared to the warm grown control leaves (Fig. 1). This is a typical response of both monocotyledonous and dicotyledonous cold tolerant annuals to growth at low temperature (Huner et al. 1993).

Figure 1. *Arabidopsis thaliana* (ecotype Col O) plants grown at 23°C (Control), shifted to 5°C for 10 days (10 d) and after development of a new rosette of leaves at 5°C (Dev). Plants shown are grown in 5 cm^2 pots at an irradiance of 150 $\mu mol\ m^{-2}\ s^{-1}$.

Cold acclimation of *Arabidopsis* also results in a marked reduction in leaf water content from 91% to 78% (Tab. 1). Furthermore, there is an almost 3-fold increase in total leaf protein, a 35% increase in chlorophyll content per unit fresh weight, and a 40% increase in specific leaf dry weight (Strand et al. 1999). A 50% increase in cell size has been shown in leaves of rye and spinach following cold acclimation (Huner et al. 1993). In contrast, following the development of new leaves at 5°C *Arabidopsis* shows a slight decrease, rather than an increase in cell size (Strand et al. 1999) but, as in spinach (Boese and Huner 1990), the number of cell layers across the leaf increases from approximately 8 to 10 (Tab. 1). Electron micrographs show that following cold acclimation by *Arabidopsis* there is also an increase in the relative volume of the cytosol at the expense of the vacuole and the cytosol is denser than in the warm grown leaves (Tab. 1; Strand et al. 1999). This may provide an important mechanism for increasing the enzymes and for the accumulation of solutes in cold acclimated leaves. Similar increases in cytoplasmic volume have been reported previously for tree species (Siminovitch et al. 1968) and winter rye (Huner et al. 1981), indicating that the changes observed in *Arabidopsis* represent a typical response to cold acclimation. However, these changes and their underlying mechanisms raise many interesting questions about the signal behind the anatomical changes, not only with respect to cold acclimation but also for increasing plant yield.

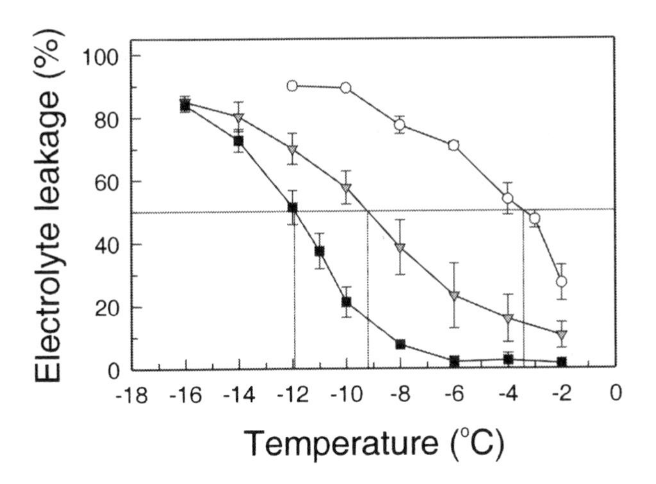

Figure 2. Freezing tolerance estimated from electrolyte leakage from control warm developed leaves (○), warm developed leaves shifted to 5°C for 10 days (▼) and cold developed leaves (■). Each value represents the mean (±SD) of at least three different leaves and all samples were exposed to freezing temperatures at the same time. Data redrawn from Hurry et al. (2000).

Electrolyte leakage has long been used as an estimate of cellular damage following freezing (Dexter et al. 1932), with the temperature that results in leakage of 50% of the cell content (T_{EL50}) being taken as a reference point to compare different genotypes or treatments. Multiple studies of freezing tolerance in *Arabidopsis* have shown that freezing tolerance is significantly increased after only a few days exposure to low temperature, from approximately –3 to –8°C (Gilmour et al. 1988; Lång et al. 1994; Uemura et al. 1995). However, these reports have suggested that *Arabidopsis* leaves are fully acclimated to the cold after only a few days of cold exposure. In contrast, we have shown that for *Arabidopsis* the new leaves that develop at low temperature show a further increase in freezing tolerance of 2-3°C compared to the shifted leaves (Fig. 2; Strand et al. 1997; Hurry et al. 2000). Thus, development of full freezing tolerance necessary for winter survival in herbaceous plants only occurs following the development of new leaves at low temperature, and therefore after the resumption of growth.

Table 1. Physical characteristics of leaves of *Arabidopsis thaliana* leaves grown at 23°C, shifted to 5°C for 10days and of leaves that developed and expanded after plants were shifted to 5°C. Leaves were fixed and sectioned for TEM. Sections were also sampled using light microscopy to measure leaf thickness and to count the cell layers.

Parameter	Control leaves	10 day shifted leaves	Cold grown leaves
Leaf thickness (μm)	265 ± 28	252 ± 7	279 ± 35
No. cell layers	7.6 ± 0.8	7.8 ± 1.0	9.6 ± 0.7
% cytoplasm	25 ± 9	44 ± 4	50 ± 12
Water content (%)	91 ± 0.4	86 ± 1.0	78 ± 0.2

3. METABOLIC ACCLIMATION TO LOW TEMPERATURE

Shifting warm grown leaves to 5°C results in a severe repression of photosynthesis in herbaceous plants (Hurry and Huner 1991; Holaday et al. 1992; Strand et al. 1997, 1999). At ambient CO_2 and the growth irradiance (150 μmol m^{-2} s^{-1}) CO_2 uptake by *Arabidopsis* leaves shifted to 5°C for 1 hour is reduced from 5 to 2 μmol CO_2 m^{-2} s^{-1} and this does not recover even in plants shifted to 5°C for up to 10 days (Fig. 3). In contrast to warm grown leaves shifted to low temperature, *Arabidopsis* leaves that develop at 5°C recover both their yield of photosynthetic electron transport (Strand et al. 1997) and their CO_2 assimilation rate at the growth irradiance, although light saturated CO_2 assimilation is somewhat reduced in cold developed *Arabidopsis* leaves compared to the warm grown control leaves (Fig. 3). Leaves of other cold hardy herbaceous plants, such as wheat, rye and oilseed rape that develop at low growth temperatures also fully recover photosynthetic capacity (Hurry et al. 1998). This recovery is important because it not only enables these plants to avoid photoinhibition (Hurry and Huner 1992; Hurry et al. 1993), which could otherwise become devastating for plants at low temperatures, but it also re-establishes the plants ability to produce sucrose. Sucrose plays a key role as a cryoprotectant but it is also the starting point for basal metabolism, and as a source for cellular energy it is essential for plant growth, maintenance and repair processes and plant survival.

We have identified three main factors that contribute to the acclimation of photosynthetic carbon metabolism to low temperature. First, the activities of a range of Calvin cycle enzymes increase. Second, there is a selective stimulation of sucrose synthesis and third, the allocation of inorganic phosphate (Pi) between different subcellular compartments is readjusted. These three factors are all aspects of the reprogramming of the carbon metabolism found to be associated with leaf development at low temperature.

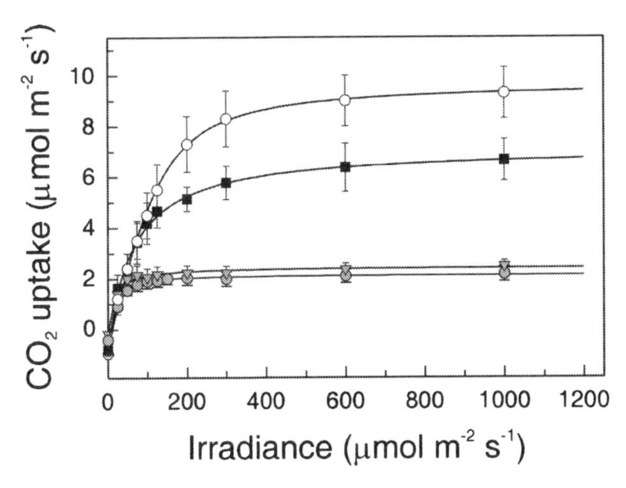

Figure 3. Photosynthetic light response curves for CO_2 uptake of Arabidopsis plants grown at 150 μmol m^{-2} s^{-1}. The light response curves were measured on attached leaves using an open gas-exchange system at ambient CO_2 concentration (35.5 Pa). Control warm developed leaves were measured at 23°C (○) and again 1 h after a shift to 5°C (●), leaves that had been shifted to 5°C for 10 days were measured at 5°C (✱) and cold developed leaves were also measured at 5°C (■).

3.1. The Calvin Cycle and Photosynthetic Gene Expression

The energy captured in photosynthetic electron transport is used to reduce CO_2 to sugars in the photosynthetic carbon reduction cycle, or the Calvin cycle. The Calvin cycle is autocatalytic and each metabolite of the cycle is both a substrate and a product. Consequently, all of the reactions must proceed at a balanced rate. The Calvin cycle is divided into three phases: *carboxylation*, *reduction* and *regeneration*. The *carboxylation* by Rubisco of the C5 sugar RuBP leads to the formation of two C3 molecules of 3-phosphoglycerate (PGA) and is followed by the *reduction* of PGA to triose-phosphate by phosphoglycerate kinase (PGK), NADP-glyceraldehyde 3-phosphate dehydrogenase (GAPDH) and triose phosphate isomerase (TPI). The *regeneration* of the CO_2 acceptor RuBP from triose-phosphate requires several steps catalyzed by aldolase (Ald), stromal fructose-1,6-bisphosphatase (sFBPase), transketolase (TK), seduheptulose-1,7-bisphosphatase (SBPase), ribulose phosphate epimerase (R5P3ep), ribose phosphate isomerase (R5PI), and phosphoribulokinase (PRK) (Fig. 4). Four enzymes in the Calvin cycle catalyze irreversible reactions; Rubisco, sFBPase, SBPase and PRK. These enzymes and GAPDH are strongly regulated by light dependent changes in the chloroplast such as reduced thioredoxin, pH and Mg^{2+} concentration. Tight regulation of the Calvin cycle is necessary to prevent depletion of the ATP and NADPH pools. PRK is particularly sensitive to changes that occur in the chloroplast and it is the step catalyzed by PRK that ultimately stops the Calvin cycle activity in the dark (Sharkey 1998).

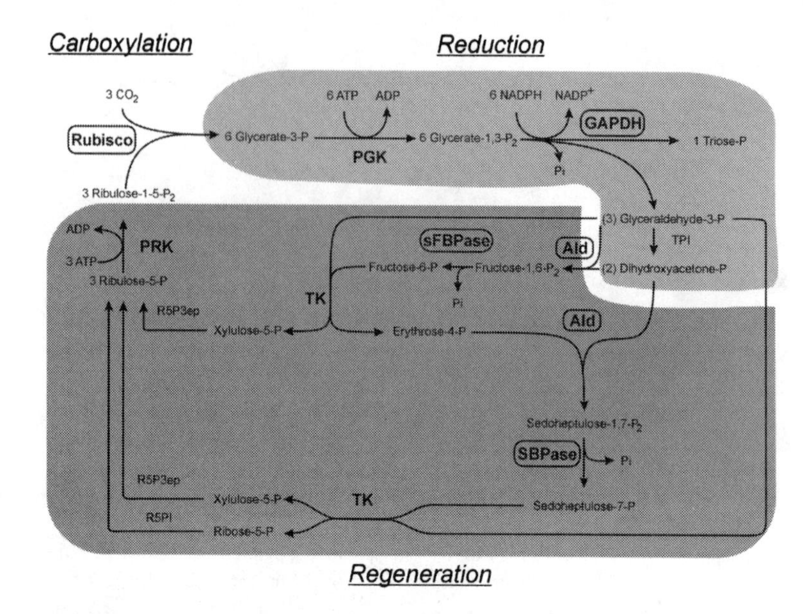

Figure 4. Schematic representation of the Calvin cycle showing the enzymes assessed to date (**bold** type face) and those that we have shown to increase more than 2-fold following leaf development at 5°C (boxed).

When we examined the activity and gene expression of eight of the Calvin cycle enzymes in leaves shifted to 5°C for 10 d we found a small increase in activity for only 3 of the Calvin cycle enzymes (Ald, GAPDH, sFBPase) (Strand et al. 1999) and a 2.5-fold increase in transcript for SBPase (Druart, Gardeström & Hurry, unpublished). However, of the four enzymes catalyzing irreversible reactions, three of these (Rubisco, sFBPase & SBPase) more than double in activity in cold developed leaves. The other light-regulated enzyme (GAPDH) also increased activity more than 2-fold following cold acclimation. Of the non-regulated enzymes catalyzing reversible reactions, Ald, which provides the substrates for both sFBPase and SBPase (Fig. 4), was the only one strongly induced by cold acclimation (Strand et al. 1999). The other enzymes assessed, PGK, TK and PRK all showed less than 50% increases and therefore became relatively much less abundant following cold acclimation. The differential regulation of these two groups of enzymes in response to low temperature may be of functional importance. GAPDH, Ald, sFBPase and SBPase are all located immediately downstream of Rubisco and the products of their reactions serve as precursors for sucrose and starch synthesis, as well as RuBP regeneration. High activity of this segment of the Calvin cycle will favor fixation of carbon dioxide and conversion of fixed carbon into sucrose and starch. On the other hand transketolase, which is not strongly induced by cold acclimation, catalyses the first reaction downstream of the steps at which substrates are removed for carbohydrate synthesis during RuBP-regeneration and low transketolase activity will restrict fluxes and decrease sequestration of Pi in metabolites in the purely regenerative part of the Calvin cycle, especially RuBP. Thus, these different responses of the reduction and regenerative phases of the Calvin cycle may function to maximize end product synthesis and Pi release under conditions that promote end product inhibition of photosynthesis and the sequestration of Pi into metabolites. PGK and PRK catalyze the reactions in which ATP is consumed in the Calvin cycle, and a decrease in their relative activity might also serve to reduce the sequestration of Pi in phosphorylated intermediates in the stroma.

Correlated with the increase in activity for the Calvin cycle enzymes in cold developed leaves is a release of the suppression of photosynthetic gene expression shown in leaves shifted to 5°C (Strand et al. 1997). Accumulation of soluble carbohydrates has been shown to repress the expression of genes that encode photosynthetic proteins. For example, the addition of glucose to *Chenopodium* cell cultures results in a rapid but reversible decrease in the steady-state transcript level of the small subunit of Rubisco (*RBCS*), of chlorophyll a/b-binding proteins (*LHC*), and of the δ-subunit of the thylakoid ATPase (*ATPD*) (Krapp et al. 1993). A rapid decrease of these transcripts also occurs following cold girdling of spinach leaves (Krapp and Stitt 1995). In both studies the reduction in mRNA was correlated with the accumulation of soluble carbohydrates. Shifting warm grown *Arabidopsis* leaves to 5°C leads to a rapid and sustained accumulation of hexose phosphates and soluble sugars (Strand et al. 1997, 1999) and to a reduction in the amount of transcript for the two *LHC* genes, *LHCB2* and *LHCB5*, and for *RBCS* (Strand et al. 1997). In contrasts, *Arabidopsis* leaves that developed in the cold show a full recovery of transcript levels for *LHCB2* and *LHCB5*, and *RBCS*. This recovery in the amount of transcript occurred even though the cold developed leaves maintained high amounts of soluble carbohydrates and even increased the partitioning into sucrose (see below). It has been reported that sugar accumulation *per se* is not the signal regulating the expression of the photosynthetic genes but that the signal may be related to the metabolism of hexose phosphates (Krapp et al. 1993; Jang and Sheen 1994; Krapp and Stitt 1995). *Arabidopsis* transformed to over-express or repress hexokinase showed that both *RBCS* and *LHC* expression was relatively more and less sensitive to sugars, respectively (Jang et al. 1997). In yeast, phosphorylation of glucose by hexokinase has been shown to be a sensor for mediating glucose repression of gene expression (Rose et al. 1991). Hexokinase itself is proposed to have a dual function as a protein kinase (Gancedo 1992) sensitive to the flux of sugars entering metabolism. Our results support the idea that the signal regulating the expression of the *LHC* genes and *RBCS* is related to hexose and sucrose metabolism rather than to the actual size of the soluble carbohydrate pools. Therefore, *Arabidopsis* leaves can acclimate their metabolism and gene regulation to the higher soluble carbohydrate levels at low temperature. However, this acclimation requires the reprogramming of carbon metabolism that only follows with leaf development at the low temperature.

3.2. Low Temperature Induces a Selective Stimulation of the Cytosolic Pathway for Sucrose Synthesis

During photosynthesis, fixed carbon is retained in the chloroplast and converted to starch or exported to the cytosol via the triose phosphate-phosphate transporter and converted to sucrose or used in glycolysis. The enzymes cytosolic FBPase (cFBPase) and SPS control the key regulatory steps in sucrose synthesis. cFBPase converts fructose-1,6-

bisphosphate (Fru1,6BP) to fructose-6-phosphate (Fru6P), and SPS converts UDP-glucose (UDPGlc) and Fru6P to sucrose-6-phosphate (Suc6P) further down stream in the same pathway (Fig. 5). These two key enzymes in the pathway for sucrose synthesis are highly regulated; AMP and the signal metabolite fructose-2,6-bisphosphate (Fru2,6BP) are allosteric regulators of cFBPase (Stitt 1990) and SPS is regulated allosterically by glucose-6-phosphate (Glc6P) and Pi, and covalently by multi-site phosphorylation (Huber and Huber 1996) (Fig. 5). Phosphorylation of SPS reduces enzyme activity by decreasing the affinity for the substrate Fru6P and the allosteric activator Glc6P. Furthermore, a binding site for a regulatory 14-3-3 protein, involving Ser229, has been identified on the SPS protein in spinach. Binding of the 14-3-3 protein appears to inhibit SPS activity (Toroser et al. 1998). However, whether *Arabidopsis* SPS is regulated according to the described mechanisms is unclear. There are indications that SPS in *Arabidopsis* is not light activated (Huber et al. 1989; Signora et al. 1998; Hurry & Stitt, unpublished data) and preliminary data indicates that there may be as many as three functionally distinct genes for SPS within the *Arabidopsis* genome (Langenkämper et al. 2001). More work is needed to elucidate the mechanisms regulating of the SPS enzyme in *Arabidopsis*.

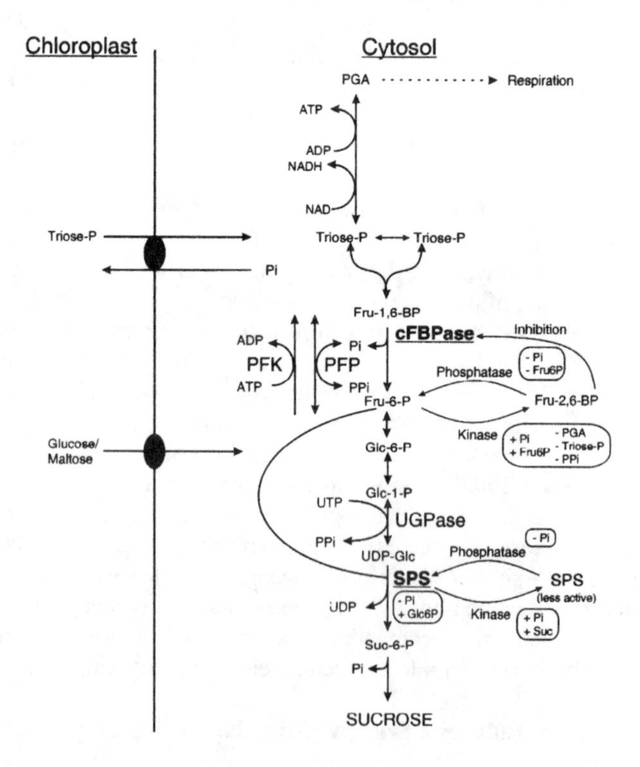

Figure 5. Schematic of the cytosolic pathways for sucrose biosynthesis showing the enzymes known to be induced by low temperatures and the factors known to regulate their activity. At this point nothing is known about the effects of low temperature on the activity of the regulatory phosphatases and kinases.

Enzymes in the pathway of sucrose synthesis have been shown to be particularly sensitive to low temperature and there is good evidence to suggest that, in the short-term, sucrose synthesis limits photosynthesis at low temperatures (Stitt and Grosse 1988). In response to this limitation, when *Arabidopsis* plants are shifted to the cold there is a rapid and strong increase in the activities and the amount of cFBPase, SPS (Strand et al. 1997, 1999) and UDP-glucose pyrophosphorylase (UGPase) (Ciereszko et al. 2001). In contrast to the regulation of the Calvin cycle enzymes, which generally show an increase in activity and protein without increased transcript levels (SBPase being the exception), the increase in these cytosolic enzymes is associated with increased gene expression. The amount of transcripts for *cFBP*, *UGP* and *SPS* increase gradually after a shift to the cold, with the highest transcript levels found in cold developed leaves (Strand et al. 1997; Ciereszko et al. 2001). A strong increase in activity of the enzymes for sucrose synthesis following cold exposure is observed in a number of other cold tolerant herbaceous species, including spinach, oilseed rape, winter wheat and winter rye, indicating that this represents a response that is crucial for cold acclimation (Guy et al. 1992; Hurry et al. 1994, 1995a). As yet, the promoter regions of *SPS* and *cFBP* have not been examined but they do not appear to contain known cold responsive elements such as the DRE/CRT (Gilmour et al. 2000) and determining the mechanisms behind this specific upregulation will be an interesting field of future research.

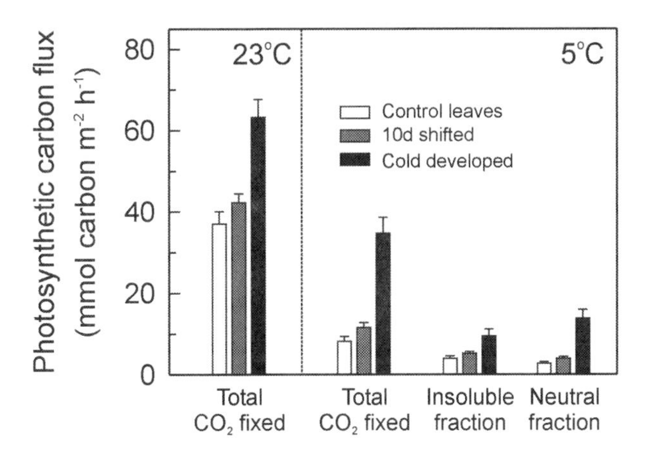

Figure 6. Photosynthetic carbon flux measured as ^{14}C incorporation at either 23°C (left panel) or 5°C (right panel). Each bar represents the mean (±SD) of at least 3 different incubations. The insoluble fraction primarily represents starch and the neutral fraction soluble sugars. Data redrawn from Strand et al. (1999).

The increase in activity of cFBPase, UGPase and SPS in the cold developed leaves results in both a recovery of total CO_2 fixation and in a change in partitioning of fixed carbon towards sucrose synthesis. To show this, photosynthetic carbon flux was measured by following ^{14}C incorporation by *Arabidopsis* leaf discs provided with CO_2 from labeled bicarbonate. Under these conditions cold developed leaves fix approximately the same amount of CO_2 at 5°C as control leaves fix at 23°C (Fig. 6). At 23°C control leaves allocate slightly more than 50% of their newly fixed carbon into starch and this proportion remains more or less constant after these leaves are shifted to 5°C for 10 days, although the total amount of CO_2 fixed by these shifted leaves at 5°C is drastically reduced. However, in the cold developed leaves more newly fixed carbon was partitioned into the neutral fraction, representing soluble carbohydrates, rather than the insoluble fraction, representing starch. 40% of the newly fixed labeled carbon was detected in the neutral fraction in the cold developed leaves, compared to only 30% in the warm grown control leaves (Fig. 6). Thus, cold acclimation in *Arabidopsis* leaves results in a specific stimulation of the pathway for sucrose synthesis that leads to the recovery of photosynthesis at low temperature and an increased production of cryoprotective carbohydrates.

3.3. Availability of Inorganic Phosphate (Pi) for Metabolism

One function of sucrose and starch synthesis is to release phosphate bound to metabolites so that it can be utilized in new reactions. Photosynthesis responds within seconds to alterations in the balance between triose phosphate production and utilization, even though end product synthesis is responsible for only one-ninth of the phosphate turnover under normal photorespiratory conditions (Sharkey 1998). The accumulation of large pools of phosphorylated intermediates found in leaves exposed to a short-term (hours) shift to low temperature (Labate and Leegood 1988) is thought to reflect a limitation in end product synthesis. Thus, accumulation of phosphorylated intermediates contributes to feedback down-regulation of photosynthesis because the stromal pool of Pi can be depleted to the point where photophosphorylation becomes Pi-limited (Labate and Leegood 1988; Foyer et al. 1990). We investigated whether the repression of photosynthesis in leaves shifted to low temperature for 10 d was due to Pi limitation by comparing the Pi status of the cold exposed (shifted leaves) and cold developed leaves. Under stable growth conditions the subcellular compartmentation of Pi is tightly regulated and 85 to 95% of the Pi may be located in the vacuole (Bieleski and Ferguson 1983). Prolonged exposure to low temperature, inducing Pi deficiency, can consequently be buffered by release of Pi from the vacuole. Our results indicate that changes in Pi compartmentation do occur during cold acclimation. Although the total pool of Pi in the cells does not change greatly after cold exposure, more Pi is available for metabolism (Strand et al. 1999). Thus, even though phosphorylated intermediates increase at low temperature, Pi availability over the long-term increases rather than decreases. This indicates that Pi is not likely to inhibit photophosphorylation and photosynthesis after long-term exposure (days) to low temperature, in contrast to the effects of short-term exposure reported previously (Labate and Leegood 1988).

3.4. Changes of Pi, a Signal Initiating the Cold Acclimation Process?

The signals that trigger changes in photosynthetic carbon metabolism during cold acclimation have not been identified. Molecular analysis of low temperature signaling has revealed a network of pathways that overlap with signaling pathways mediating responses to salt and water stress (Bohnert and Sheveleva 1998). As discussed above, short-term exposure to low temperature leads to a rapid and acute Pi-limitation in photosynthetic metabolism. We suspected that these rapid changes in Pi availability might trigger or modulate metabolic acclimation to low temperatures. To investigate whether changes in the amount of available phosphate is involved in triggering the cold acclimation process we compared the responses of the *Arabidopsis pho1-2* and *pho2-1* mutants with decreased and increased leaf phosphate, respectively. The PHO1 protein has been postulated to be involved in loading Pi into the xylem whereas PHO2 is believed to be a regulatory protein that monitors or is involved in the transduction information about changes of leaf Pi (Delhaize and Randall 1995).

The *pho* mutants showed striking differences in their ability to acclimate to low temperature (Hurry et al. 2000). When the plants were shifted to 5°C, the Pi-deficient mutant *pho1-2* experienced an especially strong inhibition of photosynthesis and showed clear signs of Pi limitation. Nevertheless, the *pho1-2* plants grew remarkably well at low temperature and showed a strong recovery of photosynthesis in cold developed leaves. Furthermore, *pho1-2* increased freezing tolerance more efficiently than did WT *Arabidopsis*. In contrast, the Pi-over accumulating mutant *pho2-1* was unable to increase freezing tolerance as much as WT. The three major factors identified to be part of the reprogramming of the carbon metabolism during cold acclimation; an increase in the Pi available for metabolism, the selective stimulation of sucrose synthesis and the increased activities of Calvin cycle enzymes, were all affected in the *pho* mutants. The increase in the availability of Pi for metabolism, which is typically seen during long-term exposure to low temperature, was blocked in the Pi-over accumulating *pho2-1* mutant. In cold developed leaves of WT and *pho1-2* the pool of accessible Pi increased several fold, whereas the *pho2-1* showed the same pool size of available Pi as when grown at 23°C. Consequently, the *pho2-1* plants had only 50% of the available Pi compared to WT. Clearly, the *pho2-1* mutant did not respond to the increased sequestering of phosphate in metabolic intermediates at low temperature.

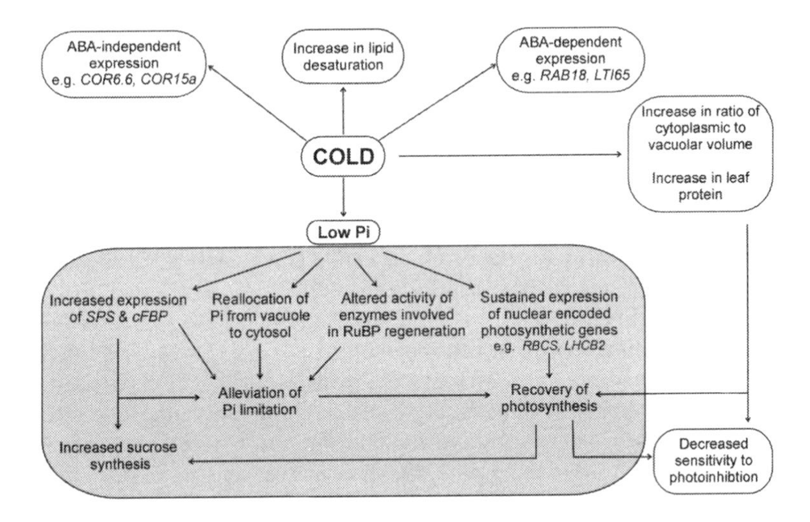

Figure 7. Schematic illustration of the interaction between phosphate availability, photosynthetic acclimation and the development of cold hardiness in *Arabidopsis thaliana*. Redrawn from Hurry et al. (2000).

The cold induced stimulation of the cytosolic enzymes, cFBPase and SPS, was also strongly affected in the *pho* mutants. In the *pho2-1* plants the increase of SPS protein and activity was attenuated, and the increase of cFBPase activity was strongly attenuated when exposed to low temperature. In *pho1-2* at 23°C, cFBPase protein and activity, and SPS protein and activity were 3-fold higher than in WT 23°C-leaves. After transfer of the *pho1-2* plants to 5°C, cFBPase and SPS increased further. These results indicate that changes in expression of the cytosolic enzymes are related to changes in Pi. The higher activities of cFBPase and SPS in *pho1-2* resulted in higher amounts of leaf sugars and a more pronounced mobilization of the carbohydrates at low temperature whereas the reduced activity of these enzymes in the *pho2-1* lead to lower amounts of leaf sugars. The response of *RBCS* and *LHC* expression at low temperature was also modified in the *pho* mutants. The release of the suppression of *RBCS* and *LHC* expression that is typically seen after cold acclimation (see above) was impaired in the *pho2-1* leaves. This suggests an important role for Pi in modulating the expression of these nuclear encoded photosynthetic genes under conditions that maintain high amounts of carbohydrates, such as low temperature. Associated with the impaired recovery of *RBCS* and *LHC* expression was the inability of the *pho2-1* plants to increase the activities of the Calvin cycle enzymes following cold acclimation.

These results establish a link between changes of Pi and specific aspects of cold acclimation related to photosynthetic carbon metabolism (Fig. 7). Two models could explain the interaction between Pi and cold acclimation in the *pho* mutants: a) lower level of Pi in the *pho1-2* leaves act to promote and even anticipate the responses during cold acclimation, whereas these processes are hindered by the higher levels of Pi in the *pho2-1* leaves b) the PHO2 protein is a regulatory protein that monitors information about changes of leaf Pi, including changes in Pi availability after exposure to low temperature. A role for PHO2 as a regulatory protein would explain the depletion of metabolically available free Pi and the large accumulation of hexose-P at 5°C in the *pho2-1* plants compared to WT. The role of Pi as a signal involved in the metabolic acclimation to low temperature will become clearer when the mutations have been identified.

4. EXPERIMENTS WITH TRANSGENIC *ARABIDOPSIS*

The results discussed so far give a clear indication that the selective stimulation of sucrose synthesis is a major factor contributing to the recovery of photosynthesis and for the reprogramming of carbon metabolism during cold acclimation. Consequently, the two regulatory enzymes in the sucrose biosynthesis pathway, cFBPase and SPS, are likely to be key enzymes in this process. To test the hypothesis that these enzymes are important not only for the acclimation of carbon metabolism but also for the development of cold tolerance, transgenic *Arabidopsis* plants were made with altered expression of cFBPase and SPS. We made transgenic *Arabidopsis* plants expressing either the cFBPase or the SPS gene in an antisense direction (anti-fbp and anti-sps, respectively) (see Strand 2000) and plants expressing the *Zea mays* SPS gene in the sense direction (over-sps) (see Signora et al. 1998; Strand et al. 2000). The degree of antisense repression was of the order of 70-80% for both enzymes and the SPS activity in the over-expressing lines was as much as 8 times higher than the WT under normal warm growth conditions. These three types of transgenic plants resulted in different metabolic backgrounds: the anti-fbp plants had a reduced rate of sucrose synthesis but a compensatory increase in starch synthesis; the anti-sps plant had a reduced rate of sucrose synthesis but no increase in starch synthesis; and the over-sps plants had an increased capacity for sucrose synthesis. The cold induced upregulation of the enzyme activity of cFBPase and SPS was blocked in the respective antisense lines (Strand 2000). SPS activity increased in the over-sps plants following cold exposure and the size of the increase correlated with the increase found in WT plants. This increase in the over-sps plants is therefore likely to be the result of the cold-induced increase in expression of the endogenous *SPS* gene, indicating that this is a specific response to low growth temperature and not a response to increased sucrose or metabolite levels.

4.1. Effect of the Introduced Transgenes on Photosynthesis at Low Temperatures

As we have shown (Fig. 3), when *Arabidopsis* plants are shifted to low temperature photosynthesis is strongly inhibited. The extent of this inhibition in leaves shifted to 5°C for 10 days was similar in WT and both antisense lines but was much less in the plants over-expressing SPS (Fig. 8). Following the development of new leaves at 5°C, photosynthesis recovered fully in both WT and the over-sps plants but under these fully acclimated conditions there was little further benefit to photosynthetic capacity of the over-expression of SPS. Plants over-expressing SPS are therefore to a large extent "pre-acclimated" to growth at low temperatures. Furthermore, the recovery of photosynthetic activity was strongly attenuated in both antisense lines (Fig. 8) even after a new rosette of leaves had developed at 5°C.

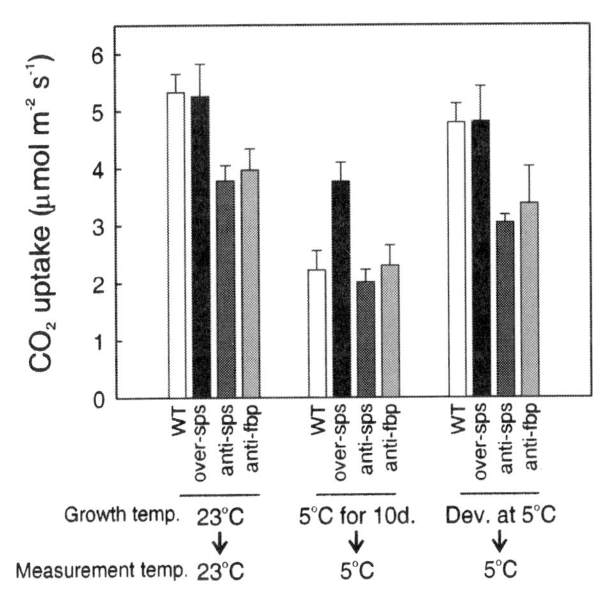

Figure 8. Effect of the introduced transgenes on photosynthetic CO_2 uptake at ambient growth conditions. Each bar represents the mean (±SD) of at least 4 different leaves.

To determine whether these changes in single leaf photosynthetic rates also translate into changes in total carbon acquisition by these plants we measured continuous whole plant CO_2-exchange over periods up to several days and have found that these changes in single leaf measurements are also reflected in the total carbon acquired by the plants (Cavaco & Hurry, unpublished). ^{14}C flux measurements also indicate that the over-sps plants show an increase in flux into the soluble sugar pools at low temperature and that both antisense lines show reduced soluble sugar synthesis, especially the anti-fbp plants (Strand 2000). These results confirm our initial hypothesis that the upregulation of the cytosolic pathway for sucrose synthesis at low temperature is a critical element for photosynthetic recovery. These results also demonstrate that the enhancement of this pathway in transgenic plants leads to greater total carbon acquisition by plants exposed to short- and long-term reductions in temperature, and to enhanced synthesis of soluble sugars at low temperature.

4.2. Effect of the Introduced Transgenes on Freezing Tolerance

The questions that remain are whether the development of freezing tolerance is also affected in the three transgenic lines, and whether this correlates with the effects we observed on photosynthetic acclimation and soluble sugar production.

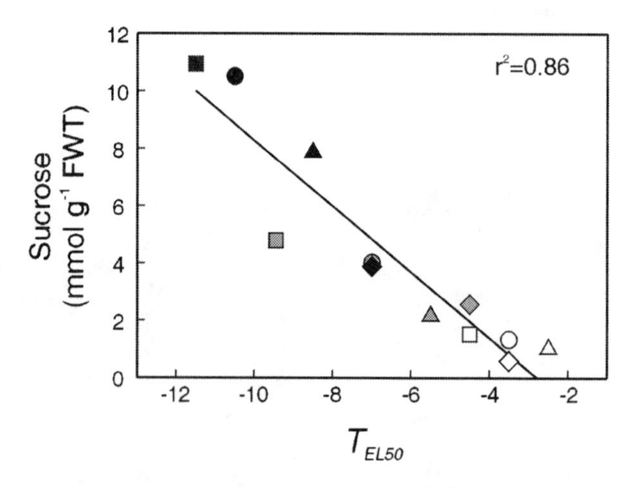

Figure 9. Correlation between sucrose content and freezing tolerance (T_{EL50}) in wild type and transgenic *Arabidopsis*. Symbols represent wild type (○, ●, ●), over-sps (□, ■, ■), anti-sps (Δ, ▲, ▲) and anti-fbp (◇, ◆, ◆) measured in control non-hardened leaves (○, □, Δ, ◇), after 10 days at 5°C (●, ■, ▲, ◆) and in leaves that developed at 5°C (●,■,▲,◆).

Correlated with the improved photosynthetic capacity of the over-sps plants at 5°C was an increased freezing tolerance in leaves shifted to 5°C for 10 days (Fig. 9; Strand 2000). WT plants increased their freezing tolerance down to temperatures around -7°C after exposure to 5°C for 10 d whereas the over-sps plants achieved a freezing tolerance around -9°C during the same exposure time. After the development of new leaves at 5°C, WT and the over-sps plants increased freezing tolerance further to around –11 to –12°C. In contrast, the development of freezing tolerance by both antisense lines was strongly attenuated with the anti-fbp and anti-sps plants only reached a freezing tolerance of around -5°C after 10 d exposure to 5°C (Fig. 9). This attenuation of the development of freezing tolerance was even more striking in the fully cold developed leaves, with the anti-fbp plants not even reach the freezing tolerance of WT leaves shifted to low temperature (Fig. 9). When we plot these changes in freezing tolerance against the sucrose content present in the leaves at the time they were harvested for the freezing tests we find a strong linear relationship, confirming that the lower activities of the sucrose biosynthetic enzymes in the antisense plants resulted in the accumulation of smaller pools of sucrose following cold development compared to WT and the over-sps plants, and that this is a strong contributing factor to the development of freezing tolerance in these leaves (Fig. 9).

5. FUTURE TARGETS

The work presented above describes only the responses of a small fraction of the complex of metabolic networks that must be coordinated in order for plants to acclimate successful to low temperatures. With the advent of high throughput sequencing and the development of a number of genomics platforms we now have the ability to look at these interacting metabolic networks as a whole, rather than as individual pathways. Chief among these new methods is large-scale DNA arrays. However, as an alternative for species where DNA arrays are either not available or not accessible, we have used PCR-based representational difference analysis (RDA) to examination differential gene expression between two mRNA populations (Hubank and Schatz, 1994).

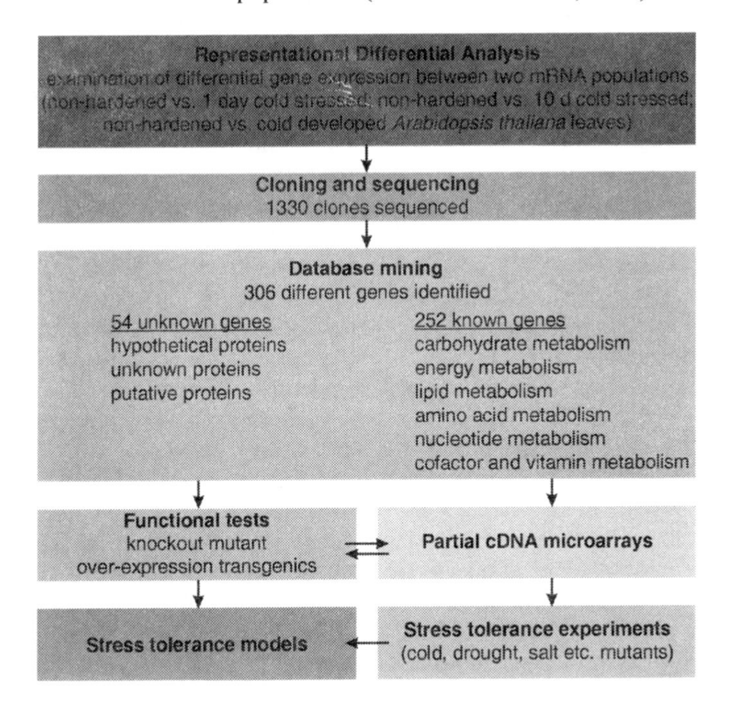

Figure 10. Schematic showing the approach being taken to identify new regulatory sites in metabolism and novel proteins critical for cold acclimation in *Arabidopsis thaliana*.

With this method we have sequenced 6 different RDA enriched libraries and identified 306 differentially expressed genes. Sequence analysis identified 252 transcripts for genes with known functions, including genes already implicated in cold stress and cold acclimation responses, and 54 transcripts for genes encoding proteins with no known function. The known genes, in addition to stress responsive genes, include enzymes involved in a number of different metabolic pathways (Fig. 10) and when a detailed analysis of their expression patterns is complete these may provide us with new targets and new target pathways for transgenic manipulation. The genes that have no significant homology to genes with known functions represent novel candidate genes for cold acclimation. However, as with all such large-scale cloning and expression studies it must be recognized that the majority of the changes in gene expression will represent trivial changes in mRNA accumulation and not significant acclimation responses. Nevertheless, functional tests using knockout mutants and large-scale expression studies (gene chips) in response to several stresses will give us more information about the role of both the known and unknown genes in *Arabidopsis* and information of this type will provide the basis for developing more comprehensive stress tolerance models in plants.

6. CONCLUSIONS

Temperate herbaceous winter annuals exploit the autumn months to establish site occupancy. This requires an acclimation strategy that involves the enhancement of pathways involved in photosynthetic carbon metabolism and coordinated changes in respiratory metabolism. Even if such acclimation is expensive, in terms of lower nitrogen or phosphate use efficiency or high respiratory costs, establishing site occupancy during the autumn may still provide a competitive advantage for over-wintering versus spring germinating annuals. We have identified three major factors that contribute to the acclimation of photosynthetic carbon metabolism to low temperature. First, there is a general increase in the activities of the Calvin cycle enzymes. This releases the limitation of enzyme activity in the Calvin cycle and is necessary for the recovery of CO_2 fixation at low temperature. Second, the allocation of Pi between different subcellular compartments is readjusted. Cold acclimation increases several-fold the amount of metabolically available Pi in the cytosol. The change in Pi distribution in the cell appears to be a critical response to low temperature necessary for initiating the changes to carbon metabolism. Third, there is a selective cold induced stimulation of sucrose synthesis. CFBPase, UGPase and SPS are strongly increased at both transcript and activity levels when plants are exposed to low temperatures and we have shown that it is possible to improve photosynthetic performance at low temperature and to increase freezing tolerance by increasing the activity of the sucrose biosynthetic pathway in transgenic plants.

These findings illustrate the importance of viewing adaptation to low temperature as the integrated acclimation of all metabolic processes in the cell, and not as the adaptive change of isolated processes. In this respect the new functional genomics, proteomics and metabolomics platforms being established will be powerful tools for building a more comprehensive picture of the interactions between the different cellular compartments during acclimation. The signal transduction pathways that mediate the acclimation response of photosynthetic, respiratory and cytosolic metabolism to low temperature remain unknown. Studies investigating the role of cellular redox poise indicate that this is one element in the puzzle (Huner et al. 1995) and, as we have shown here, this may be linked to the accumulation of sugars and changes in cellular Pi status. However, the long-term acclimation of herbaceous winter annuals to low temperature depends on *development* at the low temperature and unraveling the basis of this developmental response, and the signals that direct it, remains an exciting challenge.

REFERENCES

Bieleski, R.L., and Ferguson, I.B., 1983, Physiology and metabolism of phosphate and its compounds, in: *Encyclopedia of Plant Physiology*, New Series, vol 15A, A. Läuchli and R.L. Bieleski eds., Springer-Verlag, Berlin, pp. 422-449

Boese, S.R., and Huner, N.P.A., 1990, Effect of growth temperature and temperature shifts on spinach leaf morphology and photosynthesis, *Plant Physiol.* **94**: 1830-1836.

Bohnert, H., and Sheveleva, E., 1998, Plant adaptations - making metabolism move, *Curr Opinion Plant Biol.* **1**: 267-274.

Ciereszko, I., Johnsson, H., and Kleczkowski, L.A., 2001, Sucrose and light regulation of a cold-inducible UDP-glucose pyrophosphorylase gene via a hexokinase-independent and abscisic acid-insensitive pathway in *Arabidopsis, Biochem J.* **354**: 67-72.

Delhaize, E., and Randall, P.J., 1995, Characterization of a phosphate-accumulator mutant of *Arabidopsis thaliana, Plant Physiol.* **107**: 207-213.

Dexter, S.T., Tottingham, W.E., and Garber, L.G., 1932, Investigation of hardiness of plants by measurement of electrical conductivity, *Plant Physiol.* **7**: 63-78.

Foyer, C., Furbank, R., Harbinson, J., and Horton, P., 1990, The mechanisms contributing to photosynthetic control of electron transport by carbon assimilation in leaves, *Photosynth Res.* **25**: 83-100.

Gancedo, J.M., 1992, Glucose repression in *Saccharomyces cerevisiae* is directly associated with hexose phosphorylation by hexokinase PI and PII, *Eur J Biochem.* **206**: 297-313.

Gilmour, S.J., Hajela, R.K., and Thomashow, M.F., 1988, Cold acclimation in *Arabidopsis thaliana, Plant Physiol.* **87**: 745-750.

Gilmour, S.J., Sebolt, A.M., Salazar, M.P., Everard, J.D., and Thomashow, M.F., 2000, Over-expression of the Arabidopsis *CBF3* transcriptional activator mimics multiple biochemical changes associated with cold acclimation, *Plant Physiol.* **124**: 1854-1865.

Guy, C.L., Huber, J.L.A., and Huber, S.C., 1992, Sucrose phosphate synthase and sucrose accumulation at low temperature, *Plant Physiol.* **100**: 502-508.

Holaday, A.S., Martindale, W., Alred, R., Brooks, A., and Leegood, R.C., 1992, Changes in activities of enzymes of carbon metabolism in leaves during exposure of plants to low temperature, *Plant Physiol.* **98**: 1105-1114.

Hubank, M., and Schatz, D.G., 1994, Identifying differences in mRNA expression by representation difference analysis of cDNA, *Nucl Acid Res.* **22**:5640-5648.

Huber, S.C., and Huber, J.L., 1996, Role and regulation of sucrose-phosphate synthase in higher plants, *Annu Rev Plant Physiol Plant Mol Biol.* **47**: 431-444.

Huber, S.C., Nielson, T.H., Huber, J.L.A., and Pharr, D.M., 1989, Variation among species in light activation of sucrose-phosphate synthase, *Plant Cell Physiol.* **30**: 277-285

Huner, N.P.A., Palta, J.P., Li, P.H., and Carter, J.V., 1981, Anatomical changes in leaves of Puma rye in response to growth at cold-hardening temperatures, *Bot Gaz.* **142**: 55-62.

Huner, N.P.A., Öquist, G., Hurry, V.M., Krol, M., Falk, S., and Griffith, M., 1993, Photosynthesis, photoinhibition and low temperature acclimation in cold tolerant plants, *Photosynth Res.* **37**: 19-39.

Huner, N.P.A., Maxwell, D.P., Gray, G.R., Savitch, L.V., Laudenbach, D.E., and Falk, S., 1995, Photosynthetic response to light and temperature - PSII excitation pressure and redox signaling, *Acta Physiol Plant.* **17**: 167-176.

Hurry, V.M., and Huner, N.P.A., 1991, Low growth temperature effects a differential inhibition of photosynthesis in spring and winter wheat, *Plant Physiol.* **96**: 491-497.

Hurry, V.M., and Huner, N.P.A., 1992, Effect of cold-hardening on sensitivity of winter and spring wheat leaves to short-term photoinhibition and recovery of photosynthesis, *Plant Physiol.* **100**: 1283-1290.

Hurry, V.M., Gardeström, P., and Öquist, G., 1993, Reduced sensitivity to photoinhibition following frost-hardening of winter rye is due to increased phosphate availability, *Planta.* **190**: 484-490.

Hurry, V.M., Malmberg, G., Gardeström, P., and Öquist, G., 1994, Effects of a short-term shift to low temperature and of long-term cold hardening on photosynthesis and ribulose 1,5-bisphosphate carboxylase/oxygenase and sucrose phosphate synthase activity in leaves of winter rye (*Secale cereale* L.), *Plant Physiol.* **106**: 983-990.

Hurry, V.M., Keerberg, O., Pärnik, T., Gardeström, P., and Öquist, G., 1995a, Cold-hardening results in increased activity of enzymes involved in carbon metabolism in leaves of winter rye (*Secale cereale* L.), *Planta.* **195**: 554-562.

Hurry, V.M., Strand, Å., Tobiæson, M., Gardeström, P., and Öquist, G., 1995b, Cold hardening of spring and winter wheat and rape results in differential effects on growth, carbon metabolism, and carbohydrate content, *Plant Physiol.* **109**: 697-706.

Hurry, V., Keerberg, O., Pärnik, T., Öquist, G., and Gardeström, P., 1996, Effect of cold hardening on the components of respiratory decarboxylation in the dark and in the light in leaves of winter rye, *Plant Physiol.* **111**: 713-719.

Hurry, V., Huner, N., Selstam, E., Gardeström, P., and Öquist, G., 1998, Photosynthesis at low growth temperature, in: *Photosynthesis: A Comprehensive Treatise* A.S. Raghavendra ed., Cambridge University Press, Cambridge , pp. 238-249.

Hurry, V., Strand Å., Furbank, R., and Stitt, M., 2000, The role of inorganic phosphate in the development of freezing tolerance and the acclimatization of photosynthesis to low temperature is revealed by the *pho* mutants of *Arabidopsis thaliana, Plant J.* **24**: 383-396.

Jang, J-C., and Sheen, J., 1994, Sugar sensing in higher plants, *Plant Cell.* **6**: 1665-1679.

Jang, J-C., León, P., Zhou, L., and Sheen, J., 1997, Hexokinase as a sugar sensor in plants, *Plant Cell.* **9**: 5-19.

Krapp, A., and Stitt, M., 1995, An evaluation of direct and indirect mechanisms for the "sink-regulation" of photosynthesis in spinach: changes in gas exchange, carbohydrates, metabolites, enzyme activities and steady-state transcript levels after cold-girdling source leaves, *Planta.* **195**: 313-323.

Krapp, A., Hofmann, G., Schäfer, C., and Stitt, M., 1993, Regulation of the expression of *rbcS* and other photosynthetic genes by carbohydrates: a mechanism for the "sink regulation" of photosynthesis, *Plant J.* **3**: 817-828.

Labate, C.A., and Leegood, R.C., 1988, Limitation of photosynthesis by changes in temperature. Factors affecting the response of carbon dioxide assimilation to temperature in barley leaves, *Planta.* **173**: 519-527.

Lång, V., Mäntaylä, E., Welin, B., Sundberg, B., and Palva, E.T., 1994, Alterations in water status, endogenous abscisic acid content, and expression of *rab18* gene during the development of freezing tolerance in *Arabidopsis thaliana, Plant Physiol.* **104**: 1341-1349.

Langenkämper, G., Fung, R.W.M., Newcomb, R.D., Atkinson, R.G., Gardener, R.C., and MacRae, E.A., 2001, Sucrose phosphate synthase genes in plants belong to three different families, *J Mol Gen.* (in press).

Larcher, W., 1995, *Physiological Plant Ecology*, 3rd edn. Springer, Berlin.

Öquist, G., Hurry, V.M., and Huner, N.P.A., 1993, Low temperature effects on photosynthesis and correlation with freezing tolerance in spring and winter cultivars of wheat and rye, *Plant Physiol.* **101**: 245-250.

Rose, M., Albig, W., and Entian, K.D., 1991, Glucose repression in *Saccharomyces cerevisiae* is directly associated with hexose phosphorylation by hexokinase PI and PII, *Eur J Biochem.* **199**: 511-518.

Sharkey, T.D., 1998, Photosynthetic carbon reduction, in: *Photosynthesis: A Comprehensive Treatise* A.S. Raghavendra ed., Cambridge University Press, Cambridge , pp. 111-122.

Signora, L., Galtier, N., Skot, L., Lucas, H., and Foyer, C.H., 1998, Over-expression of sucrose phosphate synthase in *Arabidopsis thaliana* results in increased foliar sucrose/starch ratios and favours decreased foliar carbohydrate accumulation in plants after prolonged growth with CO_2 enrichment, *J Exp Bot.* **49**: 669-680.

Siminovitch, D., Rheume, B., Pomeroy, K., and Lepage, M., 1968, Phospholipid, protein, and neucleic acid increases in protoplasm and membrane structures associated with development of extreme freezing resistance in black locust tree cells, *Cryobiology.* **5**: 202-225.

Stitt, M., 1990, Fructose-2,6-bisphosphate as a regulatory molecule in plants, *Annu Rev Plant Physiol Plant Mol Biol.* **41**: 153-185.

Stitt, M., and Grosse, H., 1988, Interactions between sucrose synthesis and CO_2 fixation IV. Temperature-dependent adjustment of the relation between sucrose synthesis and CO_2 fixation, *J Plant Physiol.* **133**: 392-400.

Strand, Å., 2000, Metabolic Acclimation To Low Temperature - How To Make Metabolism Move, Ph.D. Dissertation, Umeå University, Umeå.

Strand, Å., Hurry, V., Gustafsson, P., and Gardeström, P., 1997, Development of *Arabidopsis thaliana* leaves at low temperature releases the suppression of photosynthesis and photosynthetic gene expression despite the accumulation of soluble carbohydrates, *Plant J.* **12**: 605-614.

Strand, Å., Hurry, V., Henkes, S., Huner, N.P.A., Gustafsson, P., Gardeström, P., and Stitt, M., 1999, Acclimation of Arabidopsis leaves developing at low temperatures. Increasing cytoplasmic volume accompanies increased activities of enzymes in the Calvin cycle and in the sucrose-biosynthesis pathway, *Plant Physiol.* **119**: 1387-1397.

Strand, Å., Zrenner, R., Trevanion, S., Stitt, M., Gustafsson, P., and Gardeström, P., 2000, Decreased expression of two key enzymes in the sucrose biosynthesis pathway, cytosolic fructose-1,6-bisphosphatase and sucrose phosphate synthase, has remarkably different consequences for photosynthetic carbon metabolism in transgenic *Arabidopsis thaliana, Plant J.* **23**: 759-770.

Toroser, D., Athwal, G.S., and Huber, S.C., 1998, Site-specific regulatory interaction between spinach leaf sucrose-phosphate synthase and 14-3-3 proteins, *FEBS Lett.* **435**: 110-114.

Uemura, M., Joseph, R.A., and Steponkus, P.L., 1995, Cold acclimation of *Arabidopsis thaliana*: effect on plasma membrane lipid composition and freeze-induced lesions. *Plant Physiol.* **109**: 15-30.

CHANGES IN THE PLASMA MEMBRANE FROM *ARABIDOPSIS THALIANA* WITHIN ONE WEEK OF COLD ACCLIMATION

Yukio Kawamura and Matsuo Uemura*

1. INTRODUCTION

Many plant species have obtained the capability to resist freezing and/or low temperatures and have thus extended their geographical distribution into regions with cold climates. A number of studies have been carried out to try to determine how these plants acclimate to low temperatures and survive under freezing conditions, and the results of those studies have indicated that irreversible damage in the plasma membrane that occurs during freeze-induced dehydration is the primary cause of freezing injury and, hence, plants must increase the cryostability of the plasma membrane during the course of cold acclimation.

Arabidopsis, a small *Brassicaceae* plant, has become one of the most widely used plant species for studies on cold acclimation and freezing injury because the plant cold-acclimates and shows various changes in cells, tissues and organs in a very short period. For example, enhanced freezing tolerance can be detected after only 1 day of cold acclimation, and maximum freezing tolerance is attained by cold acclimation for 1 week at 2°C (Uemura et al., 1995; Wanner and Junttila, 1999). Specific changes can be determined even earlier; changes in the behavior of membranes have been observed within 6 hours of cold acclimation (Ristic and Ashworth, 1993).

Uemura et al. (1995) reported that there are substantial changes in the lipid composition of the plasma membrane of Arabidopsis after 1 week of cold acclimation. There is an increase in the proportion of phospholipids as a consequence of an increase in the proportions of di-unsaturated species of both phosphatidylcholine and phosphatidylethanolamine. In contrast, the proportion of mono-unsaturated species of

*Yukio Kawamura and Matsuo Uemura, Cryobiosystem Research Center, Iwate University, Morioka 020-8550, Japan.

Plant Cold Hardiness, edited by Li and Palva
Kluwer Academic/Plenum Publishers, 2002

cerebrosides and free sterols occurs, while there is little change in the proportions of sterylglucosides and acylated sterylglucosides.

Although there are several reports on changes in plasma membrane proteins during cold acclimation in some plants (Yoshida and Uemura, 1984; Uemura and Yoshida, 1984; Zhou et al., 1994), it remains to be determined what role these proteins play during cold acclimation. There have only been two studies in which the proteins in the plasma membrane that change during cold acclimation have been identified. Goodwin et al. (1996) showed that the cell wall-plasma membrane linker protein increases at the mRNA level in *Brassica napus*, and Breton et al. (2000) recently reported that annexin increased at the protein level in wheat after 8 hours of cold acclimation. However, there have been no studies in which any plasma membrane proteins that change during cold acclimation in Arabidopsis have been identified.

On the other hand, many studies have shown that gene expression is considerably altered during cold acclimation in various plant species, and a number of genes that are regulated by low temperatures have been identified (Thomashow, 1998). The *COR* (cold-regulated) genes of Arabidopsis are known as one group of these genes and have been shown to encode various proteins, such as COR6.6, COR15a, and COR78 or homologs of LEA II proteins such as COR47. Many proteins encoded in these genes are hydrophilic polypeptides, but none of them have been shown to be membrane proteins with transmembrane domains. Since the amounts of mRNA of the genes encoding these proteins as well as the amounts of proteins themselves correlate positively with freezing tolerance, it has been speculated that these genes play roles in the increase in freezing tolerance (Thomashow, 1998). However, it has not been determined what roles these gene products play in order to reduce or prevent freezing injury during cold acclimation except for one case. Steponkus et al. (1998) hypothesized that the chloroplastal stroma polypeptide COR15am polypeptide, the final product encoded in the *COR15a* gene, affects the cryobehavior of the plasma membrane during freeze-induced dehydration. Since there are many proteins that are peripherally associated on the plasma membrane without a transmembrane region (Santoni et al., 1998), it is possible that some of the polypeptides encoded in the *COR* genes interact with the plasma membrane only under dehydrated conditions induced by freezing and contribute to the increase in the cryostability of the plasma membrane.

In this study, we examined changes in the plasma membrane, especially in the plasma membrane proteins, within 1 week of cold acclimation of Arabidopsis. Peptide mass finger printing, which is very useful for protein identification in Arabidopsis with the completion of the Arabidopsis genome sequencing project (The Arabidopsis Genome Initiative, 2000), was carried out using matrix-assisted laser desorption-ionization delayed extractionreflectron-time of flight (MALDI-TOF) mass spectrometry. Furthermore, recent advances in mass spectrometry techniques have enabled identification of proteins with a relatively small amount separated in two-dimensional isoelectrofocusing polyacrylamide gel electrophoresis (2D-PAGE).

2. MATERIALS AND METHODS

2.1. Isolation and Determination of Freezing Tolerance of Protoplasts

Protoplasts were enzymatically isolated from either nonacclimated or cold-acclimated leaves according to the method of Uemura et al. (1995). After protoplasts had been frozen, freezing tolerance of protoplasts was determined with fluorescein diacetate staining as described previously (Uemura et al., 1995).

2.2. Plasma Membrane Isolation and Lipid Analysis

Microsomal fractions were prepared from Arabidopsis leaves as previously reported (Uemura et al., 1995). After vacuolar membranes had been removed by the floating centrifugation as previously described (Matsuura-Endo et al., 1990), plasma membrane-enriched fractions were isolated using a two-phase partition system consisting of 5.6% (w/v) PEG 3,350 (Sigma) and 5.6% (w/w) dextran T-500 (Pharmacia) according to the methods of Uemura et al. (1995). Protein content was determined by a dye-protein binding procedure using BSA as a standard (Bradford, 1976). Lipid analysis was performed according to the method of Uemura et al. (1995).

2.3. Protein Analysis

Proteins were analyzed using 2D-PAGE. For the first dimension, isoelectrofocusing electrophoresis was carried out with 50 μg of plasma membrane proteins using gel strips forming an immobilized linear pH gradient from 4 to 7. Strips were rehydrated for 12 h at 100 V at 20°C with 7 M thiourea and 2 M urea lysis buffer, as described previously (Rabilloud et al., 1997), containing 4% (w/v) CHAPS, 20 mM dithiothreitol, 0.5% (v/v) IPG buffer (Amersham Pharmacia Biotech) and membrane proteins. Isoelectrofocusing was performed at 20°C in an IPGphor system (Amersham Pharmacia Biotech) with a successive increase in voltage: 500 V for 1 h, 1,000 V for 1 h, 2,000 V for 1 h, 4,000 V for 1 h, 6,000 V for 2 h and 8,000 V for 4 h (the total voltage hour value reaching ca. 50 kVhrs). After isoelectrofocusing electrophoresis, gel strips were equilibrated in a denaturing solution [1% (w/v) dithiothreitol, 6 M urea, 30% (w/v) glycerol, 4% (w/v) SDS and 50 mM Tris-HCl (pH 6.8)], and then SDS-PAGE on 12% (w/v) polyacrylamide gels was carried out according to the method of Laemmli (1970). Proteins in gels were visualized with silver-staining (Oakley et al., 1980) or modified silver staining for mass peptide finger printing (Shevchenko et al., 1996).

Spots on silver-stained 2-D gels were excised and were then subjected to alkylation tryptic digestion using a modified procedure of Jensen et al. (1999). The gel pieces containing proteins were washed twice with a solution containing 20 mM EDTA and 100 mM NH_4HCO_3 in 30% (v/v) acetonitrile, dried by a centrifugal concentrator (TOMY, Japan), and then rehydrated with reducing solution [10 mM EDTA, 10 mM dithiothreitol and 100 mM NH_4HCO_3] for 1 h at 60° C. After cooling to room temperature, the gel pieces were dried and then rehydrated with alkylating solution [10 mM EDTA, 10 mM iodeacetoamide and 100 mM NH_4HCO_3] for 30 min at room temperature. The gel pieces were washed twice with water and then minced. The minced gel pieces were dried and then rehydrated in a digestion solution containing 100 mM NH_4HCO_3 and 10 mM $CaCl_2$ with 0.1 to 0.2 μg of trypsin at 37°C overnight. After digestion, the minced gel pieces were separated from the digestion solution and successively washed with 0.1% (v/v) trifluoroacetic acid (TFA), 0.1% TFA in acetonitrile, 0.1% TFA in 50% (v/v) acetonitrile, and then 0.1% TFA in acetonitrile at room temperature in order to further extract the peptides from the minced gel pieces. Pooled extracts (including the digestion solution and both the aqueous and organic washes) were concentrated using a centrifugal concentrator and then desalted by a ZipTipC_{18} (Millipore).

Tryptic peptide masses were measured using a MALDI-TOF mass spectrometer equipped with a nitrogen laser (337 nm) (Voyager-DE STR, PE Biosystems). A solution of peptides was mixed with the same volume of a matrix solution consisting of saturated α-cyano-4-hydroxycinnamic acid (CHCA) prepared in 50% (v/v) acetonitrile/0.1% (v/v) trifluoroacetic acid, and the peptides were co-crystallized with CHCA by evaporating organic solvents. All MALDI spectra were internally calibrated using a standard peptide mixture. Mono-isotopic masses from all spectra recorded for a given peptide were selected and analyzed. Matching of experimental mass spectra with theoretical mass spectra obtained from various Arabidopsis databases was performed using three sequence database search programs, MS-Fit (http://prospector.ucsf.edu/), ProFound (http://prowl.rockefeller.edu/) and Pepident (http://www.expasy.ch/tools/peptident.html). Database queries were carried out for mono-isotopic masses using the following parameters: peptide mass tolerance of ±50 ppm (ppm = [experimental mass (in daltons) – theoretical mass] / theoretical mass, expressed in parts per million); the maximum number of missed tryptic cleavages of 1 or 2; and modifications, including conversion of oxidation of Met.

3. RESULTS AND DISCUSSION

3.1. Expansion-Induced Lysis and Lipid Composition

Enhanced freezing tolerance of Arabidopsis leaves, when determined by the electrolyte leakage tests, is observed after only 1 day of cold acclimation at 2°C, and maximum freezing tolerance is attained by cold acclimation for 7 days (Uemura et al., 1995). Freezing tolerance of protoplasts isolated from either nonacclimated or 7-day-cold-acclimated leaves is similar to that of respective leaves (Uemura et al., 1995). When protoplasts are frozen at temperatures ranging from -1 to -5°C, the freeze-induced osmotic contraction results in endocytotic vesiculation of the plasma membrane. Endocytotic vesiculation intrinsically is not injurious per se, but reductions in large areas of the plasma membrane are irreversible and, consequently, protoplasts lyse during osmotic expansion following thawing of a suspended medium. This form of injury is referred to as expansion-induced lysis (Steponkus et al., 1993). In protoplasts isolated from nonacclimated Arabidopsis leaves, the incidence of expansion-induced lysis has been reported to be ≤ 10% at temperatures ranging from -2 to -4°C (Uemura et al., 1995). On the other hand, expansion-induced lysis does not occur in protoplasts isolated from

7-day-cold-acclimated leaves. However, it has not been determined when the expansion-induced lysis disappears during 7 days of cold acclimation.

Freezing tolerance (LT_{50}) of protoplasts isolated from leaves nonacclimated and cold-acclimated for 1, 3 and 7 days at 2°C was determined to be -4.2, -7.1, -10.1 and -14.6°C, respectively (data not shown). Since the survival curve of protoplasts of 7-day-cold-acclimated leaves almost completely overlapped with that of protoplasts of 28-day-cold-acclimated leaves (data not shown), it is likely that maximum freezing tolerance of protoplasts was attained by 7 days of cold acclimation. After 1 day of cold acclimation, survival at temperatures ranging from -6 to -10°C rapidly increased (30-40% increase), while the increase in survival at temperatures ranging from -2 to -4°C was relatively small (0-17% increase) (Table 1). There was a large increase in survival at -2 and -3°C between 1 and 3 days of cold acclimation (16-22% increase).

To determine the temperature range over which expansion-induced lysis in protoplasts isolated from nonacclimated leaves occurs, survival of protoplasts was compared after conventional freeze/thaw treatment and freeze/hypertonic-thaw treatment according to the method of Uemura et al. (1995). As shown in Table 2, expansion-induced lysis occurred at temperatures ranging from -2 to -4°C, and its incidence was $\leq$ 28%. The incidence of expansion-induced lysis was slightly decreased by 1 day of cold acclimation, and it greatly decreased between 1 and 3 days of cold acclimation. Clearly, there was a very small incidence of expansion-induced lysis after cold acclimation for 3 days (Table 2). Thus, we concluded that the occurrence of expansion-induced lysis disappeared between 1 and 3 days of cold acclimation.

It is known that the decrease in the incidence of expansion-induced lysis is related to the increase in mono- or di-unsaturated species of phosphatidylcholine in the plasma membrane (Steponkus et al., 1993). In Arabidopsis, the proportion of phospholipids, especially the di-unsaturated species of phosphatidylcholine and phosphatidylethanolamine, increases after 7 days of cold acclimation (Uemura et al., 1995). Thus, it was expected that there is a significant change in lipid composition of the plasma membrane between 1 and 3 days of cold acclimation. In the present study, we show the lipid composition of the plasma membrane isolated after 1 day of cold acclimation (Table 3). The amount of phospholipids on the basis of proteins increased after 1 day of cold acclimation, but there were no significant changes in the phospholipid composition of the plasma membrane when expressed as mol% of the total lipids. Thus, the phospholipid composition of plasma membrane does not change substantially, at least after 1day of cold acclimation. For the other lipid classes, however, the cerebrosides were remarkably decreased after 1 day of cold acclimation even when expressed as both µmol/mg protein and mol% of the total lipids. In future, further studies are needed to analyze the lipid composition of the plasma membrane after 3 days of cold acclimation.

Table 1. Increase in % survival during cold acclimation[a]

	[ACC1d] - [NA]	[ACC3d] - [ACC1d]	[ACC7d] - [ACC3d]
(°C)	(%)	(%)	(%)
-2	0.1 ± 6	16 ± 6	5 ± 6
-3	14 ± 5	22 ± 5	
-4	17 ± 5	6 ± 5	4 ± 5
-6	44 ± 4	12 ± 4	11 ± 5
-8	43 ± 3	10 ± 4	24 ± 4
-10	31 ± 2	19 ± 3	15 ± 4

[a] To determine the increase in % survival, the survival of protoplasts isolated from leaves nonacclimated (NA) or acclimated for 1 day (ACC1d), 3 days (ACC3d) or 7 days (ACC7d) was determined after conventional freeze/thaw treatment according to the method of Uemura et al. (1995). Values are given as means ± s.e. (n=3).

Table 2. Incidence of expansion-induced lysis at temperatures ranging from -2 to -4°C[a]

	NA			ACC 1 day			ACC 3 days		
	con	hyper/thaw	EIL	con	hyper/thaw	EIL	con	hyper/thaw	EIL
(°C)	(%)	(%)	(%)	(%)	(%)	(%)	(%)	(%)	(%)
-2	80 ± 4	101 ± 7	20 ± 8	81 ± 4	101 ± 5	21 ± 6	97 ± 4	92 ± 4	-4.5 ± 5
-3	59 ± 4	87 ± 5	28 ± 6	73 ± 4	89 ± 4	16 ± 5	95 ± 4	87 ± 3	-7.6 ± 5
-4	55 ± 3	74 ± 4	19 ± 5	72 ± 4	88 ± 4	16 ± 5	78 ± 3	86 ± 3	7.3 ± 5

[a] To determine the incidence of expansion-induced lysis (EIL) in protoplasts isolated from leaves nonacclimated (NA) or acclimated (ACC) for 1 day or 3 days, survival of protoplasts after conventional freeze/thaw (con) or freeze/hypertonic-thaw (hyper/thaw) treatment was determined according to the method of Uemura et al. (1995). The incidence of EIL was calculated as the difference between survival of protoplasts after conventional freeze/thaw treatment and that after freeze/hypertonic-thaw treatment. Values are given as means ± s.e. (n=3).

Table 3. Lipid composition of plasma membrane isolated from Arabidopsis leaves nonacclimated (NA) and cold-acclimated for 1 day (ACC1d)

Lipids	mol % of total lipids			mol % of phospholipids			mol / mg protein		
	NA	ACC1d	AC/NA	NA	ACC1d	AC/NA	NA	ACC1d	AC/NA
Phospholipid									
PC	17.94	18.40	1.03	34.87	33.33	0.96	0.327	0.374	1.14
PE	19.41	21.59	1.11	37.72	39.10	1.04	0.354	0.439	1.24
PA	5.86	7.17	1.22	11.40	12.98	1.14	0.107	0.146	1.36
PI + PS	5.33	5.20	0.98	10.37	9.41	0.91	0.097	0.106	1.09
PG	2.91	2.86	0.98	5.65	5.18	0.92	0.053	0.058	1.10
Subtotal	51.45	55.22	1.07	100.01	100.00		0.938	1.123	1.20
Sterol									
Free sterol	32.39	33.42	1.03				0.592	0.679	1.15
Sterylglucoside	5.48	4.16	0.76				0.101	0.084	0.83
Acylated sterylglucoside	3.63	3.38	0.93				0.067	0.068	1.02
Subtotal	41.50	40.96	0.99				0.760	0.831	1.09
Cerebroside	5.68	2.99	0.53				0.104	0.060	0.58
Monogalactosyldiacylglyceride	0.40	0.51	1.28				0.007	0.010	1.40
Digalactosyldiacylglyceride	0.95	0.32	0.34				0.018	0.006	0.37
Total	99.98	100.00					1.827	2.030	1.11

3.2. Change in Plasma Membrane Proteins

One day of cold acclimation remarkably decreased the injury of protoplasts at temperatures ranging from -6 to -10°C (Table 1). The injury that occurred at temperatures below -5° C is thought to be a consequence of freeze-induced formation of the hexagonal II phase associated with the plasma membrane, and this injury leads to a loss of osmotic responsiveness in the protoplasts (Steponkus et al., 1993). Sugawara and Steponkus (1990) reported that an increase in the proportion of di-unsaturated species of phosphatidylcholine contributes to the decrease in propensity for freeze-induced formation of the hexagonal II phase. However, the proportion of phosphatidylcholine in total lipids was hardly changed in the plasma membrane after 1 day of cold acclimation (Table 3). What then prevents the occurrence of injury in the temperature range of -6 to -10° C? Since Wanner and Junttila (1994) reported that sugar accumulation is evident within 2 hours of cold acclimation in Arabidopsis, it is possible that the increase in intracellular sugars contributes to the decrease in the occurrence of this type of injury. However, this increase may be insufficient for the enhancement of freezing tolerance, since several single-gene mutants that accumulate sugars normally in response to low temperatures are nevertheless defective in freezing tolerance (McKown et al., 1996). Rather, it is possible that changes in the plasma membrane proteins could be associated with the decreased incidence of the loss of osmotic responsiveness. Thus, we tried to identify the changes in plasma membrane proteins during cold acclimation by a proteomics technique using 2D-PAGE and MALDI-TOF mass spectrometry.

Since the solubility of membrane proteins greatly increases following the addition of a twitterionic detergent, such as CHAPS, together with thiourea and urea (Rabilloud et al., 1997), the plasma membrane proteins were solubilized using these chemicals and were then separated into solubilized and unsolubilized fractions by ultracentrifugation. Next, the solubilized fraction was analyzed by 2D-PAGE over the pI range of 4 to 7 (Fig. 1). In this experiment, it was found that there were 12 membrane proteins that increased with cold acclimation. Most of these membrane proteins increased after 1 day of cold acclimation, and some acidic proteins seemed to be newly synthesized during cold acclimation for 1 day. Next, the membrane proteins that increased during cold acclimation were identified using a MALDI-TOF mass spectrometric system.

As shown in Table 4, these proteins were classified into 6 groups: DNA repair protein RAD23 homologue (spot No. 1), dehydrin of COR47 group (spots No. 2 and 3), carbonic anhydrase 2 (spots No. 4 - 8), outer membrane lipoprotein-like protein (spot No. 9), nodulin VfENOD18-like protein (spot No. 10), and mature rubisco small subunit (spots No. 11 and 12). Three programs, which were designed to predict membrane-spanning regions, indicated that almost all of these proteins had no transmembrane domains (Table 4). Santoni et al. (1998) also reported that the majority of Arabidopsis plasma membrane proteins separated on 2D-PAGE has no transmembrane domains. Thus, the membrane proteins solubilized with urea, thiourea and CHAPS and separated on 2D-PAGE may be peripherally and non-covalently bound on the plasma membrane, and many transmembrane proteins may be collected in the unsolubilized fraction.

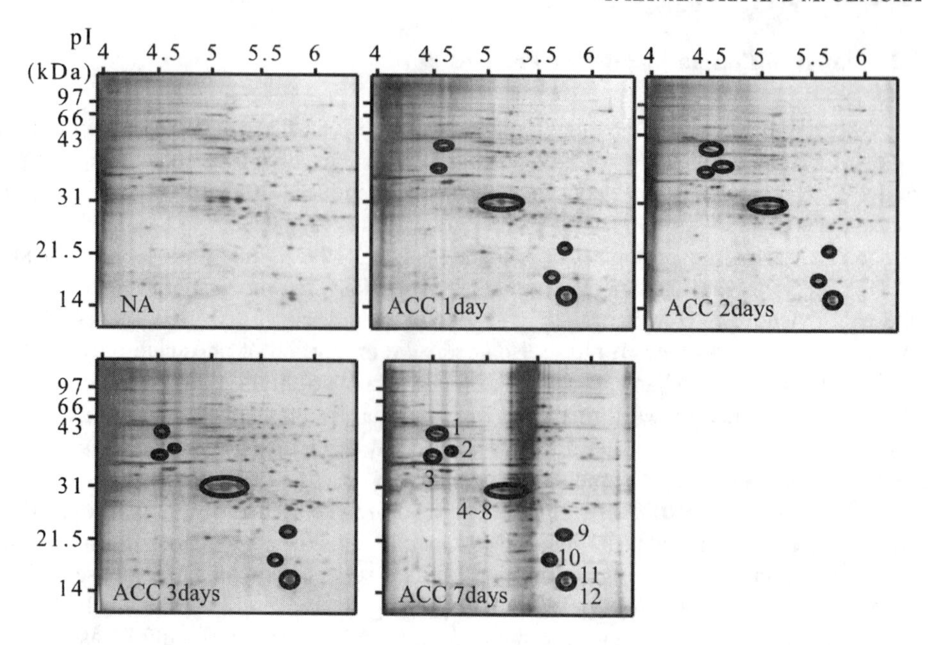

Figure 1. 2D-PAGE of the plasma membrane proteins isolated from nonacclimated (NA) and cold-acclimated (ACC) leaves of Arabidopsis. After isoelectrofocusing PAGE had been carried out in the pI range of 4 to 7, SDS-PAGE on 12% gel was performed. Some membrane proteins increased after 1 day of cold acclimation, and more-acidic proteins were newly synthesized by cold acclimation for 1 day. In this experiment, 12 spots that increased with cold acclimation were found (circled with numbers).

3.3. Cold-Induced Plasma Membrane Proteins

3.3.1. DNA Repair Protein RAD23 Homologue (spot No. 1)

Generally, RAD23 protein belongs to the ubiquitin family and has a ubiquitin-like domain in the N-terminus and two ubiquitin-associated domains in the C-terminus (Ortolan et al., 2000; Chen et al., 2001). A motif search program (http://motif.genome.ad.jp/) revealed that Arabidopsis RAD23 homologue possesses these domains as do RAD23 proteins in other species. Many studies have suggested that RAD23 is involved in nucleotide excision repair in yeast, humans and plants (Batty and Wood, 2000; Guzder et al., 1998; Jansen et al., 1998; Watokins et al., 1993; Sturm and Lienhard, 1998). It has also recently been reported that RAD23 protein may be involved in proteolysis due to 26 S proteasome. The N-terminal ubiquitin-like domain of RAD23 protein binds 26 S proteasome (Schauber et al., 1998). Interestingly, the yeast double mutant of RAD23 and RPN10, which is a component of 26S proteasome, displays a chilling-sensitive phenotype, whereas a single mutant of either RAD23 or RPN10 does not (Lambertson et al., 1999). Furthermore, an RAD23 mutant that lacks the N-terminal ubiquitin-like domain and is unable to bind 26S proteasome does not suppress the chilling-sensitive defects shown in the *rad23□ rpn10□* mutant (Lambertson et al., 1999). On the other hand, two ubiquitin-associated domains in the C-terminus of RAD23 protein

Table 4. Identification of plasma membrane proteins that changed during cold acclimation on 2D-PAGE

Spot	molecular size (kDa) / pI[a] Exp.	Theo.	Protein name	Accession[b]	Code[c]	Localization[d] (PSORT)	Transmembrane[e] (PSORT /SOSUI/TMpred)
1	44.3 / 4.5	40.1 / 4.58	RAD23 homolog	9758825	At5g38470	Cyt.	0 / 0 / 0
2	40.0 / 4.7	29.5 / 5.12	Dehydrin ERD10	556472	At1g20450	Nuc., Chl.	0 / 0 / 0
3	37.9 / 4.5	20.8 / 5.41	Dehydrin ERD14	13265523	At1g76180	Nuc., Chl.	0 / 0 / 0
4	31.3 / 5.0	28.3 / 5.36	carbonic anhydrase 2	11692920	At5g14740	Cyt.	0 / 0 / 0
5	31.3 / 5.1	28.3 / 5.36	carbonic anhydrase 2	11692920	At5g14740	Cyt.	0 / 0 / 0
6	31.3 / 5.3	28.3 / 5.36	carbonic anhydrase 2	11692920	At5g14740	Cyt.	0 / 0 / 0
7	30.4 / 5.0	28.3 / 5.36	carbonic anhydrase 2	11692920	At5g14740	Cyt.	0 / 0 / 0
8	30.4 / 5.1	28.3 / 5.36	carbonic anhydrase 2	11692920	At5g14740	Cyt.	0 / 0 / 0
9	22.7 / 5.7	21.4 / 5.98	outer membrane lipoprotein	9759532	At5g58070	Micro.	0 / 0 / 1
10	18.8 / 5.6	17.8 / 5.66	Nodulin VfENOD18-like	7630019	At3g53990	Cyt.	0 / 0 / 1
11	16.6 / 5.8	14.7 / 5.70	mature Rubisco small subunit	13926229	At1g67090	Micro.	0 / 0 / 0
12	15.9 / 5.8	14.7 / 5.70	mature Rubisco small subunit	13926229	At1g67090	Micro.	0 / 0 / 0

[a] Experimental molecular size and pI (Exp.) were determined from the mobility on 2D-PAGE gels, and the theoritical molecular size and pI (Theo.) were calculated on the basis of amino acid sequences.
[b] Accession number in NCBI database (http://www.ncbi.nlm.nih.gov/entrez/query.fcgi).
[c] Unique gene code in the MIPS Arabidopsis database (http://mips.gsf.de/proj/thal/), which is the www access to data of the Arabidopsis Genome Initiative compiled, analyzed, annotated and stored at MIPS by the MIPS Arabidopsis group.
[d] Localization in the cells predicted by the PSORT (http://psort.ims.u-tokyo.ac.jp/) program is shown as Cyt.:cytosoplasm, Nuc.:nucleus, Chl.:chloroplast, and Micro.:microbody.
[e] The number of transmembrane domains was predicted by either the PSORT, SOSUI (http://sosui.proteome.bio.tuat.ac.jp/~sosui/proteome/sosuiframe0.html) or TMpred (http://www.ch.embnet.org/software/TMPRED_form.html) programs.

bind to ubiquitin and prevent the expansion of nascent multi-ubiquitin chains (Ortlan et al., 2000; Chen et al., 2001). From these results, Chen et al. (2001) have proposed that it functions primarily as a proteasome localization signal. Thus, it is possible that the Arabidopsis RAD23 homologue identified in the present study promotes proteolysis of plasma membrane proteins during cold acclimation.

3.3.2. COR47-Like Dehydrins, ERD10 and ERD14 (spots No. 2 and 3)

ERD10 and ERD14 are very similar to COR47 and belong to the LEA group 2 proteins, which are also known as dehydrins (Thomashow, 1998). In Arabidopsis, it has been reported that dehydrins, including ERD10 and ERD14, accumulate in response to low temperatures, water deficiency and ABA treatment of both transcript and protein levels (Nylander et al., 2001). The molecular sizes of both ERD10 and ERD14 determined from the mobility on SDS-PAGE gels were substantially larger than the theoretical molecular sizes (Table 4) but were almost the same as the sizes reported by Nylander et al. (2001). It has been reported that the theoretical molecular sizes of other dehydrins are also different from the sizes determined by SDS-PAGE (Close, 1996). The Arabidopsis dehydrins can be subgrouped into three types, and ERD10, ERD14 and

COR47 belong to the family of acidic SK-type dehydrins (Nylander et al., 2001). Prediction of protein localization sites in cells using the PSORT program (http://psort.ims.u-tokyo.ac.jp/) indicated that these proteins might be localized in the nucleus or chloroplast (Table 4). However, Santoni et al. (1998) reported that ERD14 and COR47 proteins exist in the plasma membrane fraction. Furthermore, it has been reported that WCOR410 of wheat acidic dehydrin, which is very similar to Arabidopsis COR47-like proteins, accumulates at high levels in the vicinity of the plasma membrane of cells in the vascular transition area during cold acclimation (Danyluk et al., 1998). Thus, it is possible that acidic dehydrins, including ERD10 and ERD14, are peripherally bound to the plasma membrane.

3.3.3. Carbonic Anhydrase 2 (spots No. 4-8) and Mature Rubisco Small Subunit (spots No. 11 and 12)

Carbonic anhydrase plays a crucial role in the CO_2-concentrating mechanism (Badger and Price, 1994). In C_3 plants, carbonic anhydrase 2 is one of the most abundant proteins in leaves, and it has been reported that its isoforms can be classified in two types: chloroplast type and cytosolic type (Fett and Coleman, 1994). As shown in Table 4, five spots on 2D-PAGE were identified as carbonic anhydrase 2 of the cytosolic type, although it was not known whether multiple spots resulted from post-translocational modification or artifacts that occurred during 2D-PAGE. Immunolocalization studies have shown that chloroplastic carbonic anhydrase 2 is preferentially located in the stromal areas surrounding the grana and thylakoids and that some are associated with the chloroplast membranes (Rumeau et al., 1996). Thus, it is possible that carbonic anhydrase 2 of the cytosolic type is associated with the plasma membrane.

In this experiment, the mature Rubisco small subunit was found in the plasma membrane fraction and was the protein encoded in At1g67090 of the nuclear gene. The Rubisco small subunit is generally translated as a precursor from the nuclear gene including the signal peptide that is necessary to be targeted to the chloroplast, and it is then cleaved after being transporting into the chloroplast. The TargetP program (http://www.cbs.dtu.dk/services/TargetP/) predicted that the signal peptide of this Rubisco small subunit precursor may be 54 N-terminal amino acid residues (data not shown). On the other hand, the mature Rubisco small subunit may also be localized in a region other than chloroplast, because there is cDNA (NCB Accession No. 13926229) that lacks 55 N-terminal amino acid residues of the Rubisco small subunit coded in At1g67090, and the product of this cDNA clearly has no signal peptide to the chloroplast. Predictably, this protein does not localize in the chloroplast according to the PSORT program (http://psort.ims.u-tokyo.ac.jp/) (Table 4). The origin of this cDNA may be mRNA formed by alternative splicing. Thus, it is possible that the mature Rubisco small subunit is associated with the plasma membrane and contributes to the rapid increase in amounts of sugars during the early stage of cold acclimation (reported in the paper of Wanner and Junttila (1994)) together with carbonic anhydrase 2.

3.3.4. Outer Membrane Lipoprotein-Like Protein (spot No. 9)

The results of BLAST and FASTA searches showed that the outer membrane lipoprotein-like protein of Arabidopsis leaves was very similar to the bacterial lipocalin (data not shown), which is homologous to the animal apolipoprotein D (Akerstrom et al., 2000). All of these proteins belong to the lipocalin family. The lipocalin family has diverse functions, including lipid transport, retinol transport, invertebrate cryptic coloration, olfaction, pheromone transport and prostaglandin synthesis (Akerstrom et al., 2000). These roles are due to several common molecular-recognition properties of lipocalins: the ability to bind a wide range of small hydrophobic molecules, binding to specific cell-surface receptors and formation of complexes with soluble molecules (Flower, 1996). It has been thought that membrane-anchored bacterial lipocalins may play important roles in membrane biogenesis and repair (Bishop, 2000). Interestingly, it has also been reported that the bacterial lipocalin may contribute to the adaptation of cells to high osmotic stress (Bishop, 2000). Thus, the increase inArabidopsis lipoprotein during cold acclimation may be related to the resistance against osmotic stress.

3.3.5. Nodurin VfENOD18-Like Protein (spot No. 10)

Bacteria of the family Rhizobiaceae enter a symbiotic interaction with legume plants, which leads to the formation of a novel plant organ, the root nodule (Mylona et al., 1995). The root nodule is induced as a result of bacteria-plant interactions involving specific patterns of gene expression by both partners (Schultze and Kondorosi, 1998). Several plant genes that are expressed exclusively or predominantly in root nodules have been identified (Pawlowski, 1997; Sanchez et al., 1991). The encoded gene products have been designated as nodulins and are divided into early and late nodulins depending on the time of their appearance during root nodule organogenesis. The Arabidopsis nodulin-like protein that increased during cold acclimation was very similar to the early nodulin gene product VfENOD18 from *Vicia faba*. Homologues of nodulin genes have been identified in some non-leguminous plants, including Arabidopsis, suggesting that these gene products are not directly related to the formation of the nodule. The results of a motif search using the Pfam database (http://www.motif.genome.ad.jp/) showed that the nodulin-like protein identified in this study and the VfENOD18 protein belonged to the universal stress protein family (Pfam Accession PF00582). In *Escherichia coli*, the universal stress protein A contributes to the tolerance to carbon-source starvation; to toxic agents, including heavy metals, oxidants, acids and antibiotics; and to osmotic shock (Nystrom and Neidhardt, 1992; 1994). Thus, the nodulin-like protein may be associated with the tolerance to osmotic and oxidative stresses caused by the exposure to freezing and low temperature.

4. CONCLUSIONS

In the present study, we identified six groups of plasma membrane proteins that are changed during cold acclimation of Arabidopsis. Among those proteins, dehydrins and the proteins related to CO_2 fixation, proteolysis or osmotic stress increased within only 1 day of cold acclimation and then continued to gradually increase over a period of 7 days of cold acclimation. On the other hand, the lipid composition of plasma membrane does not change substantially, at least after 1 day of cold acclimation. These results indicate that changes in the plasma membrane proteins occur more rapidly than those in the lipid composition during the process of cold acclimation. Some cold-induced plasma membrane proteins may be related to changes in the incidence of freeze-induced lesions associated with the plasma membrane in the early stage of cold acclimation.

ACKNOWLEDGEMENTS

This study was partially supported by PROBRAIN. We thank A. Iwabuchi for her technical assistance and our laboratory members for their stimulating discussion.

REFERENCES

Akerstrom, B., Flower, D.R. and Salier, J.P., 2000, Lipocalins: unity in diversity, *Biochim. Biophys. Acta* **1482**: 1-8.

Badger, M.R. and Price, G.D., 1994, The role of carbonic anhydrase in photosynthesis, *Annu Rev Plant Physiol Plant Mol. Biol.* **45**: 369-392.

Batty, D.P. and Wood, R.D., 2000, Damage recognition in nucleotide excision repair of DNA, *Gene* **241**: 193-204.

Bishop, R.E., 2000, The bacterial lipocalins, *Biochim. Biophys. Acta* **1482**: 73-83.

Bradford, M.M., 1976, A rapid and sensitive method for the quantitation of microgram quantities of protein utilizing the principle of protein-dye binding, *Anal. Biochem.* **72**: 248-254.

Breton, G., Vazquez-Tello, A., Danyluk, J. and Sarhan, F., 2000, Two novel intrinsic annexins accumulate in wheat membranes in response to low temperature, *Plant Cell Physiol.* **41**: 177-184.

Chen, L., Shinde, U., Ortolan, T.G. and Madura, K., 2001, Ubiquitin-associated (UBA) domains in Rad23 bind ubiquitin and promote inhibition of multi-ubiquitin chain assembly, *EMBO Rep.* **2**: 933-938.

Close, T.J., 1996, Dehydrins: emergence of a biochemical role of a family of plant dehydrin proteins, *Physiol. Plant.* **97**: 795-803.

Fett, J.P. and Coleman, J.R., 1994, Characterization and expression of two cDNAs encoding carbonic anhydrase in *Arabidopsis thaliana*, *Plant Physiol.* **105**: 707-713.

Flower, D.R., 1996, The lipocalin protein family: structure and function, *Biochem. J.* **318**: 1-14.

Goodwin, W., Pallas, J.A. and Jenkins, G.I., 1996, Transcripts of a gene encoding a putative cell wall-plasma membrane linker protein are specifically cold-induced in *Brassica napus*, *Plant Mol. Biol.* **31**: 771-781.

Guzder, S.N., Sung, P., Prakash, L. and Prakash, S., 1998, Affinity of yeast nucleotide excision repair factor 2, consisting of the Rad4 and Rad23 proteins, for ultraviolet damaged DNA, *J. Biol. Chem.* **273**: 31541-31546.

Jansen, L.E., Verhage, R.A. and Brouwer, J., 1998, Preferential binding of yeast Rad4-Rad23 complex to damaged DNA, *J. Biol. Chem.* **273**: 33111-33114.

Jensen, O.N., Wilm, M., Shevchenko, A. and Mann, M., 1999, Sample preparation methods for mass spectrometric peptide mapping directlyfrom 2-DE gels, in: *Methods in Molecular Biology, Vol.112: 2-D Proteome Analysis Protocols,* A. J. Link, ed., Humana Press Inc., Totowa, NJ, pp. 513-530.

Laemmli, U.K., 1970, Cleavage of structural proteins during the assembly of the head of bacteriophage T4, *Nature* **227**: 680-685.

Lambertson, D., Chen, L. and Madura, K., 1999, Pleiotropic defects caused by loss of the proteasome-interacting factors Rad23 and Rpn10 of *Saccharomyces cerevisiae*, *Genetics* **153**: 69-79.

Matsuura-Endo, C., Maeshima, M. and Yoshida, S., 1990, Subunitcomposition of vacuolar membrane H^+-ATPase from mung bean, *Eur. J. Biochem.* **187**: 745-751.

McKown, R., Kuroki, G. and Warren, G., 1996, Cold responses of Arabidopsis mutants impaired in freezing tolerance, *J. Exp. Bot.* **47**: 1919–1925.

Mylona, P., Pawlowski, K. and Bisseling, T., 1995, Symbiotic nitrogen fixation, *Plant Cell* **7**: 869-885.

Nylander, M., Svensson, J., Palva, E.T. and Welin, B.V., 2001, Stress-induced accumulation and tissue-specific localization of dehydrins in *Arabidopsis thaliana*, *Plant Mol. Biol.* **45**: 263-279.

Nystrom, T. and Neidhardt, F.C., 1992, Cloning, mapping and nucleotide sequencing of a gene encoding a universal stress protein in *Escherichia coli*, *Mol. Microbiol.* **6**: 3187-3198.

Nystrom, T. and Neidhardt, F.C., 1994, Expression and role of the universal stress protein, UspA, of *Escherichia coli* during growth arrest, *Mol. Microbiol.* **11**: 537-544.

Oakley, B.R., Kirsch, D.R. and Morris, N.R., 1980, A simplified ultrasensitive silver stain for detecting proteins in polyacrylamide gels, *Anal. Biochem.* **105**: 361-363.

Ortolan, T.G., Tongaonkar, P., Lambertson, D., Chen, L., Schauber, C. and Madura, K., 2000, The DNA repair protein rad23 is a negative regulator of multi-ubiquitin chain assembly, *Nat. Cell Biol.* **2**: 601-608.

Pawlowski, K., 1997, Nodule-specific gene expression, *Physiol. Plant.* **99**: 617-631.

Rabilloud, T., Adessi, C., Giraudel, A. and Lunardi, J., 1997, Improvement of the solubilization of proteins in two-dimensional electrophoresis with immobilized pH gradients, *Electrophoresis* **18**: 307-316.

Ristic, Z. and Ashworth, E.N., 1993, Changes in leaf ultrastructure and carbohydrates in *Arabidopsis thaliana* L. (Heyn) cv. Columbia during rapid cold acclimation, *Protoplasma* **172**: 111-123.

Rumeau, D., Cuine, S., Fina, L., Gault, N., Nicole, M. and Peltier, G., 1996, Subcellular distribution of carbonic anhydrase in *Solanum tuberosum* L. leaves: characterization of two compartment-specific isoforms, *Planta* **199**: 79-88.

Sanchez, F., Padilla, J.E., Perez, H. and Lara, M., 1991, Control of nodulin gene in root-nodule developement and metabolism, *Annu. Rev. Plant Physiol. Plant Mol. Biol.* **42**: 507-528.

Santoni, V., Rouquie, D., Doumas, P., Mansion, M., Boutry, M., Degand, H., Dupree, P., Packman, L., Sherrier, J., Prime, T., Bauw, G., Posada, E., Rouze, P., Dehais, P., Sahnoun, I., Barlier, I. and Rossignol, M., 1998, Use of a proteome strategy for tagging proteins present at the plasma membrane, *Plant J.* **16**: 633-641.

Schauber, C., Chen, L., Tongaonkar, P., Vega, I., Lambertson, D., Potts, W. and Madura, K., 1998, Rad23 links DNA repair to the ubiquitin/proteasome pathway, *Nature* **391**: 715-718.

Schultze, M. and Kondorosi, A., 1998, Regulation of symbiotic root nodule development, *Annu. Rev. Genet.* **32**: 33-57.

Shevchenko, A., Wilm, M., Vorm, O. and Mann, M., 1996, Mass spectrometric sequencing of proteins silver-stained polyacrylamide gels, *Anal. Chem.* **68**: 850-858.

Steponkus, P.L., Uemura, M., Joseph, R.A., Gilmour, S.J. and Thomashow, M.F., 1998, Mode of action of the COR15a gene on the freezing tolerance of *Arabidopsis thaliana*, *Proc. Natl. Acad. Sci. U S A* **95**: 14570-14575.

Steponkus, P.L., Uemura, M. and Webb, M.S., 1993, A contrast of the cryostability of the plasma membrane of winter rye and spring ort - two species that widely differ in their freezing tolerance and plasma membrane lipid composition, in: *Advances in Low-temperature Biology, Vol. 2,* P.L. Steponkus, ed., JAI Press, London, pp. 211-312.

Sturm, A. and Lienhard, S., 1998, Two isoforms of plant RAD23 complement a UV-sensitive *rad23* mutant in yeast, *Plant J.* **13**: 815-821.

Sugawara, Y. and Steponcus, P.L., 1990, Effect of cold acclimation and modification of the plasma membrane lipid composition on lamellar-to-hexagonal II phase transitions in rye protoplasts, *Cryobiology* **27**: 667.

The Arabidopsis Genome Initiative., 2000, Analysis of the genome sequence of the flowering plant *Arabidopsis thaliana*, *Nature* **408**: 796-815.

Thomashow, M.F., 1998, Role of cold-responsive genes in plant freezing tolerance, *Plant Physiol.* **118**: 1-8.

Uemura, M., Joseph, R.A. and Steponkus, P.L., 1995, Cold acclimation of *Arabidopsis thaliana*. Effect on plasma membrane lipid composition and freeze-induced lesions, *Plant Physiol.* **109**: 15-30.

Uemura, M. and Yoshida, S., 1984, Involvement of plasma membrane alterations in cold acclimation of winter rye seedlings (*Secale cereale* L. cv. Puma), *Plant Physiol.* **75**: 818-826.

Wanner, L.A. and Junttila, O., 1999, Cold-induced freezing tolerance in Arabidopsis, *Plant Physiol.* **120**: 391-400.

Watkins, J.F., Sung, P., Prakash, L. and Prakash, S., 1993, The *Saccharomyces cerevisiae* DNA repair gene

RAD23 encodes a nuclear protein containing a ubiquitin-like domain required for biological function, *Mol. Cell Biol.***13**: 7757-7765.

Yoshida, S. and Uemura, M., 1984, Protein and lipid composition of isolated plasma membranes from orchard grass (*Dactylis glomerata* L.) and changes during cold acclimation, *Plant Physiol.* **75**: 31-37.

Zhou, B.L., Arakawa, K., Fujikawa, S. and Yoshida, S., 1994, Cold-induced alterations in plasma membrane proteins that are specifically related to the development of freezing tolerance in cold-hardy winter wheat, *Plant Cell Physiol.* **35**: 175-182.

CRYOPROTECTIN, A CABBAGE PROTEIN PROTECTING THYLAKOIDS FROM FREEZE-THAW DAMAGE
Expression of candidate genes in *E. coli*

Silke M. Schilling[1], Hany A. M. Sror[2], Dirk K. Hincha[3], Jürgen M. Schmitt[1], and Carsten A. Köhn[1]

1. ABSTRACT

We purified and partially sequenced a cryoprotective protein (cryoprotectin) from the leaves of cold-acclimated savoy cabbage (*Brassica oleracea*). Cryoprotectin protects thylakolds isolated from the leaves of non-acclimated spinach (*Spinacia oleracca*) from freeze-thaw damage. Sequencing of cryoprotectin revealed the copurification of at least two isoforms. The sequence data showed a high degree of similarity to a number of genes belonging to the class of lipid transfer proteins (LTP). The *wax9* gene family of *Brassica oleracea* was chosen to clarify the function of individual genes with respect to cryoprotective activity. The five *wax9* genes so far known were cloned and expressed in *Escherichia coli*. The preliminary data show for the first time that WAX9A and WAX9C have cryoprotective activity in an *in vitro* test assay. WAX9B and WAX9D had only low levels of cryoprotective activity while we could not detect any cryoprotective activity for WAX9E.

2. INTRODUCTION

Cryoprotective proteins in leaf extracts of cold acclimated spinach and cabbage have long been known (Volger and Heber, 1975). Sieg et al. (1996) purified a cryoprotective

[1] FU-Berlin, Institut für Pflanzenphysiologie, Königin-Luise-Str. 12-16a, 14195 Berlin, Germany.
[2] Permanent address: Biochemistry Department, Faculty of Agriculture, Ain-shams University, Cairo, Egypt.
[3] Max-Planck Institut für Molekulare Pflanzenphysiologie, 14424 Potsdam, Germany.

Plant Cold Hardiness, edited by Li and Palva
Kluwer Academic/Plenum Publishers, 2002

protein from cold acclimated cabbage leaves to homogeneity as judged by a silver-stained SDS-PAGE gel. Partial sequencing of the protein revealed the copurification of at least two isoforms (Hincha et al., 200 1). The sequence data showed a high degree of similarity to a number of lipid transfer proteins found within the *Brassicaceae.*

The *wax9* gene family of *Brassica oleracea* consists of 5 genes designated *wax9A-D* (Pyee and Kolattukudy, 1995) and *wax9E* (Hincha et al., 2001). WAX9D has been identified as the major protein in the surface wax of Brokkoli (*Brassica* oleracea var. *italica*) (Pyee et al., 1994) while WAX9E has been purified from the cuticula of savoy cabbage (*Brassica oleracea L. convar. capitata* (L.) Alef. *var. sabauda* L.) (Hincha et al., 2001). Genes encoding similar or identical proteins have been found in *Brassica napus.* The corresponding cDNAs or proteins have been identified in imbibed seeds and germinating seedlings (Soufleri et al.; 1996; Ostergaard et al., 1995) and in cold induced leaves (Sohal et al., 1999). The partial sequence of a calmodulin-binding protein from *Brassica chinensis* with 98 % similarity to the WAX9B/Q9S9GI protein has been described by Liu et al. (2001). All these proteins belong to the class of lipid transfer proteins.

LTPs are small (7-10 kDa) proteins containing eight strictly conserved cysteine residues forming four disulfide bridges (Kader, 1996; Kader, 1997). Determination of the three-dimensional structure of LTPs by X-ray crystallography and NMR led to a model predicting that the protein is composed of four a-helices and a C-terminal tail which form an internal hydrophobic cavity that runs through the molecule (Gomar et al., 1998; Lee et al., 1998; Shin et al., 1995). The hydrophobic cavity can accommodate one or two molecules of a variety of fatty acids (Douliez et al., 2001; Han et al., 2001; Lerche et al., 1997; Tassin-Moindrot et al., 2000). Another characteristic of LTPs is their remarkable stability towards denaturant, heat and proteases (Lindorff-Larscn and Winther, 2001; Hincha et al., 2001).

LTPs were originally defined by their ability to transfer lipids between membranes in *vitro.* Despite studying LTP-lipid interactions in detail as well as studying expression patterns of LTP-mRNAs and LTP-promoters the biological function of LTPS is not clear yet. As LTPs in general have a leader sequence directing the mature protein to the extraplasmatic space it has been proposed that LTPs play a role in cuticula formation (Pyee et al., 1994; Thoma et al., 1993, Sterk et al., 1991). For several LTPs an antimicrobial and/or antifungal activity has been described (Cammue et al., 1995; GarciaOlmedo et al., 1995; Kristensen et al., 2000). There is also evidence that LTPs are involved in the response to environmental stresses like cold (Dunn et al., 1998, Pearce et al., 1998), drought (Colmenero-Flores et al., 1997; Plant et al., 1991; Treviflo and O'Connell, 1998), salinity (Torres-Schumann et al., 1992) and heavy metals (Hollenbach et al., 1997).

LTPs are also interesting from a medical point of view as they are a pan-allergen in plant-derived food (Asero et al., 2000; Pastorello et al., 2001; 2000; Garcia-Casado et al., 2001). They also have been identified as pollen allergens (Colombo et al., 1998; Toriyama et al., 1998).

The sequence data obtained by Hincha et al. (2001) for the cryoprotective protein did not allow the assignment of cryoprotectivity to a specific gene. Attempts to further purify the cryoprotectin extract used for sequencing failed so far. On one hand the concentration of cryoprotectin in leaf extracts is very low and on the other hand highly purified cryoprotectin samples are active for a short time only. To elucidate the function of specific genes with respect to cryoprotective-activity and/or lipid transfer activity we decided to clone and express the members of the *wax9* gene family in *E. coli* and evaluate the recombinant proteins for cryoprotectivity and for lipid transfer activity.

3. RESULTS

3.1. Selecting Candidate Genes for In Vitro Expression

Basis for the work has been the partial sequencing of a cryoprotectin sample which appeared to be homogeneous as judged from silver stained SDS-PAGE gels (Sieg et al., 1996; Hincha et al., 2001). Sequencing of two distinct N-termini revealed the copurification of at least two isoforms. After tryptic digestion further 21 internal peptides were purified (Hincha et al., 2001). The sequencing results are summarized in the bottom lines of Figure 1. Overlapping sequences were combined into continuous sequences.

Comparison with the databases showed a high degree of similarity to 15 genes of the *Brassicaceae, 5* genes belong to the wax9 gene family of *Brassica oleracea,* 7 genes belong to the LTP-family of *Brassica napus, 1* gene is from *Brassica chinensis* and 2 genes belong to the LTP-family of *Arabidopsis thaliana.* The mature WAX9D protein is identical with the mature Q9S9F9 protein as well as the mature WAX9B protein is identical with the mature Q9S9GI protein. The mature PCBPIO protein has a similarity of 98% to the mature WAX9B/Q9S9G 1 proteins.

The first 25 amino acids in Figure I represent the leader sequence. Omission of a leader sequence means that there was no sequence data available in the database, it does not mean that the protein does not have a leader.
The *wax9* gene family was chosen for further investigations and for cloning purposes because of the following reasons:

1. The genes originate from *Brassica oleracea,* the same species we are using to purify cryoprotectin.
2. For all sequenced cryoprotectin fragments there are almost 100% identical sequences within the WAX9 proteins.
3. The use of *Brassica napus* as a source for cloning cDNAs or genomic DNA would be more laborious because of the allopolyploid nature of *Brassica napus* originating from *Brassica oleracea and Brassica campestris.* The pattern of a lipid transfer protein gene family homolog to the *wax9* gene family will be most probably much more complex.

3.2. Selecting an Expression System

Expression systems for recombinant plant proteins are available in *Escherichia coli, Pichia pastoris, and Saccharomyces cerevisiae.* The use of *P. pastoris or S. cerevisiae*

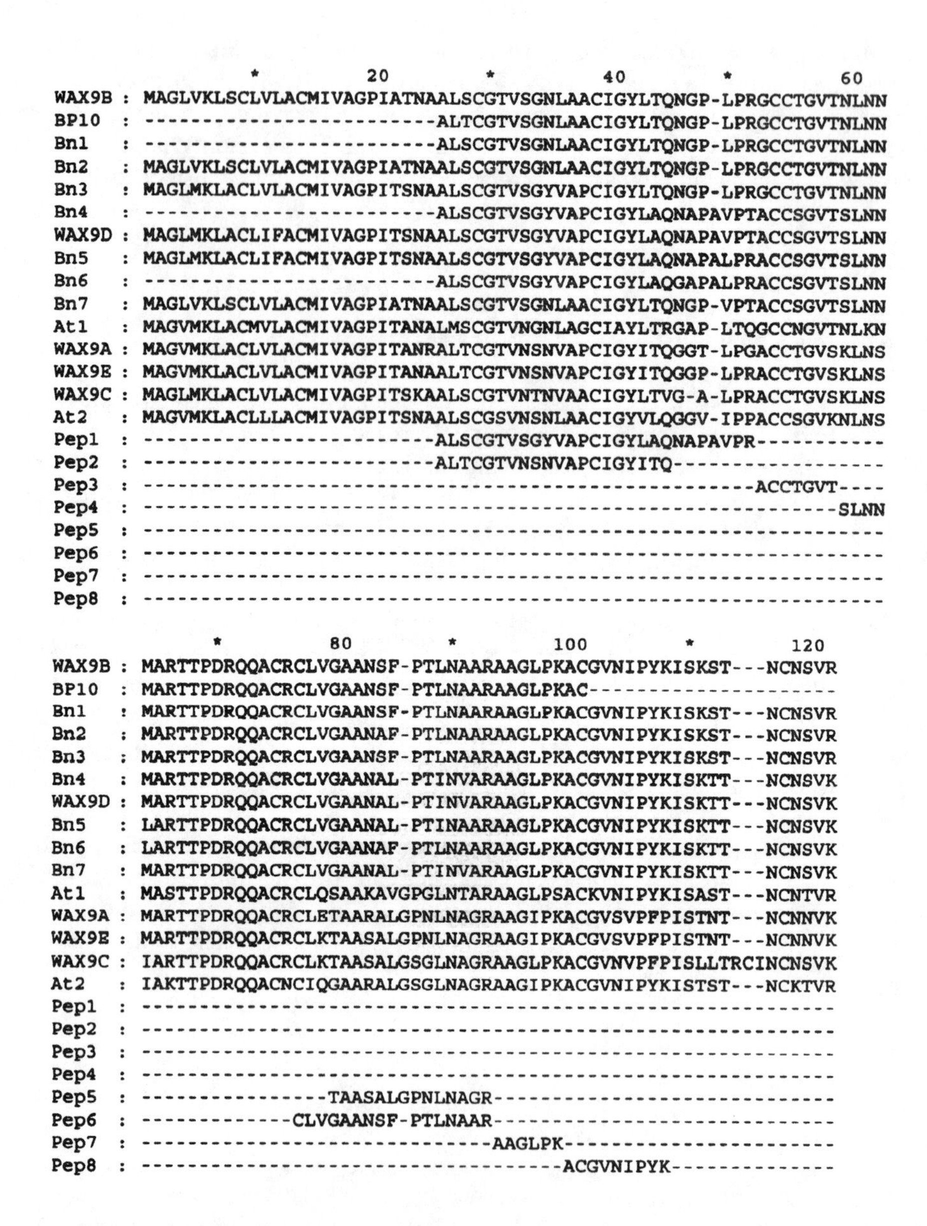

Figure 1. Alignment of 15 LTP sequences with the sequence fragments obtained from sequencing highly purified cryoprotectin samples by Hincha et al. (200 1). The sequences were aligned using Clustal X (Thompson et al., 1997). The sequence names have been abbreviated to fit on the page. Their origin and accession numbers in GenBank or Swiss-Prot/TrEMBL are: WAX9: *Brassica oleracea;* WAX9A: GenBank L33904; WAX9B: GenBank L33905; WAX9C: GenBank L33906; WAX9D: GenBank L33907; WAX9E: GenBank AF093751; *PBIO: Brassica chinensis,* Liu et al. (2001); Bn: *Brassica napus;* Bnl: TREMBL Q9S9GI; Bn2: Swiss-Prot Q42614; Bn3: Swiss-Prot Q42615; Bn4: TREMBL Q9S9F9; Bn5: TREMBL Q9SMMI; Bn6: TREMBL Q9S9GO-1 Bn7: Swiss-Prot Q42616; At: *Arabidopsis thaliana;* Atl: GenBank AY059927; At2: GenBank AY049296; Pepl-8: Sequences obtained from sequencing highly purified cryoprotectin samples by Hincha et al. (2001). Overlapping sequences were combined into continuous fragments.

has the advantage that secretion of the recombinant proteins through the secretary pathway permits posttranslational modifications like correct disulfide bond formation to occur. Successful expression of a functional wheat lipid transfer protein in *P. pastoris* has been reported by Klein et al. (1998) although they encountered problems with the correct processing at the N-terminal end of the protein.

Advantages of *E. coli* are the availability of a large collection of vectors and host strains as well as the easy handling. Lullien-Pellerin et al. (1999) and Masuta et al. (1992) purified functional lipid transfer proteins expressed in *E. coli.* In the case of LullienPellerin et al. (1999) the protein accumulated in insoluble cytoplasmatic inclusion bodies and was purified and refolded from them. Masuta et al. (1992) expressed the protein as a fusion with the maltose binding protein, purified the protein by affinity column chromatography, and subjected the protein to factor Xa cleavage to obtain the mature LTP.

We decided to use *E. coli* for expression to take advantage of the easy handling and the experience available in the laboratory. For expression the vector pET-32 (Novagen) was used. In Figure 2 the main features of the resulting recombinant protein are shown. Fusing thioredoxin (trx-tag) to the N-terminal end of the protein is supposed to increase the solubility of the recombinant protein. The protein can be purified by affinity column chromatography using the His-tag or the S-tag. After purification the leader can be cleaved using enterokinase. As the leader remains bound on the affinity column after cleavage the pure mature protein is easily recovered.

The host used was the strain Origami B(DE3) plys (Novagen). This strain carries *the trxblgor* mutations. Under physiological conditions the *Escherichia coli* cytoplasm is maintained in a reduced state that strongly disfavours the formation of stable disulfid bonds in proteins. The *trxblgor* mutations inactivate the thioredoxin reductase and the glutathione oxido-reductase allowing the fon-nation of disulfide bonds in the cytoplasm (Stewart et a]., 1998; Bessette et al., 1999). Correct folding of proteins containing disulfide bonds is enhanced, thereby improving the solubility of the proteins.

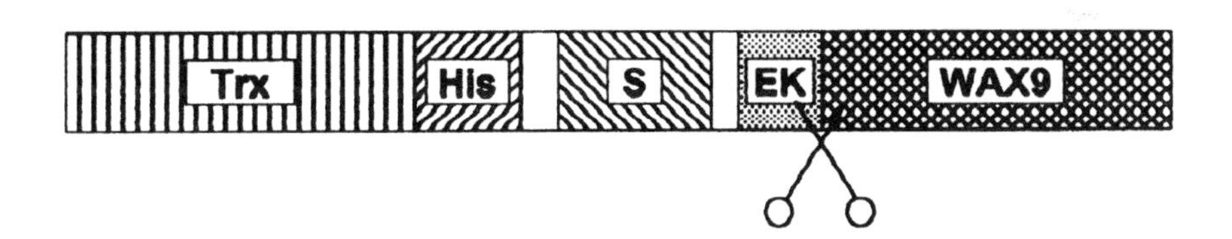

Figure 2. Schematic representation of the recombinant WAX9 protein precursor. Trx: Thioredoxin-tag, His: His-tag, S: S-tag, EK: Enterokinase cleavage site, WAX9: WAX9 protein A, B, C, D, or E.

3.3. Expression and Purification of Recombinant WAX9 Proteins

The first step was to determine the solubility of the recombinant protein. After induction the bacteria were disrupted using a French press and sonication. The lysate was then centrifuged to separate soluble and insoluble proteins. Recombinant WAX9A, B, D, and E accumulated in the soluble phase while WAX9C accumulated in form of inclusion bodies in the insoluble phase. Inclusion bodies were washed and then solubilized.

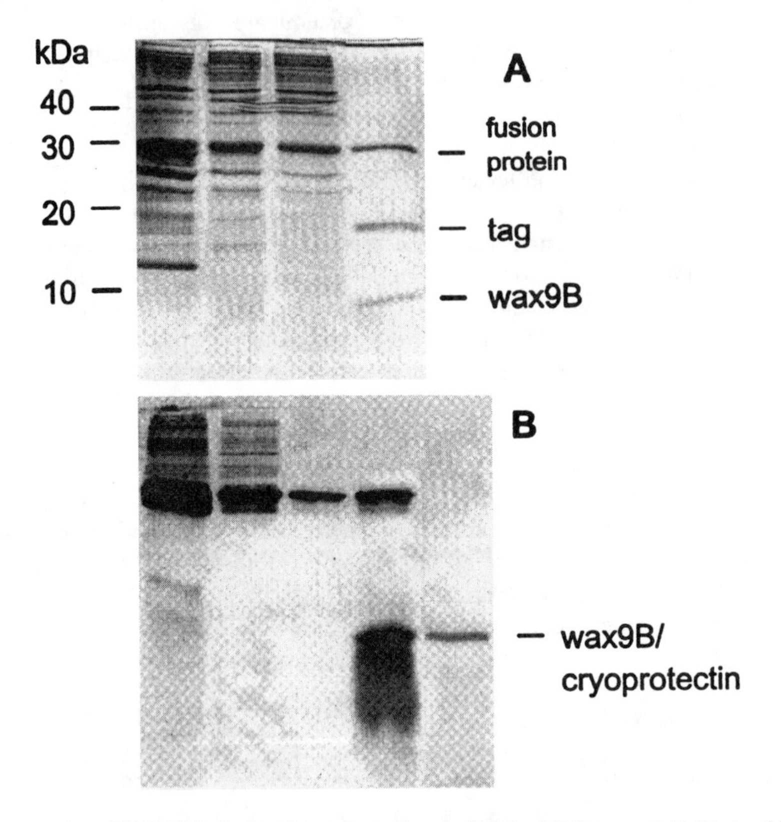

Figure 3. Expression of WAX9B in *Escherichia coli.* A: Coomassie stained SDS page gel; B: Western blot using an anti-LTP antibody; Lane 1: Total proteins *of E. coli* after induction, Lane 2: Flow through of a Ni-agarose column, Lane 3: Washing of Ni-agarose column, Lane 4: Partial digestion of the eluted recombinant protein, Lane 5: Crude cryoprotectin extract from cold acclimated *Brassica oleracea.*

The recombinant protein was purified by affinity column chromatography using Ni-agarose to bind the His-tag. The protein was then cleaved at the enterokinase cleavage site setting free the mature protein while the tag remained bound to the Ni-agarose. Alternatively the recombinant protein was eluted and cleaved in solution (Figure 3). The leader could then be removed from solution by a second affinity column chromatography step using Ni-agarose if required.

In comparison to the native WAX9 proteins the recombinant proteins have two additional amino acids (Ala, Met) at the N-terminal end. Western blots using an anti-LTP antibody as probe were used to confirm purification of the recombinant LTPs (Figure 3). An anti-his-tag antibody was used to monitor purification of WAX9C up to the step where the leader is removed by cleavage as the anti-LTP antibody did not recognize the WAX9C protein. The antigenic region of WAX9C differs in two amino acids from the synthetic peptide used to raise the antibody. The anti-LTP antibody detects a protein of similar size in crude cryoprotectin preparations as well as in samples of the purified WAX9 proteins (Figure 3).

When eluting the leader together with the mature WAX9 protein a discrepancy was observed in the amount of leader-protein in relation to mature WAX9 protein as judged by Coomassie stained SDS-PAGE gels. The amount of leader protein always was higher than the amount of mature WAX9 protein (Figure 3). This observation was confirmed by western blots using the anti-LTP antibodies comparing protein samples before and after cleavage. The signal always was considerably lower after cleavage of the protein. When using glass tubes or siliconized plastic tubes the mature WAX9 protein disappeared almost completely.

3.4. Cryoprotective activity of the recombinant WAX9 proteins

Initially the uncleaved recombinant protein was tested for cryoprotective activity. No activity was detectable for WAX9A-E. Cleavage of the precursor allowed the

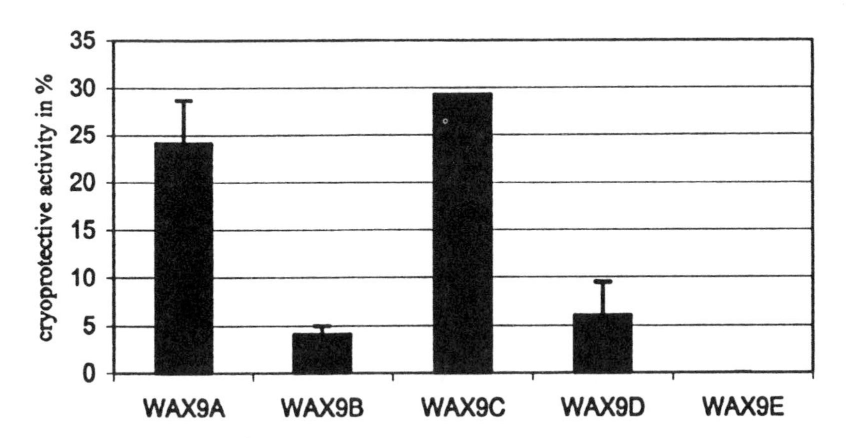

Figure 4. Cryoprotective activity of recombinant WAX9 proteins. The proteins were tested after clcavagc of the **precursor.** Differences in protein concentration& were adjusted to 10 jig/ml by calculation as there is a linear correlation between cryoprotective activity and protein concentration (Hincha and Schmitt, 1992b).

detection of cryoprotective activity (Figure 4). According to the preliminary data presented in Figure 4 WAX9A and WAX9C showed cryoprotective activity of 24% and 29% respectively. For WAX9B and WAX9D low levels of cryoprotective activity were detected. For WAX9E no cryoprotective activity could be detected at all.

4. DISCUSSION

The preliminary data presented in Figure 4 show for the first time that individual LTPs possess cryoprotective activity. WAX9A had an activity of 24% and WAX9C of 29%. The finding that WAX9A is one of the cryoprotectin candidate genes is in agreement with the sequencing results (Figure l). One of the N-terminal ends sequenced corresponds to WAX9A. WAX9B and WAX9D had low levels of cryoprotective activity which can not considered to be significant as an assay immanent variance makes it prohibitive to interpret activities below 10 %. Under this aspect it also has to be pointed out that for the results presented low protein concentrations were used and that only two repetitions were performed because of the low protein yield. Hincha and Schmitt (1992b) observed a low unspecific cryoprotective activity of proteins using BSA as control. This effect was observed for protein concentrations higher than 100 µg/ml reaching saturation at 400 µg/ml with 20 % cryoprotective activity. The protein concentrations used for the assays presented in this paper ranged between 10 and 38 µg/ml lying clearly below the 100 µg/ml value.

The main obstacle for a more accurate characterization of the WAX9 proteins has so far been the low yield of recombinant protein after cleavage. One explanation is that the mature WAX9 protein has a high affinity to a wide range of materials (glass, siliconized tubes, plastic) and is lost by adsorption. The other possibility is that contamination with a protease leads to degradation of the protein during the cleavage of the precursor with enterokinase, a serine-protease. Once this problem is solved the correlation between protein concentration and cryoprotective activity will be examined for the WAX9 proteins. Crude extracts prepared from *Brassica oleracea* leaves have cryoprotective activities of up to 100 %. In case that single proteins do not reach such high levels of activity it will be interesting to study the interaction of two or more WAX9 proteins. Possibly the effects of WAX9A and WAX9C are additive.

Hincha et al. (2001) observed that WAX9E protein purified from the cuticula of *Brassica oleracea* had lipid transfer activity while no cryoprotective activity could be detected. As we obtained the highest protein yields with WAX9E we were able to determine lipid transfer activity for it. Our results confirmed the observation of Hincha et al. (2001). The recombinant WAX9E had high lipid transfer activity (data not shown) but no cryoprotective activity (Figure 4). The LTP-activity demonstrates that the absence of cryoprotective activity is not due to incorrect protein folding but that WAX9E lacks the ability to protect thylakoids during the freeze-thaw cycle. For highly purified cryoprotectin samples, Hincha et al. (2001) showed the contrary. Cryoprotectin had cryoprotective activity but no lipid transfer activity. Due to low protein concentrations we could so far not determine lipid transfer activities for WAX9A-D.

When thylakoids are subjected to a freeze-thaw cycle in the presence of an artificial stroma medium, damage of thylakoids - measured as the release of platocyanin from the thylakoid lumen - shows biphasic kinetics (Hincha et al., 1996). During the first 30 min there is a rapid component that is directly dependent on the freezing temperature. A second slow and linearly time dependent component follows. Hincha et al. (1996) propose that the first rapid component is due to a reduced extensibility of the thylakoid membranes after the freeze-thaw cycle. The second slow phase is attributed to the solute loading of the thylakoids.

The *in vitro* freezing process can be pictured as follows (Hincha and Schmitt 1992a): Thylakoids are suspended in a medium with low solute concentration (Figure 5 A, B). During freezing water is removed from solution by ice formation. The solutes and thylakoids are concentrated in a small volume. As the solute concentration of the medium increases, water permeates out of the thylakoids driven by the concentration gradient between medium and thylakoid lumen. The thylakoids shrink (Figure 5C). At the same time solutes permeate through the membrane into the thylakoid lumen driven by the step concentration gradient (solute loading, Figure 5C). During thawing the concentration of the medium is decreased by the melting ice. The thylakoids swell osmotically. At this time the reduced extensibility of the membranes can lead to rupture of the thylakoids (Figure 5D). Rupture is enhanced by the uptake of solutes during freezing leading to increased swelling of the thylakoids. Thylakoids reseal after rupture but have a reduced volume (Figure 5E).

Adding cryoprotectin to the test assays reduces the damage of thylakoids during both phases (Hincha et al., 1990). This indicates that interaction of cryoprotectin with the thylakoid membranes modifies the characteristics of the membranes. Elasticity of the membrane after a freeze thaw cycle is increased while the permeability of the membranes is reduced during freezing.

Further analysis of a possible mode of action of cryoprotectin showed that cryoprotection and lipid transfer activity exclude each other (this paper; Hincha et al., 2001). This can be understood from the observation that cryoprotectin binds to thylakoids as well as to lipid model membranes (Sror, Schmitt, Hincha, unpublished data).

As a conclusion of the observations discussed above we propose the following model for the mode of action of cryoprotectin (Figure 5): The cryoprotective protein binds to the thylakold membrane (Figure 5F). Binding results in the modification of the membrane characteristics in so far that the elasticity of the membrane after a freeze thaw cycle is increased and that the permeability of the membranes is reduced (Figure 5F). As a result the rupture of thylakoids is reduced due to increased membrane extensibility and reduced solute loading (Figure 5G).

At the first instant it appears to be a contradiction that a LTP binds in a stable way to membranes as lipid transfer involves short and transient binding of lipid molecules during the transfer between membranes. But it has to be taken into account that lipid transfer has been the original definition of this class of proteins based on *in vitro* lipid transfer test. Until now there is no clear evidence that the biological function of this proteins is lipid transfer. For many LTPs it has been proposed that they function in pathogen defense or stress resistance. For example, the Ace-AMP1 protein extracted

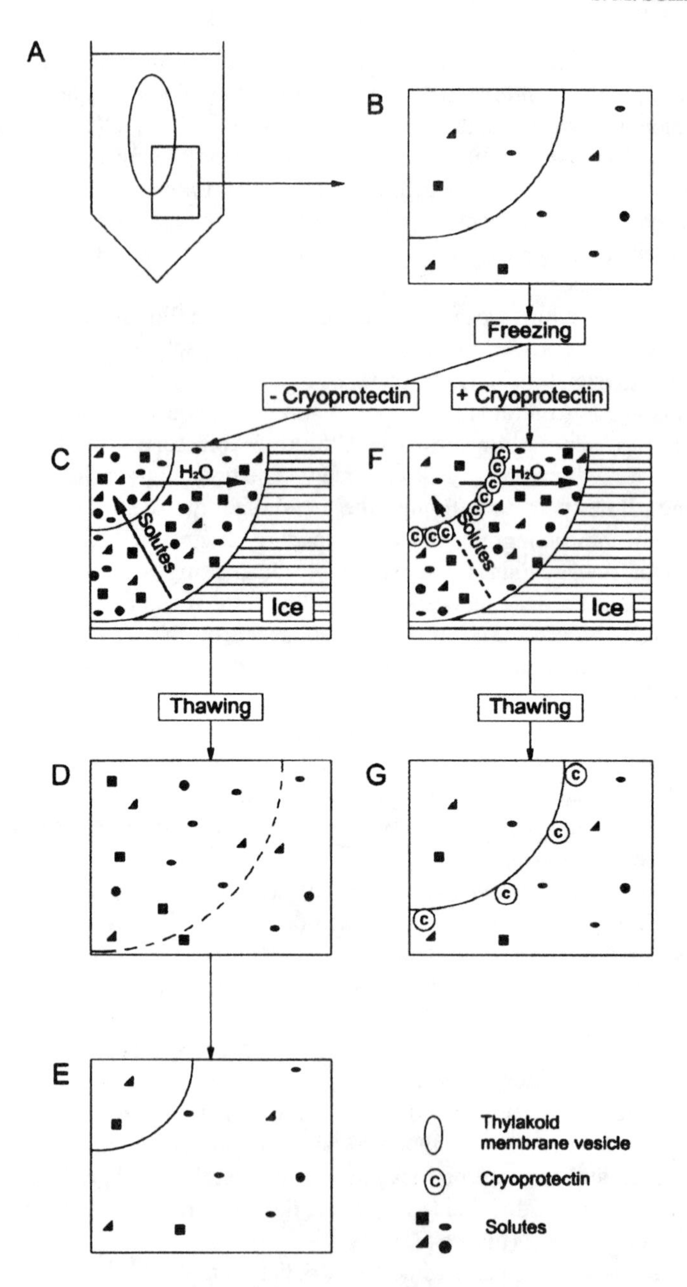

Figure 5. Schematic representation of a thylakoid suspension during an in vitro freeze thaw cycle with and without cryoprotectin added. A: General view of the experiment, B-G: Thylakoid sector in detail, B: Thylakoid before freezing, C: Thylakoid during the freezing process without cryoprotectin added, D: Thylakoid after thawing - the thylakoid swells and ruptures, E: Thylakoids reseal after rupture their volume is reduced, F: Thylakoid during the freezing process with cryoprotectin added to the medium, G: Thylakoid after thawing the thylakoid remains intact. For further details see the text.

from onion seeds is structurally related to the LTPs but does not possess lipid transfer activity. Ace-AMP1 has strong antifungal and antibacterial activity. The typical cavity, which is supposed to accommodate lipid molecules during transfer, is blocked in this protein (Tassin et al., 1998). Nevertheless it can interact with lipid membranes (Tassin et al., 1998).

For most plant species cited in the literature LTP gene families with a varying number of genes are reported. At least 26 putative LTP genes have been detected in *Arabidopsis thaliana* (Database at Salk Institute Genomic Analysis Laboratory, http://signal.salk.edu/cgi-bin/tdnaexpress). Although the main characteristics of LTPs are conserved within the gene families there is considerable variation between single genes suggesting that different members of the gene families might have different functions.

A range of LTPs has been purified from plants and the complete amino acid sequence has been determined. Comparison with the precursor protein deduced from the corresponding nucleotide sequences reveals the existence of a N-terminal leader at the end of the gene products. The leader sequence is supposed to direct the precursor protein to the endoplasmic reticulum. Because the proteins lack an endoplasmic reticulum retention signal they are expected to enter the secretary pathway (Kader, 1996).

In contrast, our working hypothesis for cryoprotectin would require that the protein is localized intracellularly. It could be shown indeed that gold labeled cryoprotectin binds to chloroplast and the cytoplasm in cabbage leaf sections fixed and embedded for electron microscopy. Using the anti-LTP antibody for immunogold labeling of sections from cold acclimated cabbage leaves revealed the localization of LTPs in chloroplasts and nuclei (Tischendorf, Schmitt, Hincha, unpublished data).

Also some other authors report evidence that LTPs are located intracellularly. LTPs have been reported to be associated with glyoxysomes in castor bean cotyledons (Tsuboi et al., 1992) and mitochondria and microsomes from maize seedlings (Douady et al., 1986). Osafune et al. (1996) suggest that LTPs in developing castor bean fruits are transported by two different routes - one is to the vacuole by vesicles and the other to the extraplasmic space, probably by the Golgi route. Nielsen et al. (1996) also reported that an antifungal LTP from sugar beet is located intra- and extracellularly.

However, it remains to be explained how a LTP is addressed to these compartments. Cammue et al. (1995) reported an additional C-terminal extension of 12 amino acids for Ace-AMP1 which has similarities to vacuolar sorting signals (Nakamura and Matsuoka, 1993). Another possibility would be alternative splicing as it had been suggested for the C-terminal end of a maize LTP by Arondel et al. (1991). Differentially spliced mRNAs could result in products having different sorting signals.

5. MATERIALS AND METHODS

5.1. Materials

Savoy cabbage *(Brassica oleracea L. convar. capitata* (L.) Alef. *var. sabauda L. cv* Imposa) was grown in the field. For cold acclimation plants stayed in the field during winter until they were exposed to subzero temperatures for several days. Broccoli (*Brassica oleracea L. convar. botrytis* (L.) Alef *var. cymosa Duch.* cv Monterey) and spinach (*Spinacia oleracea cv* Monnopa) were grown in growth chambers (16 h, 25'C day and 8 h, 15' night cycle).

5.2. Cloning of wax9 genes into an expression vector

Total DNA was extracted according to Rogers and Bendich (1985) using Cetyl-trimethylammoniumbromide (CTAB) to precipitate the DNA. DNA fragments encompassing the particular wax9 genes were amplified by a PCR reaction (60 °C annealing temperature) using the primers published by Pyee and Kolattukudy (1995) for *wax9A-D. For wax9E* new primers were designed.

wax9A:	Forward:	5'- CAACAGAAACACTAATAG - 3'
	Reverse:	5'- CTCATCCATATTATAATATTCGGAACGTCCCCTAAA - *3'*
wax9B:	Forward:	5'- CTCATTTGCAAACAGTCAAAAGCTCC - 3'
	Reverse:	5'- TCATACATTATATTAGATCGTACGTCCGCTTGAGTT - 3'
wax9C:	Forward:	5'- CATCCGCAAAAACGCCTAAGAGAAA - 3'
	Reverse:	5'- CTCATCCATTTCATAGCATGGAAAGTGTCCACTCAA - *3'*
wax9D:	Forward:	5'- CAAAAAAAATCTAAGAGAG - 3'
	Reverse:	5'- GCTCATCCATTACATAATACTTGAACGTCTGCATTA - 3'
wax9E:	Forward:	5'-ACCAACAGAAACACTAATAGA-3'
	Reverse:	5'- CTAAAGCTTCATCACACTGTC - 3'

The PCR-product obtained from the first PCR reaction was reamplified (PCR reaction, 60 °C annealing temperature) to obtain fragments suitable for cloning the *wax9* genes into pET-32. The reverse primer was designed so that the intron at the C-terminal end was omitted and that the 10 nucleotides missing at the C-terminal end were included into the primer. In addition the restriction sites needed for cloning into pET-32 were linked to the *wax9* gene by integrating the corresponding recognition sites into the primer sequence.

wax9A:	Forward:	5'- CATGCCATGGCTCTGACCTGTGGCACCG - 3'
	Reverse:	5'- CCGCTCGAGTCATTTCACGCTTGTTGCAGT - 3'
wax9B:	Forward:	5'- CATGCCATGGCTCTAAGCTGTGGAACCG - 3'
	Reverse:	5'- CCGGAATTCTCATCTAACGCTGTTGCAGTTGGTGCTT- 3'
wax9C:	Forward:	5'- CATGCCATGGCTTTGAGCTGTGGCACGG - 3'
	Reverse:	5'- CCGGAATTCTTATTTCACGCTGTTGCAGTTGATGCAC- 3'
wax9D:	Forward:	5'- CATGCCATGGCTCTGAGTTGTGGCACCG - 3'
	Reverse:	5'- CCGCTCGAGTCATTTCACACTGTTGCAGTT - 3'
wax9E:	Forward:	5'- CATGCCATGGCTCTGACCTGTGGCACCG - 3'
	Reverse:	5'- CCGCTCGAGTCATTTCACGTTGTTGCAGTT - 3'

The resulting fragments were gel purified and cloned into pET-32 using the Ncol and Xhol restriction sites (*wax9A, D, E*) or the Ncol and EcoRl restriction sites (*wax9B, C*). In frame integration of the fragments was confirmed by sequencing. The resultant vectors were then used to transform Origami B (DE3) pLys cells according to the manufacturers manual.

5.3. Expression of WAX9A-E in *E. coli* and Purification of the Recombinant Protein

Protein expression was performed according to the manufacturers manual (Novagen). For induction IPTG with a final concentration of 1 mM was added. Cells were grown for 4 h after induction, harvested and stored at -20 'C. Cells were disrupted by two passes through a French Press. DNA was disrupted by sonication to increase fluidity of the solution. Recombinant protein was purified by affinity chromatography using Ni-agarose (Qiagen) according to the manufacturers manual. The protein was then cleaved using enterokinase (Invitrogen) according to the manufacturers manual. The mature protein was stored at -20 °C.

5.4. Protein Gel Electrophoresis and Western Blotting

Electrophoresis of proteins was performed in polyacrylamide gels under reducing conditions in the presence of SDS as described by Laemmli (1970). Proteins were stained with Coomassic or transferred to nitrocellulose membranes by electroblotting (Towbin et al., 1979). The filters were blocked with MPT-buffer (10 mM Tris (pH 7.4), 150 mM NaCl, 0,05 % Tween 20, 0,03 % Casein) and probed with the anti-LTP antiserum (Hincha et al., 2001). Bound antibodies were visualized by probing with goat-anti-rabbit antibodies linked to alkaline phosphatase and subsequent calorimetric detection as described in Sambrook et al. (1989).

The anti-LTP antibody was raised using a synthetic peptide with the following amino acid sequence derived from the N-terminal end of the WAX9A, E proteins: ALTCGTVNSNVAPCI (Hincha et al., 2001).

5.5. Cryoprotection Assay

Thylakoids were isolated from spinach leaves as described by Hincha and Schmitt (1992b). The membranes were washed three times with 10 mM NaCl. Chlorophyll content was determined according to Arnon (1949) and adjusted to approximately to 1 mg/ml by dilution with 10 mM NaCl. Protein samples were transferred to 10 mM sucrose, 1 mM $MnCl_2$, 1 mM $CaCl_2$ by gel filtration (Sephadex G25). 100 µl of protein sample were then mixed with 100 µl of thylakoid suspension and the mixture was incubated in a freezer at -20 °C for 3 h. The samples were rapidly thawed in a water bath at room temperature. Under these conditions thylakoids rupture during an *in vitro* freeze thaw cycle. They reseal after rupture but have a reduced volume. Thylakoid volume can therefore be used to measure the damage incurred by the thylakoids during the freeze thaw cycle (Hincha and Schmitt, 1992b).

To determine the thylakoid volume samples were diluted with 100 µl of 10 mM $MgCl_2$, filled into haematocrit capillaries and centrifuged in a haematocrit centrifuge at 12000 rpm for 15 min. Thylakoid volume was determined with the help of a magnifying glass as the height of the thylakoid pellet at the bottom of the haematocrit capillary.

6. ACKNOWLEDGMENTS

We would like to thank Felicitas Wronski and Hilde K6th for excellent technical assistance. This research was supported by grants to J. M. Schmitt and D. K. Hincha from the Deutsche Forschungsgemeinschaft. H. A. M. Sror is supported by a stipend from the Egyptian government.

REFERENCES

Arnon, D. J., 1949, Copper enzymes in isolated chloroplasts. Polyphenoloxidase in *Beta vulgaris, Plant Physiol.* **24**:1-15.

Arondel, V., Tchang, F., Baillet, B., Vignols, F., Grellet, F., Delseny, M., Kader, J. C., Puigdomenech, P., 1991, Multiple MRNA coding for phospholipid-transfer protein from *Zea mays* arise from alternative splicing, *Gene* **99**:133-136.

Asero, R., Mistrello, G., Roncarolo, D., de Vries, S. C., Gautier, M.-F., Ciurana, C. L. F., Verbeek, E., Mohammadi, T., Knul-Brettlova, V., Akkerdaas, J. H., Bulder, I., Aalberse, R. C., and van Ree, R., 2000, Lipid transfer protein: a pan-allergen in plant-derived foods that is highly resistant to pepsin digestion, *Int. Arch. Allergy Immunol.* 122:20-32.

Bessette, P. H., Aslund, F., Beckwith, J., and Georgiou, G., 1999, Efficient folding of proteins with multiple disulfide bonds in the *Escherichia coli* cytoplasm, PNAS **96**:13703-13708.

Cammue, B. P. A., Thevissen, K., Hendriks, M., Eggermont, K., Goderis, 1. J., Prost, P., Van Damme, J., Osbom, R. W., Guerbette, F., Kader, J.-C., and Broekaert, W. F., 1995, A potent antimicrobial protein from onion seeds showing sequence homology to plant lipid transfer proteins, *Plant Physiol.* **109**:445-455.

Colmenero-Flores, J. M., Campos, F., Garciarrubio, A., and Covarrubias, A. A., 1997, Characterization of *Phaseolus vulgaris* EDNA clones responsive to water deficit: identification of a novel late embryogenesis abundant-like protein, *Plant Mol. Biol.* **35**:393-405.

Colombo, P., Duro, G., Costa, M. A., Izzo, V., Mirisola, M., Locorotondo, G., Cocchiaral R., and Geraci, D., 1998, An update on allergens. Parietaria pollen allergens, Allergy **53**:917-921.

Douady, D., Grosbois, M., Guerbette, F., and Kader, J.-C., 1986, Phospholipid transfer protein from maize seedlings is partly membrane-bound, *Plant Sci.* **45**:151-156.

Douliez, J.-P., J6gou, S., Pato, C., Mol]6, D., Tran, V., and Marion, D., 2001, Binding of two mono-acylated lipid monomers by the barley lipid transfer protein, LTPI, as viewed by fluorescence, isothermal titration calorimetry and molecular modelling, *Eur. J. Biochem.* **268**:384-388.

Dunn, M. A., White, A. J., Vural, S., and Hughes, M. A., 1998, Identification of promoter elements in a low-temperature-responsive gene *(blt4.9)* from barley *(Hordeum vulgare L.), Plant Mol. Biol.* **38**:551-564.

Garcia-Casado, G., Crespo, J. F., Rodriguez, I., and Saicedo, G., 2001, Isolation and characterization of barley lipid transfer protein Z as beer allergens, *J. Allergy Clin. Immunol.* **108**:647-649.

Garcia-Olmedo, F., Molina, A., Segura, A., and Moreno, M., 1995, The defensive role of nonspecific lipid-transfer proteins in plants, *Trends Microbiol.* **3**:72-74.

Gomar, J., Sodano, P., Sy, D., Shin, D. H., Lee, J. Y., Suh, S. W., Marion, D., Vovelle, F., and Ptak, M., 1998, Comparison of solution structures of maize nonspecific lipid transfer protein: a model for a potential in *vivo* lipid carrier protein, *PROTEINS* **31**:160-171.

Han, G. W., Lee, J. Y., Song, H. K., Chang, C., Min, K., Moon, J., Shin, D. H., Kopka, M. L., Sawaya, M. R., Yuan, H. S., Kim, T. D., Choe, J., Lim, D., Moon, H. J., and Suh, S. W., 2001, Structural basis of nonspecific lipid binding in maize lipid-transfer protein complexes revealed by high-resolution X-ray crystallography, J *Mol. Biol.* **308**: 263-278.

Hincha, D. K., Heber, U., and Schmitt, J. M., 1990, Proteins from frost-hardy leaves protect thylakoids against mechanical freeze-thaw damage *in vitro, Planta 180:416-419*

Hincha, D. K., Neukamm, B., Sror, H. A. M., Sieg, F., Weckwarth, W., Rackels, M., Lullien-Pellerin, V., Schrbder, W., and Schmitt, J. M., 2001, Cabbage cryoprotectin is a member of the nonspecific plant lipid transfer protein gene family, *Plant Physiol.* **125**:835-846.

Hincha, D. K., and Schmitt, J. M., 1992a, Freeze thaw injury and cryoprotection of thylakoid membranes, in: *Water and Life,* G. N. Somero, C. B. Osmond, and C. L. Bolis, eds., Springer, Berlin Heidelberg New York, pp. 316-337.

Hincha, D. K., and Schmitt, J. M., 1992b, Cryoprotective leaf proteins: assay methods and heat stability, J. *Plant Physiol.* **140**:236-240.

Hincha, D. K., Sieg, F., Bakaltechva, I., K6th, H., and Schmitt, J. M., 1996, Freeze-thaw damage to thylakoid membranes: specific protection by sugars and proteins, in: *Advances in Low-Temperature Biology,* P. L. Steponkus, ed., JAI Press, London, pp. 141-183.

Hollenbach, B., Schreiber, L., Hartung, W., and Dietz, K.-J., 1997, Cadmium leads to stimulated expression of the lipid transfer protein genes in barley: implications for the involvement of lipid transfer proteins in wax assembly, *Planta* **203**:9-19.

Kader, J.-C., 1996, Lipid-transfer proteins in plants, *Annu. Rev. Plant Physiol. Plant Mol. Biol.* **47**:627-654.

Kader, J.-C., 1997, Lipid-transfer proteins: a puzzling family of plant proteins, *Trends Plant Sci.* **2**:66-70.

Klein, C., Lamotte-Gu6ry, F., Gautier, F., Moulin, G., Boze, H., Joudrier, P., and Gautier, M.-F., 1998, Highlevel secretion of a wheat lipid transfer protein in *Pichia pastoris, Protein Express. Purif.* **13**:73-82.

Kristensen, A. K., Brunstedt, J., Nielsen, K. K., Roepstorff, P., Mikkelsen, J. D., 2000, Characterization of a new antifungal non-specific lipid transfer protein (nsLTP) from sugar beet leaves, *Plant Sci.* **155**:31-40.

Laemmli, U. K., 1970, Cleavage of structural proteins during the assembly of the head of bacteriophage T4, *Nature* **227**:680-685.

Lee, J. Y., Min, K., Cha, H., Shin, D. H., Hwang, K. Y., and Sub, S. W., 1998, Rice nonspecific lipid transfer protein: The 1.6 A crystal structure in the unliganded state reveals a small hydrophobic cavity, *J. Mol.* Biol. **276**:437-448.

Lerche, M. H., Kragelund, B. B., Bech, L. M., and Poulsen, F. M., 1997, Barley lipid-transfer protein complexed with palmitoyl CoA: the structure reveals a hydrophobic binding site that can expand to fit both large and small lipid-like ligands, *Structure* **5**:291-306.

Lindorff-Larsen, K., and Winther, J. R., 2001, Surprisingly high stability of barley lipid transfer protein, LTP I, towards denaturant, heat and proteases, *FEBS Lett.* 488:145-148.

Liu, H., Xue, L., Li, C., Zhang, R., and Ling, Q., 2001, Calmodulin-binding protein BP-10, a probable new member of plant nonspecific lipid transfer protein superfamily, *Biochem. Biophys. Res. Com.* **285,** 633638

Lullien-Pellerin, V., Devaux, C., Ihorai, T., Marion, D., Pahin, V., Joudrier, P., and Gautier, M.-F., 1999, Production in *Escherichia coli* and site-directed mutagenesis of a 9-kDa nonspecific lipid transfer protein from wheat, Eur. *J Biochem.* **260**:861-868.

Masuta, C., Furuno, M., Tanaka, H., Yamada, M., and Koiwai, A., 1992, Molecular cloning of a CDNA clone for tobacco lipid transfer protein and expression of the functional protein in *Egeherichia coli, FEBS Lett.* **311**:119-123.

Nakamura,K. and Matsuoka, K., 1993, Protein targeting to the vacuole in plant cells, *Plant Physiol.* **101**: 1-5.

Nielsen, K. K., Nielsen, J. E., Madrid, S. M., and Mikkelsen, J. D., 1996, New antifungal proteins from sugar beet *(Beta vulgaris* L.) showing homology to non-specific lipid transfer proteins, *Plant Mol. Biol.* **31**:539-552.

Osafune, T., Tsuboi, S., Ehara, T., Satoh, Y., and Yamada, M., 1996, The occurrence of non-specific lipid transfer proteins in developing castor bean fruits, *Plant Sci.* **113**:125-130.

Ostergaard, J., Hojrup, P., and Knudsen, J., 1995, Amino acid sequences of three acyl-binding/lipid-transfer proteins from rape seedlings, *Bioehim. Biophyv. Acta* **1254:** 169-179.

Pastorello, E. A., Farioli, L., Pravettoni, V., Ispano, M., Scibola, E., Trambaioli, C., Giuffrida, M. G., Ansaloni, R., Godovac-Zimmermann, J., Conti, A., Fortunato, D., and Ortolani, C., 2000, The maize major allergen, which is responsible for food-induced allergic reactions, is a lipid transfer protein, J *Allergy Clin. Immunol.* **106**:744-751.

Pastorello, E. A., Pompei, C., Pravettoni, V., Brenna, O., Farioli, L., Trambaioli, C., and Conti, A., 2001, Lipid transfer proteins and 2S albumins as allergens, *Allergy* 5(Suppl. 67):45-47.

Pearce, R. S., Houlston, C. E., Atherton, K. M., Rixon, J. E., Harrison, P., Hughes, M. A., and Dunn, M. A., 1998, Localization of expression of three cold-induced genes, *blil0l, blt4.9, and blt14* in different tissues of the crown and developing leaves ofcold-acclimated cultivated barley, *Plant Physiol.* **117**:787-795.

Plant, Á. L., Cohen, A., Moses, M. S., and Bray, E. A., 1991, Nucleotide Sequence and spatial expression pattern of a drought- and abscisic acid-induced gene oftomato, *Plant Physiol.* **97**:900-906.

Pyee, J., and Kolattukudy, P. E., 1995, The gene for the major cuticular wax-associated protein and three homologous genes from broccoli *(Brassica oleracea)* and their expression patterns, *Plant J.* 7:49-59.

Pyee, J., Yu, H., and Kolattukudy, P. E., 1994, Identification of a lipid transfer protein as the major protein in the surface wax of Broccoli *(Brassica oleracca) leaves, Arch. Biochem. Biophys.* **311**:460-468.

Rogers, S. 0. and Bendich, A. J., 1985, Extraction of DNA from milligram amounts of fresh, herbarium and mummified plant tissues, *Plant Mol. Biol.* **5**:69-76.

Sambrook, J., Fritsch, E. F., and Maniatis, T., 1989, *Molecular Cloning: A Laboratory Manual,* Vol 1-3, Cold Spring Harbor Laboratory Press, Cold Spring Harbor, New York.

Shin, D. H., Lee, J. Y., Hwang, K. Y., Kim, K. K., Sub, S. W., 1995, High-resolution crystal structure of the non-specific lipid-transfer protein from maize seedlings, *Structure* **3**: 189-199.

Sieg, F., Schr6der, W., Schmitt, J. M., and Hincha, D. K., 1996, Purification and characterization of a cryoprotective protein (Cryoprotectin) from the leaves of cold-acclimated cabbage, *Plant Physiol.* **111**:215-221.

Sohal , A. K., Pallas, J. A., and Jenkins, G. I., 1999, The promoter of a *Brassica napus* lipid transfer protein gene is active in a range of tissues and stimulated by light and viral infection in transgenic *Arabidopsis, Plant Mol. Biol.* **41**:75-87.

Soufleri, I. A., Vergnolle, C., Miginiac, E., and Kader, J.-C., 1996, Germination-specific lipid transfer protein cDNAs in *Brassica napus L., Planta* **199**:229-237.

Sterk, P., Booij, H., Schellekens, G. A., Kammen, A. V., and De Vries, S. C., 1991, Cell-specific expression of the carrot EP2 lipid transfer protein gene, *Plant Cell* **3**:907-921.

Stewart, E. J., Aslund, F., and Beckwith, J., 1998, Disulfide bond formation in the *Escherichia coli* cytoplasm: an *in vivo* role reversal for the thioredoxins, *EMBO J.* **17**:5543-5550.

Tassin, S., Broekaert, W. F., Marion, D., Acland, D. P., Ptak, M., Vovelle, F., and Sodano, P., 1998, Solution structure of *Ace-AMPI,* a potent antimicrobial protein extracted from onion seeds. Structural analogies with plant nonspecific lipid transfer proteins, *Biochemistry* **37**:3623-3637.

Tassin-Moindrot, S., Caille, A., Douliez, J.-P., Marion, D., and Vovelle, F., 2000, The wide binding properties of a wheat nonspecific lipid transfer protein, *Eur. J Biochem.* **267**:1117-1124.

Thoma, S., Kaneko, Y., and Somerville, C., 1993, A nonspecific lipid transfer protein from *Arabidopsis* is a cell wall protein, *Plant J.* **3**:427-436.

Thompson, J. D., Gibson, T. J., Plewniak, F., Jeanmougin, F. and Higgins, D. G., 1997, The ClustalX windows interface: flexible strategies for multiple sequence alignment aided by quality analysis tools, *Nucleic Acid Res.* **24**:4876-4882.

Toriyama, K., Hanaoka, K., Okada, T., and Watanabe, M., 1998, Molecular cloning of a CDNA encoding a pollen extracellular protein as a potential source of a pollen allergen in *Brassica rapa, FEBS Left.* **424**:234-238.

Torres-Schumann, S., Godoy, J. A., Pintor-Toro, J. A., 1992, A probable lipid transfer protein gene is induced by NaCl in stems of tomato plants, *Plant Mol. Biol.* **18**:749-757.

Towbin, H., Stachclin, T., Gordon, J., 1979, Electrophoretic transfer of proteins from polyacrylamide gels to nitrocellulose sheets: procedure and some applications, *Proc. Nati. Acad. Sci. USA* **76**:4350-4354.

Trcvifio, M. B., and O'Connell, M. A., 1998, Three drought-responsive members of the nonspecific lipidtransfer protein gene family in *Lycopcrsican pcnncllii* show different developmental patterns of expression, *Plant Physiol.* **116**:1461-1468.

Tsuboi, S., Osafune, T., Tsugeki, R., Nishimura, M., and Yamada, M., 1992, Nonspecific lipid transfer protein in castor bean cotyledon cells: subcellular localization and a possible role in lipid metabolism, *J. Biochem.* **111**:500-508.

Volger, H. G., and Heber, U., 1975, Cryoprotective leaf proteins, *Biochim. Biophys. Acta* **412**:335-349.

15

EXTRINSIC ICE NUCLEATION IN PLANTS
What are the factors involved and can they be manipulated?

Michael Wisniewski, Michael Fuller, D. Michael Glenn, Lawrence Gusta, John Duman, and Marilyn Griffith*

1. INTRODUCTION

In the early 1980's a considerable amount of research focused on the role of extrinsic ice nucleation and its= role in inducing plants to freeze at warm sub-zero temperatures. The working hypothesis was that by controlling extrinsic nucleation events, plants could supercool well below 0 °C and thus avoid freezing (Lindow, 1995). It was felt that such a strategy could provide a significant level of frost protection to frost sensitive plants or plant parts. While the majority of reports dealt with the role of ice-nucleating-active (INA) bacteria (e.g. *Pseudomonas syringae*), related research focused on the role of other extrinsic nucleating agents and whether or not plants could actually supercool to temperatures several degrees below 0 °C due to the presence of intrinsic nucleating agents which induced the plants to freeze at warm temperatures (Ashworth and Kieft, 1995). The identification of a wide range of both extrinsic and intrinsic ice nucleating agents made the practical application of blocking extrinsic ice nucleation complex. Since that time, research emphasis has switched to identifying genes that impart cold tolerance and the transcriptional activators that regulate cold hardiness genes (Thomashow, 1998; Jaglo, et al., 2001). The hypothesis here is that by the overexpression of these types of genes, a non-acclimated or freeze-sensitive plant could be made freezing tolerant. While great progress has been made in understanding the genetic basis of cold hardiness, manipulation of this trait by molecular biology has also demonstrated itself to be complicated due to the "additional" effects of the overexpression of several cold

*Michael Wisniewski, U.S. Department of Agriculture, Agricultural Research Service, 45 Wilthsire Road, Kearneysville, WV 25430. Michael Fuller, Seale-Hayne Faculty of Agriculture, Plymouth University, Newton Abbot, UK. Michael Glenn, USDA-ARS, Kearneysville, WV 25430. Larry Gusta, University of Saskatchewan, Saskatoon, Canada. John Duman, Department of Biology, Notre Dame University, IN. Marilyn Griffith, University of Waterloo, Waterloo, Canada.

Plant Cold Hardiness, edited by Li and Palva
Kluwer Academic/Plenum Publishers, 2002

hardiness genes on the physiology and development of the target plant. Therefore, blocking extrinsic ice nucleation, although complicated, may still be a valuable approach to providing protection to frost sensitive plants.

In order for ice to form on or within a plant, ice nucleation must first occur. Although the melting point of ice is 0 °C, the freezing temperature of water is not as defined (Ashworth, 1992). In fact, although it is not commonly recognized, pure water has a low probability of freezing at temperatures warmer than -40 °C (Franks, 1985). This is because a small ice crystal embryo is necessary in order for ice to form and grow to any substantial size. The probability of forming such an ice crystal embryo in pure water, as well as the half-life of such a crystal, is low until temperatures approaching -40 °C. This temperature is referred to as the homogeneous ice nucleation point.

In nature, it is rare for water to exist in a pure state but it rather exists as an ionic or colloidal solution. In such solutions heterogeneous ice nucleation is initiated on the surface of objects or on suspended particles (Ashworth, 1992). Heterogeneous ice nucleators are very effective in inducing ice formation and are very abundant. As a consequence, freezing occurs in nature at much warmer temperatures than the homogeneous nucleation temperature.

The role of heterogeneous ice nucleators in inducing ice formation in plants is important because if methods can be developed for regulating ice nucleation, significant advances could be made in limiting frost injury to both freezing-sensitive and cold adapted plants. A major question concerns the relative importance of extrinsic ice nucleation agents, such as ice-nucleation-active (INA) bacteria (e.g. *Pseudomonas syringae*), and intrinsic nucleation agents synthesized by plants (Ashworth and Kieft, 1995). While all plants can supercool (i.e., have tissues below 0 °C without freezing) to some degree (Ashworth and Kieft, 1995; Burke, et al., 1976; Lindow, 1995; Lindow, et al., 1978), the extent of supercooling varies between plant species and is influenced by the presence of ice nucleating agents which may be of plant (Ashworth, 1992; Fuller, et al., 1994; Gross, et al., 1988) or bacterial (Lindow, 1978; Hirano, et al., 1985; Lindow, 1983) origin.

The abundance of ice nuclei on plants can be estimated by freezing of droplets of plant macerates or small portions of plant tissue (Ashworth and Kieft, 1995; Lindow, 1983) but these procedures are destructive and do not provide information on where ice formation was initiated. Ice formation in intact plants can be readily detected by measuring, with thermocouples, the heat that is released upon the freezing of the water in the plant (Ashworth, 1992; Cary and Mayland, 1970; Proebsting, et al., 1982; Quamme, et al., 1972). Nevertheless, even when arrays of temperature measuring devices are attached to plants, the actual site of ice initiation and the temperature at the site where ice nucleation occurred can only be inferred (Ashworth, et al., 1985). This is a significant technical limitation and more details of the freezing process are required in order to accurately predict freezing patterns and determine under what conditions the reduction or interference of extrinsic ice nuclei would provide significant frost protection. Recently, the ability to use infrared video thermography to directly observe ice nucleation (i.e,, initial ice formation) and propagation in plants has been demonstrated (Carter, et al., 2001, 1999; Ceccardi, et al., 1995; Fuller and Wisniewski, 1998; LeGrice, et al., 1993; Wisniewski, et al., 1997; Wisniewski, 1998; Wisniewski and Fuller, 1999; Workmaster et al., 1999). The temperature and spatial resolution of the device used in these studies has enabled the researchers to clearly define the initial site of ice nucleation as well as monitor the ice front as it spread into the surrounding tissues. Using infrared thermography it is possible to determine the role of extrinsic and intrinsic ice nucleating agents in the freezing process, rates of ice propagation, the effect of plant structure on the freezing process, and how the specific pattern of freezing relates to visual patterns of injury. It is also possible to clearly evaluate if the reduction of ice nuclei or inhibiting their activity is a feasible approach to frost protection. The present report will provide an overview of these various studies and detail the factors that apparently play a significant role in determining when a plant will freeze and how ice will propagate through a plant.

2. ROLE OF MOISTURE AND EXTRINSIC ICE NUCLEATING AGENTS

One of the critical factors in determining when a plant will freeze is the presence or absence of surface moisture (Ashworth, 1992). Dry plants will always supercool to a lower temperature than wet plants. Secondly, if ice nucleating agents, such as INA bacteria, are present, they will induce plants to freeze at a warmer temperature than just the moisture alone (Wisniewski, et al., 1997, Fuller and Wisniewski, 1998). The presence of nucleators on the surface without moisture is not effective because nucleators are only active in aqueous solutions.

In order for the presence of external ice (frozen moisture on a leaf surface) to induce ice formation in a plant, the ice must physically grow through a break in the surface of the cuticle (eg. crack or broken hair cells) or through a stomatal opening (Wisniewski and Fuller, 1999). A thick cuticle, such as found on evergreen leaves (eg. azalea, cranberry) serves as an effective barrier to external nucleation (Wisniewski and Fuller, 1999; Workmaster, et al., 1999). Water can freeze on the upper surface of these plants and the plant will continue to supercool. When external ice does induce the plant to freeze it is through the growth of ice through a stomatal opening on the abaxial surface. In many herbaceous plants, the cuticle is not an effective barrier, or there are sufficient avenues of ingress that allow ice to readily propagate from either the upper or lower surface. Providing a barrier of silicone grease sufficiently prevents external ice from inducing herbaceous plants to freeze (Wisniewski and Fuller, 1999).

3. HYDROPHOBIC BARRIERS APPLIED TO HERBACEOUS PLANTS CAN BLOCK ICE NUCLEATION

Our previous research indicated that by somehow blocking the activity of extrinsic nucleating agents one may allow plants to supercool to a lower temperature and thereby provide some frost protection. In a subsequent study we used infrared thermography to examine freezing in young tomato (*Lycopersicon esculentum*) plants and determine if a hydrophobic barrier on the plant surface could prevent the action of extrinsic nucleating agents such as Ice-Nucleating-Active (INA or Ice$^+$) bacteria (*Pseudomonas syringae,* strain Cit7) from initiating freezing within a plant. To provide a barrier to the action of extrinsic ice-nucleating agents, M-96-018, a hydrophobic kaolin particle film (Engelhard, Islin, NJ, USA) was applied to the plant surface before applying an extrinsic nucleating agent (Wisniewski, et al., In Press).

Tomato plants were grown in a greenhouse in individual pots and used when they were 4-6 weeks old. Freezing tests were conducted in a programmable freezing chamber, a radiative frost chamber, and outdoors. Freezing was visualized and recorded on videotape using an infrared radiometer. Freezing of the plants was extrinsically induced by the application of droplets (5 μl) of water containing Cit7. To provide a barrier to the action of extrinsic ice nucleating agents, an emulsion of hydrophobic kaolin (Engelhard, Inc.) was applied to the plant surface prior to application of an extrinsic nucleating agent. Results indicate that dry, young tomato plants can supercool to as low as -6 °C whereas plants having a single droplet of Cit7 would freeze at -1.5 to -2.5 °C. Application of the hydrophobic barrier blocked the effect of Cit7 and allowed whole plants to also supercool to -6 °C, despite the presence of frozen droplets on the surface of leaves and stems (Fig. 1).

Fig. 1. Infrared thermography (**A** - **C**) and a photograph (**D**) of M96-018 coated and uncoated tomato plants subjected to freezing temperatures. Each plant was sprayed with water containing Ice^{+} bacteria using a hand-operated aerosol sprayer. The coated plant is on the left in **A-C** and on the right in **D**. Large drops of water can be seen as black areas on the uncoated plant in **A**. The black areas are a result of these areas being colder than the set temperature range of the camera (-0.2 to 1.7° C). The lower temperature of these areas is due to evaporative cooling. In **B**, the uncoated plant is in the process of freezing, as evidenced by the exothermic reaction of ice formation, while the coated plant is unfrozen at approximately -2.5 °C. In **C**, the uncoated plant has almost completely frozen, except along the stem, petiole, and mid-vein. The coated plant in **C** is still unfrozen at approximately -6.0 ° C. In **D**, uncoated (left) and coated (right) plants can be seen after exposure to -6.0 °C, removal from the chamber, and thawing. The uncoated plant is completely killed while the coated plant is uninjured.

When whole plants were sprayed with water and Cit7 using an aerosol sprayer and exposed to -3 °C, plants coated with the hydrophobic particle film exhibited a significant increase in survivability over untreated plants (Fig. 2). Similar results were obtained using a radiative frost chamber (Fig. 3). Experiments conducted under natural frost conditions also resulted in less injury in the coated plants, although due to the small sample size the difference in injury between the coated and uncoated plants was not significant. The hydrophobic kaolin particle film was better at preventing plants from freezing due to extrinsic ice nucleation than normal kaolin alone or anti-transpirants with putative frost protection properties.

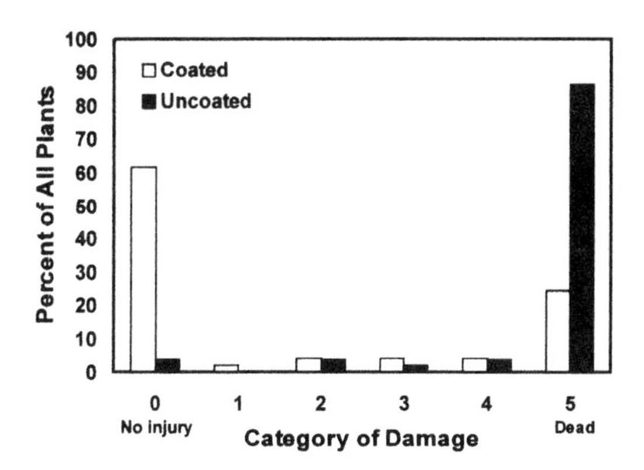

Fig. 2. Percent of M96-018-coated and uncoated tomato plants in different injury classes after exposure to -3.0 °C. Each plant was sprayed with water containing Ice^+ bacteria prior to being placed in the environmental chamber. The damage levels, combined for all classes, was analyzed by a t-test for coated vs. uncoated. The probability of a difference based on the t-test is $P > 0.0001$. $n = 51$ for uncoated plants and $n = 49$ for coated plants.

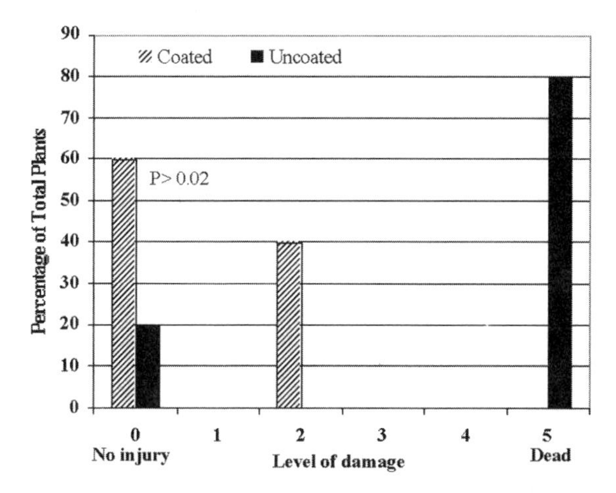

Fig. 3. Percent of M96-018-coated and uncoated tomato plants in different injury classes after exposure to -1.6 ° C for one hour in a radiative frost chamber. All plants were sprayed with water containing Ice^+ bacteria prior to being placed in the frost chamber. A t-test indicated that there was a significant difference in the level of injury between coated and uncoated plants ($P > 0.02$; $n = 6$).

The material used in the study consists of a proprietary formulation of kaolin particles that have been coated to impart hydrophobicity. It is our hypothesis that if moisture can be prevented from forming on the plant surface and activating epiphytic ice nucleators, or itself freezing and acting as source of nucleation activity, then there would be a higher probability of the plant expressing its intrinsic ability to supercool. The results of the study supported this hypothesis on both individual leaves and whole plants of tomato. It appeared that the particle film, due to its hydrophobicity, prevented moisture from being deposited on the plant surface (in many cases water would simply roll off) and lowered the amount of contact that any individual droplet had with the plant surface, effectively raising the droplet above the leaf surface (Fig. 4). Additionally, it is assumed that the moisture barrier prevented the wetting of epiphytic ice nucleators present on the leaf surface and hence their activation. Collectively, this reduced the number of potential extrinsic nucleation events, and diminished the ability of ice crystals that did form from propagating into the internal portion of the leaf and inducing the leaf to freeze. Access of ice crystals to the internal portion of the leaf is believed to occur through stomates, cracks in the cuticle, broken epidermal hairs, etc. (Wisniewski and Fuller, 1999).

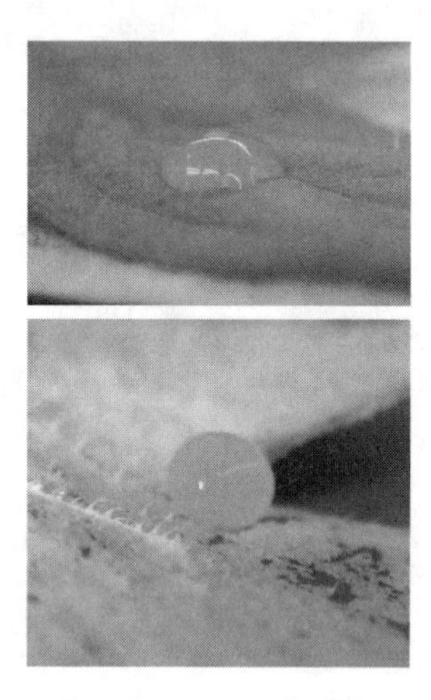

Fig. 4. Photographs of droplets of water containing Ice$^+$ bacteria, strain Cit7 of *Pseudomonas syringae*, on the surface of an uncoated tomato leaflet (upper panel) and a tomato leaflet coated with the M96-018 hydrophobic-kaolin-particle-film (lower panel). Note how the droplet on the coated leaflet is more spherical and has less contact with the leaf surface than the droplet on the uncoated leaflet.

Relevant to that study, is the work on ice nucleation in tomato plants reported by Anderson and Ashworth (1985). Using thermocouples and monitoring populations of ice-nucleation-active bacteria *(Pseudomonas syringae)*, they observed both a size and time dependency on the extent of supercooling. As plant mass increased, the extent of supercooling decreased. Additionally, at any particular sub-zero temperature the percentage of samples that would freeze increased with time. They also noted that a prime determinant of the extent of supercooling was the activity of extrinsic nucleating agents (INA bacteria, moisture). The results reported by Wisniewski, et al. (In Press) differ only in the extent to which whole plants have the potential to supercool. We have definitely observed that plant parts (leaves and leaflets) supercool to a greater extent than whole plants but have commonly seen supercooling of whole plants to temperatures of at least -6.0 °C using cooling rates of -2.0 to -2.5 °C · h^{-1}. Their inability to see supercooling below -2.0 °C in whole plants may have been due to the nucleation activity of the thermocouples themselves. Our inability to prevent some plants from freezing at -3.0 °C, however, may reflect both the size and time dependency aspects of intrinsic nucleation rather than our failure at blocking extrinsic nucleation.

The ability of the hydrophobic particle film to protect plants from frost damage has also been demonstrated for potatoes, grapevines, and citrus plants under both convective and radiative frost conditions where plants were exposed to -3 °C for two hours (Fuller, et al., submitted). In these studies, the presence of the hydrophobic particle film consistently led to less damage than that observed in uncoated plants. The particle film delayed ice crystal growth from a frozen droplet present on leaf surfaces for an average of one hour, and in some cases for the whole duration of the frost test. This time delay is significant in that it is representative of the duration of transient radiation frosts under field conditions (Fuller and Le Grice, 1998). Large scale studies under field conditions will be needed, however, to determine if the hydrophobic particle film (or a similar type of compound) can be used to provide frost protection under the complex freezing conditions that are present during natural frost episodes.

4. FREEZING OF WOODY STEMS AND BARRIERS TO ICE PROPAGATION

As previously documented and reviewed by Ashworth and Kieft (1995), the presence of effective, intrinsic nucleators, appears to be common in woody plants. These nucleators appear to be as effective as external ice nucleators, such INA bacteria, and induce stems to freeze at warm, subzero temperatures. Barriers appear to exist, however, that prevent ice propagation into lateral appendages such as buds, or newly extended primary tissues (flowers, inflorescences, etc.) (Carter et al., 2001; Wisniewski, 1997; Workmaster, 1999). These barriers are most effective if the initial freezing event occurs at a relatively warm temperature. Barriers have been observed in the propagation of ice into the strigs of Ribes and grapevines, the pedicel of cranberry fruits, and flowers of peach indicating that the ability of buds, flowers, and inflorescences to supercool in the presence of frozen stem material may be an active mechanism of freeze avoidance.

5. INFLUENCE OF COLD ACCLIMATION AND ANTIFREEZE PROTEINS ON ICE NUCLEATION

When plants are cold acclimated, they develop a greater ability to supercool. This has been demonstrated in canola and barley, and rye plants (unpublished data). In one experiment, cellular extracts of canola (*Brassica napus)* were placed on long, rectangular strips of filter paper, and the rate of freezing from one end of the strip to the other was determined using infrared thermography. Uniform times and temperatures of ice nucleation were controlled by placing a drop of ice-nucleating-active bacteria (*Pseudomonas syringae,* strain Cit7) at the bottom of each strip. Results indicated that both the sugars and proteins present in acclimated plants may play a role in regulating the rate of ice crystal growth (Fig. 5). Cellular extracts from acclimated plants delayed ice crystal growth more than extracts from non-acclimated plants. Additionally, extracts from a hardy cultivar (Express) had a greater effect than extracts from a less hardy cultivar (Quest). An effect of unidentified proteins could be seen over and above that of sugars by boiling the samples and also comparing the response of cellular extracts to pure sugar solutions with similar osmolality.

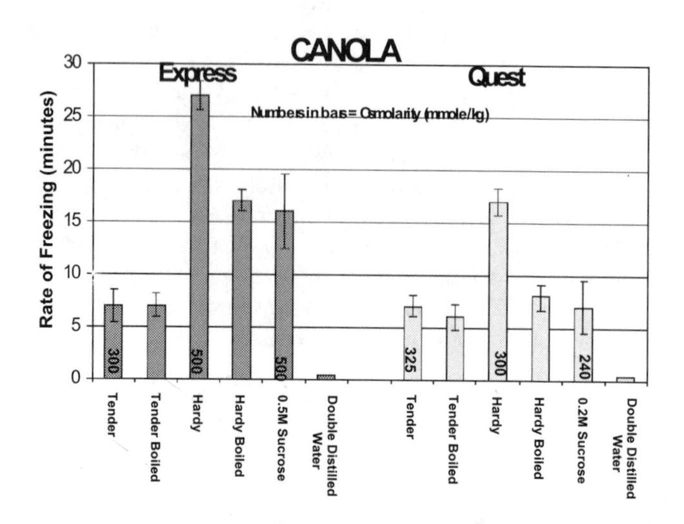

Fig. 5. Effect of cellular extracts from acclimated and non-acclimated plants of two cultivars (Express and Quest) of canola on the rate of freezing of strips of filter paper. The effect of the fresh cellular extracts was compared to solutions of sucrose with similar osmolality, boiled samples of the extracts, and deionized water.

Antifreeze proteins (AFPs), also known as hysteresis proteins (THPs), inhibit ice crystal growth by a non-colligative mechanism, lowering the freezing point of water below the melting point, thereby producing a so-called thermal hysteresis (DeVries, 1971). First described in fish, they have also been widely reported in insects (Duman, et al., 2001), but insect AFPs are structurally different than those of fishes and have higher activities. Antifreeze protein activity has also been identified in many plants (Griffith, et al., 1992; Hon, et al, 1995; Duman, 1994) but the levels of thermal hysterisis activity are comparatively low (generally 0.2-0.5 °C), as compared to fishes (genrally 0.7-1.5 °C), and insects (generally 3-6 °C). Transgenic Arabidopsis plants expressing an insect antifreeze gene derived from *Dendroides canadensis* exhibited an enhanced ability to supercool (Fig. 6). The enhanced ability to supercool was only present in the transformed lines in which the construct also coded for a signal peptide that allowed for extracellular secretion of the protein (340 lines). Transformants (270 lines) that expressed the protein but did not have the coding for the signal peptide did not supercool to a greater extent than the wild-type plants.

When acclimated canola plants were allowed to supercool to low temperatures (-12 to -15 °C) and then frozen, they exhibited no injury despite the rapid rate of ice formation and propagation. This indicates that the acclimated plants have the ability to rapidly lose water in order to prevent intracellular ice formation (unpublished data). Distinct differences in the freezing response of acclimated and non-acclimated rye plants have also been observed that may be attributed to the presence or absence of antifreeze proteins (see Chapter by Griffith, et al., this volume).

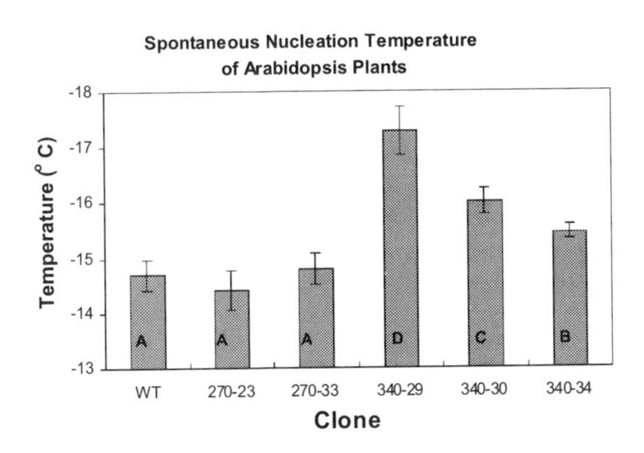

Fig. 6. Spontaneous nucleation temperature of wild-type (WT) and transgenic Arabidopsis plants expressing a gene coding for an insect antifreeze protein derived from *dendroides Canadensis*. Lines 270-23 and 270-33 are transformants that express the protein intracellularly while 340-29, 340-30, and 340-34 have a signal peptide that allows for extracellular secretion of the protein.

6. SUMMARY

The factors that determine when and to what extent a plant will freeze are complex. In herbaceous plants, it appears that extrinsic nucleating agents play a key role in initiating ice formation. Ice-nucleating-active bacteria and moisture are two major extrinsic agents, however, their influence on the freezing process can be moderated by the presence of natural or applied hydrophobic barriers. Evidence suggests that ice crystals formed on the surface of plants must physically grow into the interior of the plant in order to initiate freezing of the plant. This can occur through stomates, or cracks in the cuticle. Thick cuticles, as found on many evergreen plants, and the application of synthetic, hydrophobic materials appear to serve as effective barriers and can inhibit the effect of extrinsic ice nucleating agents by preventing moisture from collecting on the surface of the plant and/or inhibiting the growth of ice crystals into the interior of the plant. In this regard, the application of hydrophobic barriers may provide a new approach to frost protection.

The situation in woody plants is different than in herbaceous plants. In general, woody plants appear to possess native, intrinsic nucleating agents that are just as active as many extrinsic ice nucleating agents. The exact identification of the intrinsic nucleating agents in woody plants, however, is unkown. Despite the presence of internal nucleating agents that are active at warm temperatures, barriers exist in woody plants that inhibit growth of ice from older stems into primary, lateral appendages. This is important because many of tissues in woody plants that are frost sensitive are flowers and primary, elongating, shoot tissues that arise from buds attached to older stems. The barriers that prevent ice propagation into the lateral appendages are most effective when ice forms in the main stem tissues at warm temperatures (-1.0 to -2.5 °C). If significant supercooling occurs in the main stem prior to ice formation, the rate of ice crystal growth is so rapid that it easily penetrates any existing barrier between the main stem and lateral appendages. While the identification of the barrier is undefined, vascular continuity (as well as the size and amount of the vascular elements) are believed to play a key role.

Cold acclimation appears to influence the freezing response of plants. In general, acclimated plants will supercool to a greater extent than non-acclimated plants and also have the ability to lose water very rapidly, without being injured, when ice formation does occur. This can be true even when acclimated plants are supercooled to -10 to -15 °C. Evidence suggests that native antifreeze proteins can affect the freezing process in plants and also that plants transformed to express insect antifreeze proteins will supercool to a greater extent than wild-type plants.

REFERENCES

Anderson, J. A. and Ashworth, E., 1985, Ice nucleation in tomato plants, *J. Amer. Hort. Sci.* **110:** 291-296.

Andrews, P. K., Sandridge, C. R. and Toyama, T. K., 1984, Deep supercooling of dormant and deacclimating *Vitis* buds, *Amer. J. Enol. Viticul.* **35:** 175-177.

Ashworth, E. N., 1992, Formation and spread of ice in plant tissues, *Horti. Rev.* **13:** 215-255.

Ashworth, E. N. and Kieft, T. L., 1995, Ice nucleation activity associated with plants and fungi, in: R. E. Jr. Lee, G. J. Warren, and L. V. Gusta, eds, *Biological Ice Nucleation and Its Applications*, APS Press, pp. 137-162.

Ashworth, E. N., Anderson, J. A., Davis, G. A. and Lightner, G. W., 1985, Ice formation in *Prunus persica* under field conditions. *J. Amer. Soc. Hort. Sci.* **110:** 322-324.

Burke, M. J., Gusta, L. V., Quamme, H. A., Weiser, C. J. and Li, P. H., 1976, Freezing injury in plants. *Annu. Rev. Plant Physiol.* **27:** 507-528.

Carter, J., Brennan, R. and Wisniewski, M., 2001, Patterns of ice formation and movement in blackcurrant, *HortSci.* **36:** 1027-1032.

Carter, J., Brennan, R. and Wisniewski, M., 1999, Low-temperature tolerance of blackcurrant flowers. *HortSci.* **34:** 855-859.

Cary, J. W. and Mayland, H. F., 1970, Factors influencing freezing of supercooled water in tender plants. *Agro. J.* **62:** 715-719.

Ceccardi, T. L., Heath, R. L. and Ting, I. P., 1995, Low-temperature exotherm measurement using infrared thermography, *HortSci.* **30:** 140-142.

DeVries., 1971, Glycoproteins as biological antifreeze agents in Antarctic fishes, *Science* **172:** 1152-1155.

Duman, J. G., 2001, Antifreeze and ice nucleator proteins in terrestrial arthropods, *Ann. Rev. Physiol.* **63:** 327-357.

Duman, J. G., 1994, Purification and characterization of thermal hysteresis proteins from a plant, the bittersweet nightshade, *Solanum dulcamara*, *Biochim. Biophys. Acta* **1206:** 129-135.

Franks, F., 1985, Biophysics and Biochemistry at Low Temperatures. Cambridge University Press; 210 pp.

Fuller, M. P and LeGrice, L. P., 1998, A chamber for the simulation of radiation freezing of plants, *Ann. Appl. Biol.* **133:** 589-595.

Fuller, M. P., Hamed, F., Glenn, D. M. and Wisniewski, M., (submitted). Protection of crops from frost using a hydrophobic particle film and an acrylic polymer, *HortSci.*

Fuller, M. P. and Wisniewski, M., 1998, The use of infrared thermal imaging in the study of ice nucleation and freezing in plants, *J. Therm. Biol.* **23:** 81-89.

Fuller, M. P., White, G. G. and Charman, A., 1994, The freezing characteristics of cauliflower curd, *Ann. Appl. Biol.* **125:** 179-188.

Griffith, M., Ala, P., Yang, D. S. C., Hon, W. C. and Moffatt, B. A., 1992, Antifreeze protein produced endogenously in winter rye leaves, *Plant Physiol.* **100:** 593-596.

Gross, D. C., Proebsting, E. L., Jr, MacCrindle-Zimmerman, H., 1988, Development, distribution, and characteristics of intrinsic, non-bacterial ice nuclei in *Prunus* wood, *Plant Physiol.* **88:** 915-922.

Gross, D. C., Proebsting, E. L., Jr, Andrews, P. K., 1984, The effects of ice-nucleation-active bacteria on the temperatures of ice nucleation and low temperature susceptibilities of *Prunus* flower buds at various stages of development, *J. Amer. Soc. Horti. Sci.* **109:** 375-380.

Hirano, S. S., Baker, L. S. and Upper, C. D., 1985, Ice nucleation temperature of individual leaves in relation to population sizes of ice nucleation active bacteria and frost injury, *Plant Physiol.* **77:** 259-265.

Hon, W. C, Griffith, M., Mlynarz, A., Kwok, Y. C. and Yang, D. S., 1995, Antifreeze proteins in winter rye are similar to pathogenesis-related proteins, *Plant Physiol.* **109:** 879-889.

Jaglo, K. R., Kleff, S., Amundsen, K. L., Zhang, X., Volker, H, Zhang, J. Z., Deits, T. and Thomashow, M. F., 2001, Components of the Arabidopsis C-repeat/dehydration-responsive element binding factor cold-responsive pathway are conserved in *Brassica napus* and other species. *Plant Physiol.* **127:** 910-917.

LeGrice, P., Fuller, M. P. and Campbell, A., 1993, An investigation of the potential use of thermal imaging technology in the study of frost damage to sensitive crops, in: *Proc Int Conf Biol Ice Nucleation.* Univ Wyoming, Laramie: 4.

Lindow, S. E., 1995, Control of epiphytic ice-nucleation-active bacteria for management of plant frost injury, in: R. E Lee, G. J. Jr, Warren, L. V. Gusta, eds, *Biological Ice Nucleation and Its Applications.* APS Press, pp. 239-256.

Lindow, S. E., Arny, D. C. and Upper, C. D., 1978, Distribution of ice- nucleation-active bacteria on plants in nature, *Appl. Environ. Microbiol.* **36:** 831-838.

Lindow, S. E., 1983, The role of bacterial ice nucleation in frost injury to plants, *Annu. Rev. Phytopathol.* **21:** 363-384.

Proebsting, E. L., Jr, Andrews, P. K. and Gross, D., 1982, Supercooling in young developing fruit and flower buds in deciduous orchards, *HortSci.***17:** 67-68.

Quamme, H. A., Stushnoff, C. and Weiser, C. J., 1972, The relationship of exotherms to cold injury in apple stem tissues, *J. Amer. Soc. Hort. Sci.* **97:** 608-613.

Tao, H., Wisniewski, M., Zarka, D., Thomashow, M. and Duman, J., 2000, Expression of insect, *Dendroides canadensis*, antifreeze protein in a plant, *Arabidopsis thaliana*, enhances freezing survival and depresses the freezing temperature. Proceedings Symposium Insect and Plant Cold Hardiness, Victoria, British Columbia, Canada, May 2000.

Thomashow, M. F., 1998, Role of cold-responsive genes in plant freezing tolerance, *Plant Physiol.* **118:** 1-8.

Thomashow, M. F., 2001, So what's new in the field of plant cold acclimation? Lots!, *Plant Physiol.* **125:** 89-93.

Wisniewski, M., Lindow, S. E. and Ashworth, E. N., 1997, Observations of ice nucleation and propagation in plants using infrared video thermography, *Plant Physiol.* **113:** 327-334.

Wisniewski, M., 1988, The use of infrared video thermography to study freezing in plants, in: P. H. Li and T. H. H. Chen, eds, *Plant Cold Hardiness.* Plenum Press, pp. 311-316.

Wisniewski, M. and Fuller, M., 1999, Ice nucleation and deep supercooling in plants: New insights using infrared thermography, in: R. Margesin and F. Schinner, eds, *Cold Adapted Organisms: Ecology, Physiology, Enzymology and Molecular Biology.* Springer-Verlag. Berlin.

Wisniewski, M., Glenn, D. M. and Fuller, M., (In Press). The use of a hydrophobic particle film as a barrier to extrinsic ice nucleation in tomato (*Lycopersicon esculentum* L.) plants, *J. Amer. Soc. Hort. Sci.*

Workmaster, B. A., Palta, J. and Wisniewski, M., 1999, Ice nucleation and propagation in cranberry uprights and fruit using infrared thermography, *J. Amer. Soc. Hort. Sci.***124:** 619-625.

ATTENUATION OF REACTIVE OXYGEN PRODUCTION DURING CHILLING IN ABA-TREATED MAIZE CULTURED CELLS

Wen-Ping Chen and Paul H. Li*

1. INTRODUCTION

Plants encounter a wide range of environmental stresses, such as low/high temperatures, high light, drought and high salinity, during a typical life cycle. A key sign of plant stress at a molecular level is the increased production of reactive oxygen species (ROS), which result from an imbalance in the accumulation and removal of ROS during aerobic metabolism. In plants, mitochondria and chloroplasts are considered as two major sites of ROS production. Despite the role of ROS as a phytotoxin at high concentrations, recent evidence suggests that relatively low levels of ROS can serve as a signaling for plant stress acclimation (Bowler and Fluhr, 2000; Dat et al., 2000). This novel finding indicates that ROS are not simply toxic by-products of metabolism but also function as signaling molecules. Therefore, the control of ROS levels in plants during their responses to external stimuli is extremely important, because it may decide whether plants become adapted to stress or are injured by insults.

Chilling injury has a profound impact on many aspects of crop growth, development, production and storage. In the past decades, extensive attention has been paid to the mechanisms of the chilling sensitivity of plants because of the agricultural demands for improvements in the chilling tolerance of food crops. One of the proposed mechanisms to explain how chilling injury occurs in plants is ROS toxicity (Wise and Naylor, 1987; Purvis and Shewfelt, 1993). To date, improving plant cold tolerance by increasing ROS scavenging capability through genetic engineering has been successful in several crop species (Allen et al., 1997; McKersie and Bowley, 1998). In addition to defense by detoxification, recent studies have illustrated that plant mitochondria employ several strategies to avoid or minimize ROS production under normal growth conditions or stress (Møller, 2001). In this chapter, we will briefly review those newly suggested mechanisms

* Wen-Ping Chen and Paul H. Li, Laboratory of Plant Hardiness, Department of Horticultural Science, University of Minnesota, St. Paul, MN 55108.

Plant Cold Hardiness, edited by Li and Palva
Kluwer Academic/Plenum Publishers, 2002

(Figure 1) in combination with our recent studies on the mitochondrial ROS production in maize cultured cells under chilling stress. ABA treatment has been shown to induce antioxidant enzyme activities, such as superoxide dismutase, catalase and ascorbate peroxidase, in plants (Zhu and Scandalios, 1994; Guan and Scandalio, 1998; Bueno et al., 1998; Guan et al., 2000). In comparison of the control with ABA-treated maize cells, which have improved chilling tolerance (Xin and Li, 1992), we suggested that the reduced ROS production in ABA-treated maize cells during chilling might be due in part to (1) prevention of Ca^{2+} overload during chilling, and (2) induction of alternative oxidase activity and superoxide dismutase activity in mitochondria. However, each mechanism may function at different stages of chilling stress to avoid or reduce ROS damage.

2. PRODUCTION OF ROS AND ITS REGULATION IN PLANT MITOCHONDRIA

Unlike the membrane leakage that has been attributed to membrane phase transition (Lyons and Raison, 1970) or uneven contraction of membrane under chilling (Zsoldos and Karvaly, 1979), Purvis and Shewfelt (1993) suggested that the loss of membrane integrity in chilling-sensitive plants under chilling is likely mediated by ROS. Chilling causes enhanced ROS production under chilling in susceptible plants (Purvis and Shewfelt, 1993; Prasad et al., 1994). ROS, including superoxide (O_2^-), hydrogen peroxide (H_2O_2) and the hydroxyl radicals ($^{\cdot}OH$), are normal metabolites generated during metabolic processes involving electron transport. If not readily removed, ROS could cause lipid peroxidation, DNA damage, and protein denaturation, leading to cell injury or possibly death (Scandalios, 1993; Møller, 2001). Mitochondria and chloroplasts are considered the two major sites of ROS production. In heterotrophically grown plants, mitochondria are however the major site of superoxide and H_2O_2 production (Puntarulo et al., 1988; Maxwell et al., 1999) and a major source for cytosolic H_2O_2 (Puntarulo et al., 1988).

In mitochondria, superoxide anion is produced during respiration, primarily by the autooxidation of reduced mitochondrial electron-transport components, mainly at complex I and ubisemiquinone in complex III, in the presence of molecular O_2 (Figure 1). The generally accepted mechanism for the increased ROS production in mitochondria under chilling has been attributed to the inhibition of restriction of the cytochrome pathway activity by low temperatures, leading to increased ROS generation by over-reduced electron transport components (Purvis and Shewfelt, 1993).

Many strategies have been evolved in plants to keep ROS at low, steady-state levels or to prevent their attack on cellular macromolecules if produced. These include the detoxification by antioxidants such as ascorbate and glutathione, and/or by scavenging enzymes such as superoxide dismutase, catalase and peroxidase (Scandalios, 1993, Møller, 2001), as well as protection by compatible osmolytes, such as proline and glycinebetaine (Smirnoff and Cumbes, 1989; Hare and Cress, 1997; Chen et al., 2000). Moreover, plant mitochondria also possess the first line of defense mechanisms to avoid or minimize ROS production by preventing over-reduction of the electron transport chain. Recently, the alternative oxidase (AOX) pathway, downstream the ubiquinone, along with uncoupling protein (UCP) (Figures 1 and 2) in plant mitochondria are believed to act as electron transport uncouplers to allow continuous respiration when the cytochrome pathway is restricted by stress and thus keep the electron transport chain at more oxidized status (Kowaltowski et al., 1998; Møller, 2001).

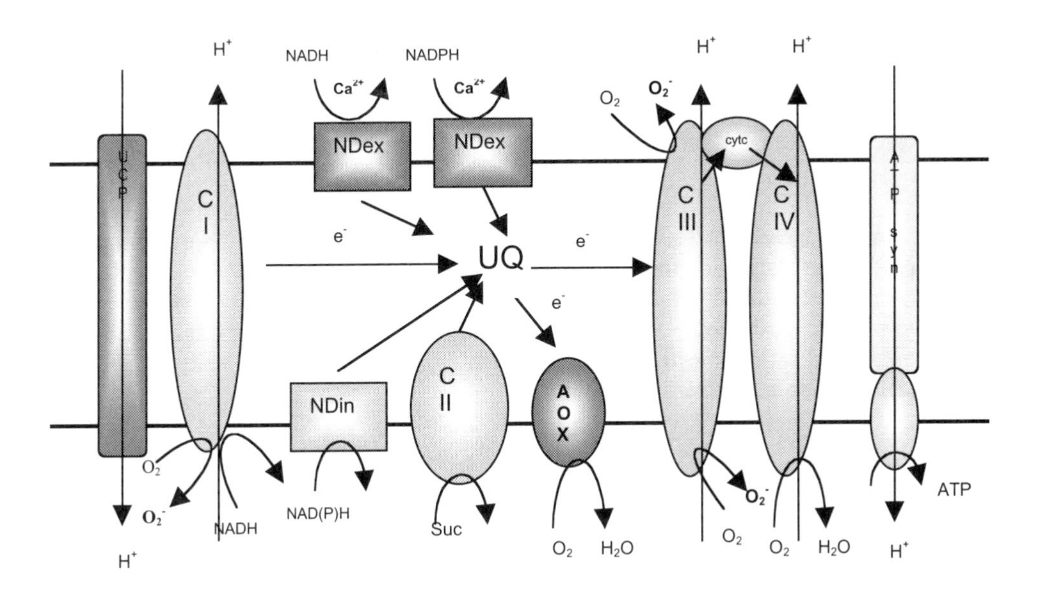

Figure 1. The electron transport chain in the inner membrane of plant mitochondria. Complex I and III are the two main sites of ROS production. Abbreviations: CI-CIV, respiratory complexes; ATP syn, ATP synthetase; AOX, alternative oxidase; UCP, uncoupling protein; NDex, external NADH or NADPH dehydrogenases; NDin, internal NADH dehydrogenase.

Based upon our understandings of the removal and reduction of ROS in plant mitochondria, attempts have been made to manipulate antioxidant defenses, with the hope of increasing plant tolerance to cold stress. Transgenic plants that overexpress genes for superoxide dismutases, ascorbate peroxidase and glutathione reductase have shown increases in tolerance to oxidative and/or cold stress despite of some exceptions (Allen et al., 1997; McKersie and Bowley, 1998). Experiments with transgenic tobacco plants that overexpress or underexpress AOX clearly indicate AOX activity can help reduce ROS production in mitochondria under normal or cytochrome pathway-restricted conditions. (Maxwell et al., 1999). No studies on cold tolerance in plants that overexpress genes for AOX or uncoupling protein have been reported, although it has been shown that AOX and uncoupling genes can be induced by low temperatures (Vanlerberghe and McIntosh, 1992; Maia et al., 1998).

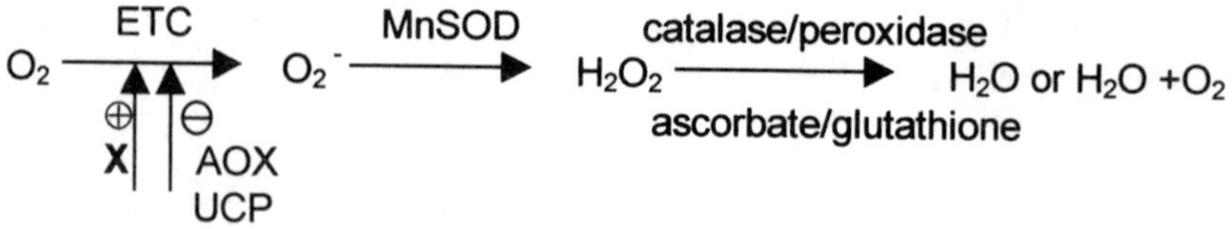

ETC MnSOD catalase/peroxidase
O_2 → O_2^{-} → H_2O_2 → H_2O or $H_2O + O_2$
⊕ ⊖ ascorbate/glutathione
X AOX
UCP

Figure 2. Defense strategies against oxidative stress in plant mitochondria. Avoidance of ROS production by activation of AOX and UCP could be the first line of defense. Detoxification of ROS by antioxidants and/or scavenging enzymes are the second line of defense. X represents any factor that disturbs respiratory pathways, resulting in over-reduction of electron transport components. Examples of those factors are respiratory inhibitors, cold stress or high concentrations of cytosolic Ca^{2+}.

Despite those mechanisms described above, we have recently demonstrated that regulation of intracellular Ca^{2+} levels may also play a role in controlling mitochondrial ROS production during chilling exposure in maize cultured cells. Our study of AOX activity in maize mitochondria suggested that the increased AOX activity in mitochondria isolated from ABA-treated cells may not be able to help reduce ROS production during 4°C chilling but rather during the periods when temperatures are gradually dropping to 4°C from 26°C or during recovery at 26°C.

3. CALCIUM OVERLOAD DURING CHILLING ENHANCED ROS PRODUCTION IN MAIZE MITOCHONDRIA

The role of Ca^{2+} as a second messenger in signaling pathways of low temperature in plants has been well documented (Monroy and Dhindsa, 1995; Knight et al., 1996). However, attentions have not been drawn enough on the Ca^{2+} cytotoxicity under cold stress although it has long been known that toxic levels of Ca^{2+} indeed occur in chilling-sensitive plants under the stress (Minorsky, 1985). A high level of cytosolic Ca^{2+} (Ca^{2+} overload) has been suggested to cause injury through (1) an inhibition of phosphate-based energy metabolism (Hepler and Wayne, 1985), (2) inducing disassembly of microtubules (Bokros et al., 1996), (3) loss of membrane integrity by activation of lipid-degrading enzymes (Ramesha and Thompson, 1984), and (4) inhibition of several key enzymes in the Calvin cycle by high levels of Ca^{2+} *in vitro* (Charles and Halliwell, 1980). Minorsky (1985) suggested that the inhibition of the key Calvin cycle enzymes may lead to a decrease in the levels of $NADP^{+}$, a shortage of which could cause photoinhibition and accumulation of ROS.

Numerous studies have shown that changes in intracellular redox and Ca^{2+} homeostasis are unifying consequences of biotic and abiotic stress. In addition, it has been suggested that a so-called 'crosstalk' may occur between Ca^{2+} and ROS. For examples, application of H_2O_2 can stimulate increase in cytosolic Ca^{2+} (Price et al., 1994), and when Ca^{2+} channel is blocked, burst of ROS production as a plant response to elicitors is prevented (Chandra and Low, 1997). Chilling often causes a prolonged cytosolic Ca^{2+} overload (Minorsky 1985; Rincon & Hanson, 1986; Jian et al., 1999, Chen and Li, 2001) as well as an increased ROS production (Scandalios, 1993; Purvis and

Shewfelt, 1993; Prasad et al., 1994; Chen et al., 2000) in chilling-sensitive plants. A question as to whether Ca^{2+} stimulates ROS production under chilling as it does in plant defense response was thus arisen. We used Ca^{2+} ionophore, A23187, to mimic the effect of chilling on intracellular Ca^{2+} levels in combination with determination of lipid peroxidation and ROS production under A23187 or chilling treatment. We concluded that chilling–induced Ca^{2+} influx and overload, as indicated by the increased $^{45}Ca^{2+}$ uptake, enhances production of ROS in maize mitochondria (Chen and Li, 2001), because (1) intracellular localization of ROS production under confocal microscopy using 2',7'-dichlorofluorescin diacetate, a ROS-specific dye, in combination with MitoTracker Red, a mitochondria-specific dye, reveals that Ca^{2+} overload in chilled maize cells was associated with increased ROS production in maize mitochondria, and (2) Ca^{2+} indeed enhanced superoxide production in isolated mitochondria, as shown in Figure 3. Under chilling, we have estimated that maize cultured cells may accumulate as high as 0.5-1 mM of cytosolic Ca^{2+}, which should have potential to stimulate mitochondrial ROS production *in vivo*, as observed by the fluorescent confocal microscopy.

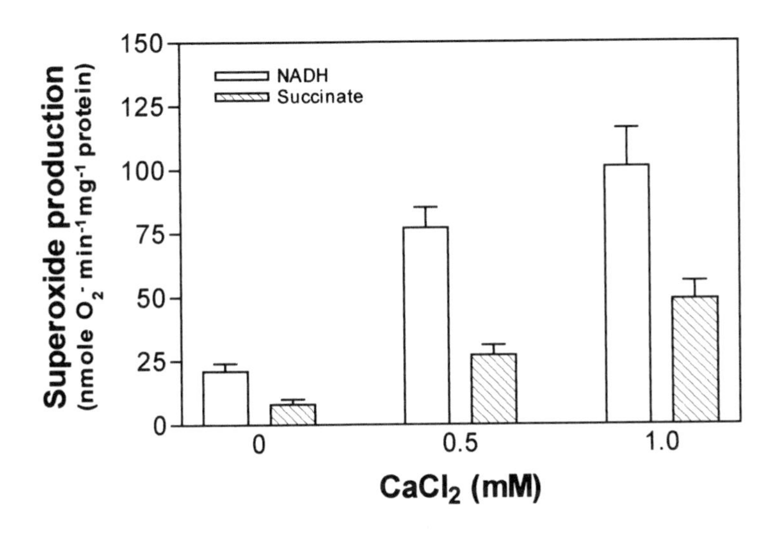

Figure 3. Ca^{2+} enhanced superoxide production in mitochondria isolated from maize cultured cells. Superoxide production was measured with either 1mM NADH or 10 mM succinate under state 4 conditions.

Ca^{2+}-induced mitochondrial ROS production in animal systems is at the level of reduced coenzyme Q where electrons are leaked as a result of the neutralization of membrane surface charge by Ca^{2+} (Kowaltowski et al., 1995). We found that (1) 2 mM Ca^{2+} caused precipitation of isolated mitochondria and (2) the Ca^{2+}-enhanced superoxide production occurred regardless of substrates used. It was thus suggested that Ca^{2+} enhanced ROS production in maize mitochondria mainly by changing the properties of mitochondrial membrane and disturbing the respiratory pathways. Unlike animal mitochondria, plant mitochondria contain additional external NADH dehydrogenases, which can be activated by Ca^{2+} (Møller, 2001). The dramatic increase in ROS production with NADH in the presence of Ca^{2+} was likely because of the activation of the external NADH dehydrogenases by Ca^{2+}, rendering the components of the Ca^{2+}-disturbed electron transport chain further over-reduced.

3.1. ABA Pretreatment Reduced Ca^{2+} Overload in Maize Cultured Cells during Prolonged Chilling

Interestingly, ABA-treated maize cultured cells did not respond to chilling with increased Ca^{2+} accumulation and ROS production, as evidence by our examinations with $^{45}Ca^{2+}$ uptake and dichlorofluorescein fluorescence (Chen and Li, 2001). *In vitro* assay indicated that Ca^{2+} still stimulated ROS production in mitochondria isolated from ABA-treated cells. We thus hypothesized that the avoidance of chilling-induced Ca^{2+} overload might help ABA-treated cells reduce Ca^{2+}-enhanced ROS production in maize mitochondria.

How ABA-treated cells reduced chilling-induced Ca^{2+} overload is an interesting question. In plants, low cytosolic Ca^{2+} concentrations are maintained by the activity of high-affinity Ca^{2+}-ATPases (pumps) and low-affinity H^+/Ca^{2+} antiporters (Bush, 1995; Sanders et al., 1999). The measurement of $^{45}Ca^{2+}$ influx and $^{45}Ca^{2+}$ accumulation in the presence of A23187 (Ca^{2+} channel-independent Ca^{2+} accumulation) suggested that the reduced Ca^{2+} accumulation in ABA-treated cells under chilling is likely achieved by increasing Ca^{2+} efflux rather than by decreasing Ca^{2+} influx in the plasma membrane. In fact, the activity of the plasma membrane Ca^{2+}-ATPase enhanced by ABA has been reported in barley aleurone protoplasts (Wang et al., 1991). A recent study with an *Arabidopsis* mutant, which is defective in vacuolar H^+-ATPase, also suggests that the regulation of cytosolic Ca^{2+} concentrations under cold stress most likely rely on the Ca^{2+} pump instead of the Ca^{2+}/H^+ antiporter (Allen et al., 2000). On the other hand, the failure of maintaining a low cytosolic Ca^{2+} level in maize cells during chilling is likely due to the dysfunction of Ca^{2+} transporters. Studies have shown that maize plasma membrane Ca^{2+}-ATPase can be impaired by prolonged chilling exposure (Jian et al., 1999).

Increases in cytosolic Ca^{2+} concentrations have been noted for a range of abiotic stresses. Our results seem to be in line with earlier findings that regulation of Ca^{2+} efflux pathways in plants are important for the function of Ca^{2+} signaling and the adaptation of plants to environmental stress, including salinity (Geisler et al., 2000), oxidative stress (Allen et al., 2000) and cold (Puhakainen et al., 1999; Jian et al., 1999; Allen et al., 2000).

4. ABA INDUCED CYANIDE-RESISTANT ALTERNATIVE OXIDASE ACTIVITY IN MAIZE CULTURED CELLS

In addition to the cyanide-sensitive cytochrome pathway that is coupled to ATP synthesis, plants possess a cyanide-resistant non-phosphorylating alternative pathway (Vanlerberghe, 1997; Møller, 2001). The alternative oxidase (AOX) is the only component in the alternative pathway, which has been suggested to function as a first line of defense to limit mitochondrial ROS production by keeping the electron transport chain more oxidized during either normal metabolism (Millenaar et al., 1998; Maxwell et al., 1999) or under cold stress (Purvis and Shewfelt, 1993). However, its role at low temperatures has been challenged by the fact that the electrons partitioning to the alternative pathway decreased as the ambient temperature was lowered (Gonzàlez-Meler et al., 1999). Ribas-Carbo et al. (2000) suggested that the alternative pathway might participate in reducing ROS production during recovery from chilling stress when cytochrome pathway activity is slowly resumed.

We examined whether cyanide-resistant AOX activity is involved in reducing mitochondrial ROS production during chilling stress in ABA-treated maize cells. As shown in Table 1, mitochondria isolated from 1-d ABA-treated cells displayed two-fold higher AOX activity than the control. ABA treatment did not affect the cytochrome pathway activity. The increased AOX activity was positively correlated with the increased AOX protein level (data not shown). How ABA treatment increased AOX protein remained unknown. We speculate that it may be due to the gene induction by H_2O_2 generated in maize during ABA treatment, because ABA treatment has been shown to induce H_2O_2 production in *Arabidopsis* guard cells (Pei et al. 2000) and application of H_2O_2 can induce AOX expression in *Petunia* and tobacco (Wagner, 1995; Maxwell et al., 1999).

Table 1. The capacity of the cytochrome pathway (CP) and the alternative pathway (AP) in mitochondria isolated from maize cells with or without 1-d ABA treatment at 26°C. Values were obtained with succinate as substrate under state 3 conditions in the presence of either 1 mM KCN or 0.1 mM n-propyl gallate. Total capacity (Total) was the sum of CP, AP and residual activity.

Treatment	Respiratory Pathway Capacity			
	CP	AP	Residual	Total
Control	114±8	36±3	0	150±6
ABA	115±9	74±5	0	189±8

The involvement of AOX activity in maize in reducing ROS production was demonstrated by measuring superoxide production in the presence of disulfiram, an AOX inhibitor. We found that superoxide production in isolated mitochondria was increased about 300% with the inhibitor at room temperature. This result appears to support earlier suggestion (Popov et al., 1997; Millenaar et al., 1998; Maxwell et al., 1999) that AOX activity plays a role in lowering mitochondrial ROS production at warm temperatures. However, *in vitro* assays indicated that AOX activity was undetectable at 4°C, consistent with an earlier report (Gonzàlez-Meler et al., 1999). We questioned whether AOX activity would have any effectiveness in reducing ROS production in maize cells when chilled at 4°C or below. There are three stages when maize cells encounter 4°C chilling exposure: (1) temperature decreasing period (from 26°C to 4°C), (2) 4°C exposure period, and (3) recovery period at 26°C. Our findings and others (Prasad et al., 1994; Kingston-Smith et al., 1999; Gonzàlez-Meler et al., 1999) have shown that cytochrome pathway activity is sensitive to mild chilling temperatures and is regaining its activity very slowly during recovery. Since the first and third periods of chilling exposure were within AOX-functioning range in maize cells, higher AOX activities in ABA mitochondria before 4°C and right after 5-d chilling exposure might have helped reduce more ROS production than the control during (1) and (3) stages. We also found that MnSOD activity in mitochondria isolated from ABA-treated maize cells was increased two-fold as compared to the control, consistent with a previous report (Zhu and Scandalios, 1994). The MnSOD activity, which can function at a wider range of chilling temperatures (Thomas et al., 1999), might help remove ROS faster in ABA-treated cells than that of the control during chilling as well as recovery. As described earlier, ABA-treated cells also benefit from the avoidance of Ca^{2+} overload when exposed to chilling, which may act as a first line defense to potentially reduce ROS production.

5. IS CHILLING INJURY IN MAIZE CELLS A PROGRAMMED CELL DEATH?

Chilling induces Ca^{2+} overload and increased ROS production in maize cultured cells. In addition, we observed ultrastructural changes in the chilled maize cells that are similar to those of the so-called programmed cell death (PCD) under pathogen infection, wounding, and H_2O_2 treatment, or during plant growth and development (Kratsch and Wise, 2000; Beers and McDowell, 2001). These changes include disappearance of starch granules, fusion and disruption of the vacuole, swelling of mitochondria, shrinkage of the cytoplasm, and chromatin fragmentation (Figure 4). PCD is defined as cell death by the expression of unique proteins or enzymes involved in the systematic degradation of the cell, which differs from necrotic cell death due to severe cell injury. Typically, PCD initially requires an increased Ca^{2+} influx (Xu and Heath, 1998) and the production of ROS (Mittler et al., 1998). Studies have shown that during PCD the activation of lytic enzymes, such as proteases responsible for protein degradation (D'silva et al., 1998) and endonucleases responsible for DNA fragmentation (Mittler and Lam, 1995), is Ca^{2+}-dependent. The loss of membrane integrity during PCD may be also due to the activation of lipid-degrading enzymes (Ramesha and Thompson, 1984) by the increased cytosolic Ca^{2+} or simply as a result of ROS-induced lipid peroxidation (Scandalios, 1993). Taken together, our results seem to suggest that chilling injury of maize may belong to PCD category (Kratsch and Wise, 2000).

6. CONCLUDING REMARKS

Literature has well documented that the imbalance of ROS status in plant systems, especially under stresses including chilling, leads to physiological dysfunction and cell injury. A better understanding of the regulation of ROS production in plants will help design better strategies to improve plant performance under stress. Mitochondria contain a number of proteins known to detoxify ROS or minimize ROS production (Figures 1 and 2), however, only the role of MnSOD in plant cold tolerance has reasonably been established. Our studies on the ABA-improved chilling tolerance in maize cultured cells have revealed that a reduced Ca^{2+} overload under prolonged chilling exposure and increased AOX and MnSOD activities may all contribute to the less ROS production in ABA-treated cells during chilling. Investigations on how ABA treatment is able to reduce the chilling-induced cytosolic Ca^{2+} overload may provide clues for attempts to improve plant chilling tolerance. To elucidate whether AOX really participates in alleviating chilling injury, transgenic plants that over- or down-express AOX gene seem to be one of the approaches.

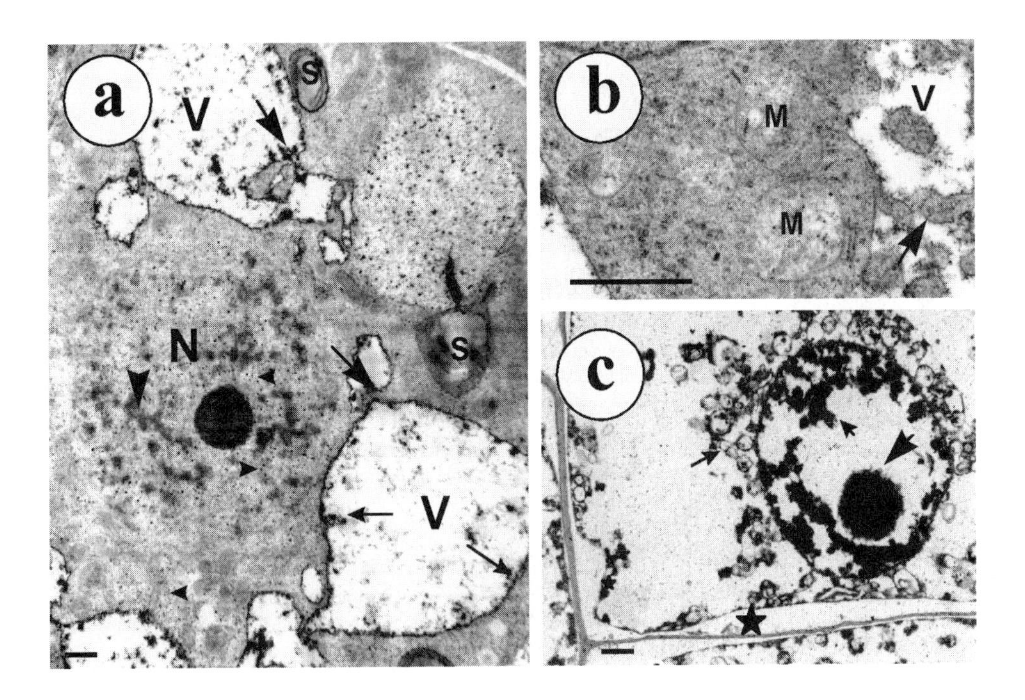

Figure 4. Ultrastructural changes in maize cultured cells after chilling at 4°C for 5 days (a, b) and 10 days (c). (a) Fusion of the vacuoles (large arrow) and accumulation of amorphous electron-dense materials (small arrow) in the vacuoles were seen. Loss of starch granules (S) and chromatin condensation (arrowhead) in the nucleus were often seen. A large amount of electron dense antimonate-Ca^{2+} deposits (small arrowheads), an indication of Ca^{2+} overload after 5 days at 4°C, were located in the cytosol and nucleus. (b) Swollen mitochondria (M) and fragments of the cytoplasm (arrowhead) inside the vacuole. (c) Cellular structures were disrupted with shrinkage of the plasma membrane (star) and many vesicle-like structures (arrow) surrounding the nucleus that contained the fragmented chromatin (small arrowhead) and nucleoli (large arrowhead). Bars=1 μm.

REFERENCES

Allen, G. J. and Sanders, D., 2000, Alteration of stimulus-specific guard cell calcium oscillations and stomatal closing in *Arabidopsis det3* mutant, *Science* **289**: 2338-2342.

Allen, R. D., Webb, R. P. and Schake S. A., 1997, Use of transgenic plants to study antioxidant defenses, *Free Rad. Biol. Med.* **23**: 473-479.

Auh, C. K. and Murphy, T. M., 1995, Plasma membrane redox enzyme is involved in the synthesis of O_2^- and H_2O_2 by *phytophthora* elicitor-stimulated rose cells, *Plant Physiol.* **107**: 1241-1247.

Beers E. P. and McDowell J. M., 2001, Regulation and execution of programmed cell death in response to pathogens, stress and developmental cues, *Curr. Opin. Plant Biol.* **4**: 561–567.

Bokros, C. L., Hugdahl, J. D., Blumenthal, S. S. D. and Morejohn L. C., 1996, Proteolytic analysis of polymerized maize tubulin: regulation of microtubule stability to low temperature and Ca^{2+} by the carboxyl termius of beta-tubulin, *Plant Cell Environ.* **19**: 539-548.

Bowler, C. and Fluhr, R., 2000, The role of calcium and activated oxygens as signals for controlling cross-tolerance, *Trends Plant Sci.* **5**: 241-246.

Bueno, P., Piqueras, A., Kurepa, J., Savoure, A., Verbruggen, N., Van Montagu, M. and Inze, D., 1998, Expression of antioxidant enzymes in response to abscisic acid and high osmoticum in tobacco BY-2 cell cultures, *Plant Sci.* **138**: 27-34.

Bush D. S., 1995, Calcium regulation in plant cells and its role in signaling, *Annu. Rev. Plant Physiol. Plant Mol. Biol.* **46**: 95-122.

Chandra, S. and Low, P. S., 1997, Measurement of Ca^{2+} fluxes during elicitation of the oxidative burst in aequorin-transformed tobacco cells, *J. Biol.Chem.* **272**: 28274–28280.

Charles, S. A. and Halliwell, B., 1980, Action of calcium ion on spinach (*Spinacia oleracea*) chloroplast fructose bisphosphatase and other enzymes in the Calvin cycle, *Biochem J.* **188**: 775-779.

Chen, W. P. and Li P. H., 2001, Chilling-induced Ca^{2+} overload enhances production of active oxygen species in maize (*Zea mays* L.) cultured cells: the effect of abscisic acid treatment, *Plant Cell Environ.* **24**: 791-800.

Chen, W. P., Li, P. H. and Chen, T. H. H., 2000, Glycinebetaine increases chilling tolerance and reduces chilling-induced lipid peroxidation in *Zea mays* L, *Plant Cell Environ.* **24**: 609-618.

D'Silva, I., Poirier, G. G. and Heath, M. C., 1998, Activation of cysteine proteases in cowpea plants during the hypersensitive response. A form of programmed cell death, *Exp. Cell Res.* **245**: 389-399.

Dat, J., Vandenabeele, S., Vranová, E., Van Montagu, M., Inzé, D. and Van Breusegem, F., 2000, Dual action of the active oxygen species during plant stress responses, *Cell. Mol. Life Sci.* **57**: 779–795.

Geisler, M., Frangne, N., Gomès, E., Martinoia, E. and Palmgren, M. G., 2000, The ACA4 gene of *Arabidopsis* encodes a vacuolar membrane calcium pump that improves salt tolerance in yeast, *Plant Physiol.* **124**: 1814–1827.

Gonzàlez-Meler, M. A., Ribas-Carbo, M., Giles, L. and Siedow, J. N., 1999, The effect of growth and measurement temperature on the activity of the alternative respiratory pathway, *Plant Physiol.* **120:** 765-772.

Guan, L. and Scandalios, J. G., 1998, Two structurally similar maize cytosolic superoxide dismutase genes, Sod4 and Sod4A, respond differentially to abscisic acid and high osmoticum, *Plant Physiol.* **117**: 217-224.

Guan, L. M., Zhao, J. and Scandalios, J. G., 2000, Cis-elements and trans-factors that regulate expression of the maize Cat1 antioxidant gene in response to ABA and osmotic stress: H_2O_2 is the likely intermediary signaling molecule for the response, *Plant J.* **22**: 87-95.

Hare, P.D. and Cress, W.A., 1997, Metabolic implications of stress-induced proline accumulation in plants, *Plant Growth Regul.* **21:** 79-102.

Hepler, P. K. and Wayne, R. O., 1985, Ca^{2+} and plant development, *Annu. Rev. Plant Physiol.* **36**: 397-439.

Jian, L. C., Li, J. H., Chen, W. P., Li, P. H. and Ahlstrand, G. G. 1999, Cytochemical localization of calcium and Ca^{2+}-ATPase activity in plant cells under chilling stress: a comparative study between the chilling-sensitive maize and the chilling-insensitive winter wheat, *Plant Cell Physiol.* **40**: 1061-1071.

Kingston-Smith, A. H., Harbinson, J. and Foyer, C. H., 1999, Acclimation of photosynthesis, H_2O_2 content and antioxidants in maize (*Zea mays*) grown at sub-optimal temperature, *Plant Cell Environ.* **22**: 1071-1083.

Knight, H., Trewavas, A. J. and Knight, M. R., 1996, Cold calcium signaling in Arabidopsis involves two cellular pools and a change in calcium signature after acclimation, *Plant Cell* **8**: 489-503.

Kowaltowski, A. J., Castilho, R. F. and Vercesi, A. E., 1995, Ca^{2+}-induced mitochondrial membrane permeabilization: role of coenzyme Q redox state, *Amer. J. Physiol.* **269**: 141-147.

Kowaltowski, A. J., Costa, A. D. T. and Vercesi, A. E., 1998, Activation of the potato plant uncoupling mitochondrial protein inhibits reactive oxygen species generation by the respiratory chain, *FEBS Lett.* **425**: 213-216.

Kratsch, H. A. and Wise, R. R., 2000, The ultrastructure of chilling stress, *Plant Cell Environ.* **23:** 337-350.

Levine, A., Tenhaken, R., Dixon, R. and Lamb, C. 1994, H_2O_2 from the oxidative burst orchestrates the plant hypersensitive disease resistance response, *Cell* **79**: 583-593.

Lyons, J. M., 1973, Chilling injury in plants, *Annu. Rev. Plant Physiol.* **24**: 445-466.

Maiaa, I. G., Benedettia, C. E., Leitea, A., Turcinellia, S. R., Vercesic, A. E. and Arruda P., 1998, AtPUMP: an Arabidopsis gene encoding a plant uncoupling mitochondrial protein, *FEBS Lett.* **429**: 403-406.

Maxwell, D. P., Wang, Y. and McIntosh, L. 1999, The alternative oxidase lowers mitochondrial reactive oxygen production in plant cells, *Proc. Natl. Acad. Sci. USA* **96**: 8271-8276.

McKersie, B. D. and Bowley, S. R., 1998, Active oxygen and freezing tolerance in transgenic plants, in: *Plant Cold Hardiness: Molecular Biology, Biochemistry, and Physiology*, P. H. Li and T. H. H. Chen, eds, Plenum Press, New York, pp. 203-214.

Millenaar, F. F., Benschop, J. J., Wagner, A. M. and Lamber, H., 1998, The role of the alternative oxidase in stabilizing the in vivo reduction state of the ubiquinone pool and the activation state of the alternative oxidase, *Plant Physiol.* **118**: 599-607.

Minorsky, P. V., 1985, A heuristic hypothesis of chilling injury in plants: A role for Ca^{2+} as the primary physiological transducer in injury, *Plant Cell Environ.* **8**: 75-94.

Mittler, R. and Lam, E., 1995, Identification, characterization, and purification of a tobacco endonuclease activity induced upon hypersensitive response cell death, *Plant Cell* **7**: 1951-1962.

Mittler, R., Feng, X. Q. and Cohen, M., 1998, Post-transcriptional suppression of cytosolic ascorbate peroxidase expression during pathogen-induced programmed cell death in tobacco, *Plant Cell* **10**: 461–473.

Møller, I. M., 2001, Plant mitochondria and oxidative stress: Electron transport, NADPH turnover, and metabolism of reactive oxygen species, *Annu. Rev. Plant Physiol. Plant Mol. Biol.* **52**: 561–591.

Monroy, A. F. and Dhindsa, R. S., 1995, Low-temperature signal transduction: Induction of cold acclimation-specific genes of alfalfa by Ca^{2+} at 25°C, *Plant Cell* **7**: 321-331.

Pei, Z. M., Murata, Y., Benning, G., Thomine, S., Klüsener, B., Allen, G. J., Grill, E. and Schroeder, J. I., 2000, Calcium channels activated by hydrogen peroxide mediate abscisic acid signaling in guard cells, *Nature* **406**: 731-734.

Popov, V. N., Simonian, R. A., Skulachev, V. P. and Starkov, A. A., 1997, Inhibition of the alternative oxidase stimulates H_2O_2 production in plant mitochondria, *FEBS Lett* **415**: 87-90.

Prasad, T. K., Anderson, M. D., Martin, B. A. and Steward, C. R., 1994, Evidence for chilling-induced oxidative stress in maize seedlings and a regulatory role for hydrogen peroxide, *Plant Cell* **6**: 65-74.

Price, A. H., Taylor, A., Ripley, S. J., Griffiths, A., Trewavas, A. J. and Knight, M. R., 1994, Oxidative signals in tobacco increase cytosolic calcium, *Plant Cell* **6**: 1301–1310.

Puntarulo, S., Galleano, M., Sanchez, R. A. and Boveris, A., 1991, Superoxide anion and hydrogen peroxide metabolism in soybean embryonic axes during germination, *Biochim. Biophys. Acta* **1074**: 277-283.

Purvis, A. C. and Shewfelt, R. L., 1993, Does the alternative pathway ameliorate chilling injury in sensitive plant tissues? *Physiol. Plant.* **88**: 712-718.

Ramesha, C. S. and Thompson, G. A., 1984, The mechanism of membrane response to chilling: Effect of temperature on phospholipids deacylation and reacylation reactions in the cells surface membrane, *J. Biol. Chem.* **259**: 8706-8712.

Ribas-Carbo, M., Aroca, R., Gonzàlez-Meler, M. A., Irigoyen, J. J. and Sánchez-Díaz, M., 2000, The electron partitioning between the cytochrome and alternative respiratory pathways during chilling recovery in two cultivars of maize differing in chilling sensitivity, *Plant Physiol.* **122**: 199-204.

Sanders, D., Brownlee, C. and Harper J. F., 1999, Communication with calcium, *Plant Cell* **11**: 691-706.

Scandalios, J. G., 1993, Oxygen stress and superoxide dismutases, *Plant Physiol.* **101**: 7-12.

Smirnoff, N. and Cumbes, Q. J., 1989, Hydroxyl radical scavenging activity of compatible solutes, *Phytochemistry* **28**: 1057-1060.

Thomas, D. J., Thomas, J. B., Prier, S. D., Nasso, N. E. and Herbert, S. K., 1999, Iron superoxide dismutase protects against chilling damage in the *cyanobacterium synechococcus* species PCC7942, *Plant Physiol.* **112**: 275–282.

Vanlerberghe, G. C., 1997, Alternative oxidase: from gene to function, *Annu. Rev. Plant Physiol. Plant Mol. Biol.* **48**: 703-734.

Vanlerberghe, G. C. and McIntosh, L., 1992, Lower temperature increases alternative pathway capacity and alternative oxidase protein in tobacco, *Plant Physiol.* **100**: 115-119.

Wagner, A. M., 1995, A role for active oxygen species as second messengers in the induction of alternative oxidase gene expression in *Petunia hybrida* cells, *FEBS Lett.* **368**: 339-342.

Wang, M., Van Duijn, B. and Schram, A.W., 1991, Abscisic acid induces a cytosolic calcium decrease in barley aleurone protoplasts, *FEBS Lett.* **278**: 69-74.

Wise, R. R. and Naylor, A. W., 1987, Chilling-enhanced photooxidation. The peroxidative destruction of lipids during chilling injury to photosynthesis and ultrastructure, *Plant Physiol.* **83**: 272-277.

Xin, Z. G. and Li, P. H., 1992, Abscisic acid-induced chilling tolerance in maize suspension-cultured cells, *Plant Physiol.* **99**: 707-711.

Xu, H. X. and Heath, M. C., 1998, Role of calcium in signal transduction during the hypersensitive response caused by basidiospore-derived infection of the cowpea rust fungus, *Plant Cell* **10**: 585–597.

Zhu, D. and Scandalios J. G., 1994, Differential accumulation of manganese-superoxide dismutase transcripts in maize in response to abscisic acid and high osmoticum, *Plant Physiol.* **106**: 173-178.

Zsoldos, F. and Karvaly B., 1979, Cold-shock injury and its relation to ion transport by roots, in: *Low temperature Stress in Crop Plants*, J. M. Lyons, D. Graham and J. K. Raison, eds, Academic Press Inc, New York, pp. 123-138.

Part III

Genetic Engineering

GENETIC ENGINEERING OF CULTIVATED PLANTS FOR ENHANCED ABIOTIC STRESS TOLERANCE

Lawrence V. Gusta, Nicole T. Nesbitt, Guohai Wu, Ximing Luo, Albert J. Robertson, Doug Waterer, and Michael L. Gusta*

1. INTRODUCTION

It is generally acknowledged abiotic stress tolerance is induced in response to an environmental stimulus resulting in the upregulation of multiple stress genes. Stimulus perception governs the extent of upregulation of stress associated proteins. This upregulation dictates when a plant initiates acclimation and thus the extent of stress tolerance the plant acquires. Often cultivated crops do not perceive an environmental stress, such as an episodic frost, early enough to acclimate. As a result, cultivated crops are often lethally injured by an unseasonal -3 to -4 °C frost whereas if given sufficient time they can acclimate to withstand much lower temperatures (e.g. -9 to -12 °C). Drought tolerance is generally induced as a consequence of plants experiencing a wet-dry cycle. During the day, plants lose turgor due to a water shortage; however, during the night plants regain turgor due to the reduced evapo-transpiration rate. As a result of the cycle, drought associated genes are upregulated. In many cool season crops, leaf temperatures exceeding 30 °C induce the formation of heat shock proteins that protect the cell. Maximum frost and drought tolerance is attained after several weeks of acclimating conditions; in contrast, enhanced heat tolerance can be attained in hours.

Abiotic stress tolerance is a complex quantitatively inherited trait with many genes involved. Therefore it is not expected that overexpression of a single gene would have a significant impact on stress tolerance. However, several recent reports have demonstrated significant enhancement of stress tolerance for plants transformed with a single gene construct. For example, Zhang and Blumwald (2001) reported transgenic tomato plants overexpressing a vacuolar Na^+/H^+ antiport protein had enhanced salinity tolerance. Jaglo et al. (2001), presented evidence that canola (*Brassica napus*) seedlings overexpressing

* Lawrence V. Gusta, Nicole T. Nesbitt, Guohai Wu, Ximing Luo, Albert J. Robertson, Doug Waterer, and Michael L. Gusta, Crop Development Centre/Plant Sciences Dept., University of Saskatchewan, 51 Campus Drive, Saskatoon, SK, S7N 5A8, Canada.

the C-repeat/dehydration responsive element binding factor (CBF) increased the freezing tolerance of both non-acclimated and cold-acclimated plants. Constitutive expression of CBF genes resulted in an increased freezing tolerance of Arabidopsis seedlings at non-acclimating temperatures (Gilmour et al., 1998). The biosynthesis of the phytohormone abscisic acid (ABA) is controlled in part by 9-cisexpoxycarotenoid dehydrogenase whose expression is induced by a drought stress (Lauchi et al., 2001). In transgenic Arabidopsis, overexpression of its gene resulted in an endogenous ABA increase and promoted transcription of drought and ABA-inducible genes (Lauchi et al., 2001). Transgenic Arabidopsis overexpressing vacuolar H^+-pyrophosphatase had enhanced tolerance to both salt and drought stresses (Gaxiola et al., 2001). Thus, there is accumulating evidence that single genes, either coding for a key biosynthetic enzyme or a transcriptional factor, can have a dramatic effect on abiotic stress tolerance.

In this study, we transformed canola (*Brassica napus)* cv. DH-12075, potato (*Solanum tuberosum*) cv. Desiree, and flax (*Linum usitatissimum*) cv. CDC Normandy with mitochondial manganese superoxide dismutase (SOD3.1) isolated from wheat (*Triticum aestivum*) accession No. U72212; G. Wu, R.W. Wilen, A.J. Robertson and L.V. Gusta; CBF1 isolated from canola accession No. AF084185; N. Zhao, G. Wu, Y-P, Gao, R.W. Wilen and L.V. Gusta; dehydrin 4 (DHN4) (Close et al., 1989); and ROB5 a stress associated gene upregulated in response to ABA, cold, heat, drought, and salinity we previously isolated from a bromegrass culture (Robertson et al., 1994). These three plant species were tested both in the field and in controlled environments for tolerance to frost, heat, and drought stresses at the seedling and flowering stages. We also determined the effect of these genes on growth and development.

2. EXPERIMENTAL PROTOCOL

The vector constructs used for transformation of canola, potato, and flax are shown in Table 1. In 2000, we produced 38, 104, and 84 transformants of canola, potato, and flax, respectively. In 2001, 93, 57, and 56 transformants of canola, potato, and flax were produced respectively. Plants generated were tested at both the seedling and maturation stage of growth for frost, heat, and drought tolerance. Plants were grown either in the field or in controlled environment chambers.

Table 1. Transformation Vector Constructs.

Construct (Promoter)	Vector	Promoter (Restriction Sites)	Gene (Restriction Sites)
35S:SOD3.1	PHS737	35S (Hind III, Xba I)	SOD3.1 (BamH I, Kpn I)
COR78: SOD3.1	Bin 19	COR78 (Sal I, BamH I)	SOD3.1 (Kpn I, EcoR I)
COR78:DHN4	Bin 19	COR78 (Sal I, BamH I)	DHN4 (Kpn I, EcoR I)
COR78:CBF1	Bin 19	COR78 (Sal I, BamH I)	CBF1 (BamH I, Kpn I)
COR78:ROB5	Bin 19	COR78 (Sal I, BamH I)	ROB5 (BamH I, Kpn I)

2.1. Frost Tolerance

Frost tolerance was determined by either a controlled freeze test in the laboratory or by subjecting plants to a natural frost in the field. For controlled freeze tests, plants were sprayed with ice nucleating active bacteria (*Pseudomonas syringae*, strain Cit7) which generally initiated freezing between -2.5 to -3 °C. Moist plant samples were placed in 2.5 x 15 cm test tubes containing a thin bottom layer of moist tissue paper. Plant samples were held at -2.5 °C for 60 min., -3 °C for 60 min., and then cooled at 2 °C h^{-1}. Generally test temperatures were separated by 1 °C intervals. Following the freeze protocol, the plant samples in the test tubes were held at 4 °C overnight. The LT_{50} °C (lowest temperature that resulted in less than 50% injury) was evaluated by the electrolyte leakage test.

2.2. Heat Tolerance

Heat tolerance was determined on both intact plants and excised stems and leaves. Plants growing in 4L pots were watered to field capacity prior to being subjected to a heat stress. For plants growing in the field, excised leaves were placed in capped test tubes (2.5 x 15 cm) containing 2 to 3 ml of water. The plant tissue, either the whole plant or excised leaves, was heated linearly from 22 to 42 °C over a twelve hour period and then held isothermal at 42 °C. The duration at this temperature varied with the plant species. Viability was determined either by regrowth, seed yield and seed quality, or by the electrolyte leakage test.

2.3. Drought Tolerance

For canola and flax, drought tolerance was determined with plants growing outside in 4L pots containing a mixture of loam-clay and peat (1:1). At the three to five leaf stage, the plants were watered to capacity, allowed to drain for one day, and then weighted each day after. Water was withheld from the plants at this time. After fourteen days the plants were re-watered to assess regrowth potential. The plants were rated visually by four independent evaluators. In the field, drought tolerance was determined by 1000 kernel weight of the harvested seed. The 2001 growing season was the driest on record, with Saskatoon receiving only 30% of its normal effective precipitation.

3. STRESS TOLERANCE

3.1. Frost Tolerance

Episodic frosts in early spring or during crop maturation in the fall, result in either plant death or delayed maturity leading to yield reduction and loss in crop quality. For example, canola experiencing a frost of -3 to -5 °C at the pod filling stage inhibits the clearing of chlorophyll during seed ripening. As a consequence, the crop is devalued due to the expense of removing chlorophyll from the extracted oil used for cooking purposes. Potato, with little or no inherent frost tolerance, is a very sensitive crop to spring and fall frosts. Fall frosts at the potato tuber bulking stage results in a dramatic loss in yield. In actively growing plants, an increase in frost tolerance of 2 or 3 °C would make a dramatic difference in crop production.

As stated previously, marginal gains in frost tolerance have been reported for canola and Arabidopsis plants transformed with CBF driven by the constitutive 35S promoter (Jaglo et al., 2001; Gilmour et al., 1998) and for alfalfa transformed with Mn-SOD (McKersie et al., 1993). In our studies, several lines of canola plants transformed with 35S:SOD3.1 had over a 90% survival rate (visual estimate of three independent ratings) versus 50% for the controls following a natural frost of -9 °C over a two night period (see Fig. 1). These plants were grown in a glasshouse at 22 °C, then transferred outside just prior to the frost. In a controlled environment freeze test control canola plants at the three to four leaf stage, subjected to a cold acclimating regime of 7 °C day and 5 °C dark for six days, were killed at -7 °C compared to -13 °C for plants transformed with COR78:SOD3.1.

In a controlled environment freeze test, potato plants grown at 10 °C day and 8 °C dark were subjected over a two day period to -9 °C for two hours. Excised leaves from the control plants had over 50% ion leakage versus 30% for lines transformed with 35S:SOD3.1 and COR78:ROB5 (see Fig. 2). In this study, lines transformed with COR78:SOD3.1, COR78:DHN4, and COR78:CBF1 were found to more frost tolerant than the control potato plants.

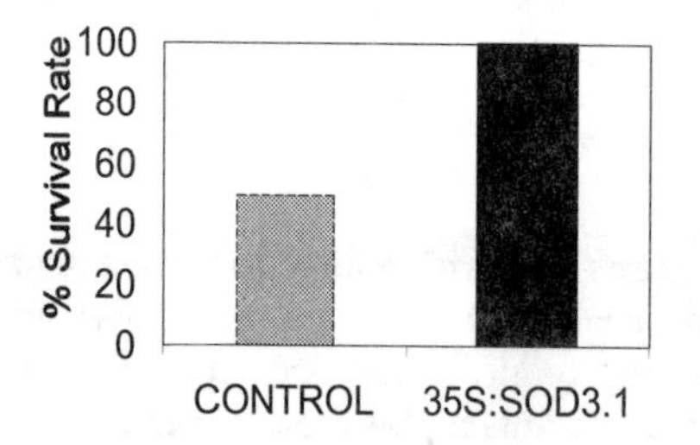

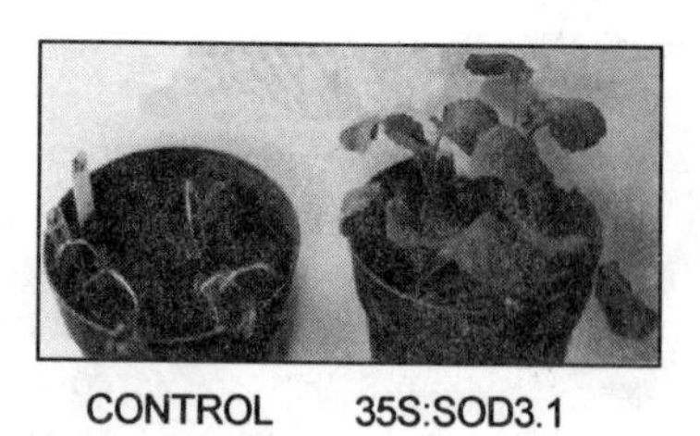

Figure 1. Frost tolerance of canola seedlings transformed with 35S:SOD3.1, following a natural frost of -9°C over a period of two nights.

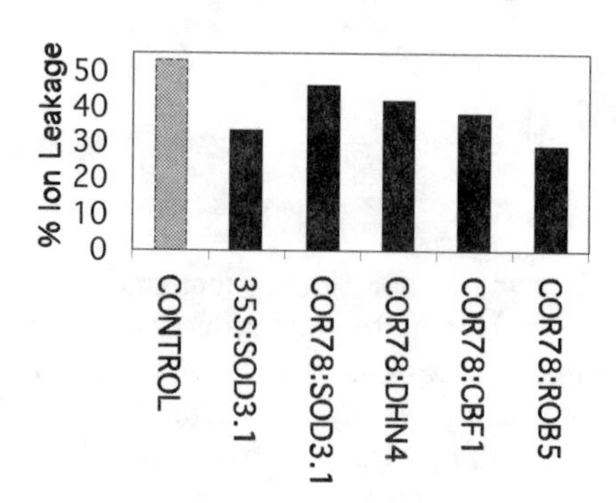

Figure 2. Frost tolerance of transgenic canola seedling subjected to a controlled freeze test of -9°C.

The freezing tolerance of flax lines transformed with 35S:SOD3.1, COR78:SOD3.1, COR78:DHN4, COR78:CBF1, and COR78:ROB5 was determined. The flax plants grown in pots under natural conditions; the time of sampling was early July. Excised stems and leaves were cooled to -9 °C at 2 °C h^{-1}, held at this temperature for two hours, then warmed to 2 °C overnight and then subjected to the same freeze thaw cycle again. After the freeze test, the control leaves had over 60% electrolyte leakage compared to 10% for the 35S:SOD3.1, COR78:CBF1 and COR78:ROB5 transformants (see Fig. 3). Leaves of flax plants growing the field were collected at the end of July and subjected to a controlled freeze test over the range of -3 to -8 °C (see Fig. 4). Control plants suffered injury between -5 to -6 °C (inflection range) compared to -8 °C for lines transformed with 35S:SOD3.1.

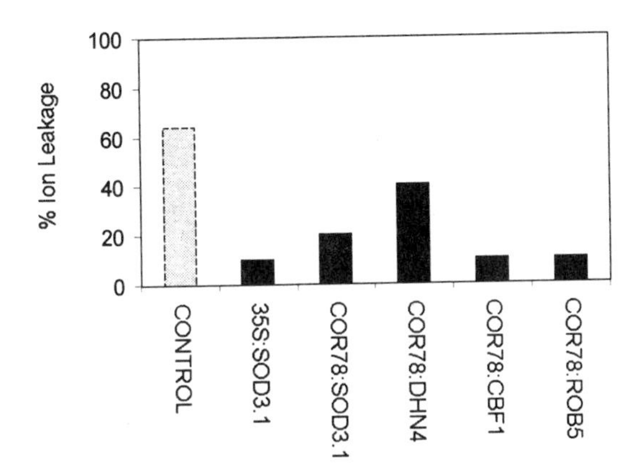

Figure 3. Frost tolerance of transgenic flax seedlings subjected to a controlled freeze test of -9°C. The plants were cooled at 2°C h^{-1}, held isothermal for two hours, warmed to 2°C overnight , and the cycle was repeated.

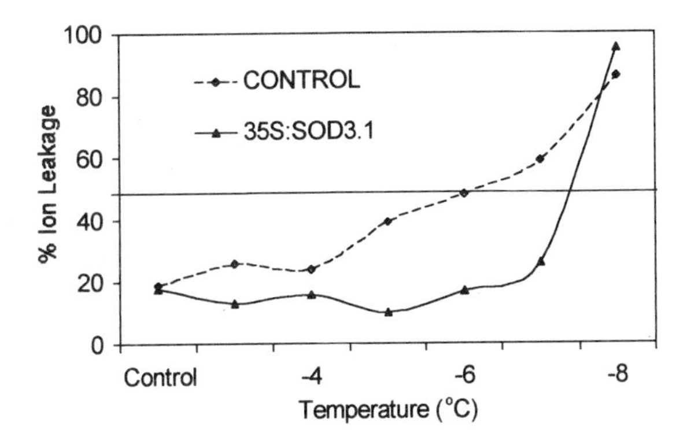

Figure 4. Frost tolerance of flax seedlings collected from the field at the end of July. The controlled freeze test ranged from -3 to -8°C.

3.2. Heat Tolerance

Both canola and flax are cool season crops and suffer high temperature stress particularly at the flowering stage (Hall, 1992; Paulsen, 1994). Yield is reduced due to a reduction in the number of pods or bolls, number of seeds per pod or boll, and seed weight. Thus yield losses can be large particularly when combined with a drought stress, however, large losses can also be observed in spite of an adequate supply of water (Angadi et al., 2000). Many cool season crops produce heat shock proteins when tissue temperature exceeds 32 to 33 °C (Vierling, 1991). Several classes of heat shock proteins have been identified that are thought to confer heat tolerance in plants. In our studies, we preconditioned the plants by slowly heating to induce natural heat shock proteins. The logic was to determine if the transformed lines had acquired either additional heat tolerance or were able to repair the induced heat injury.

For canola, dramatic increases in heat tolerance were measured in 35S:SOD3.1 transformants at the seedling stage. Control plants subjected to a heat stress at 42 °C for sixteen hours succumbed whereas several transformant lines suffered little or no visual symptoms of injury (see Fig. 5). These heat-treated plants continued to grow, flower, and set normal pod development. A similar heat stress at the flowering stage resulted in flower abortion in the case of the control plants. In 2000, 35S:SOD3.1 canola transformants heated to 42 °C for fourteen hours had less than 8% electrolyte leaf leakage versus over 60% for the control (results not shown). In 2001, we tested over 93 transformant lines and obtained positive results for the following constructs; COR78:SOD3.1, 35S:SOD3.1, COR78:DHN4, COR78:CBF1, and COR78:ROB5. The COR78 promotor has been shown to be induced by cold, desiccation, high-salt conditions, and ABA (Yamaguchi-Shinozaki and Shinozaki, 1993). Horvath et al. (1993) reported that the threshold temperature at which COR78 transcripts accumulated is between 10 and 12 °C. The seed test weight was significantly higher for these transformants compared to the control (results not shown).

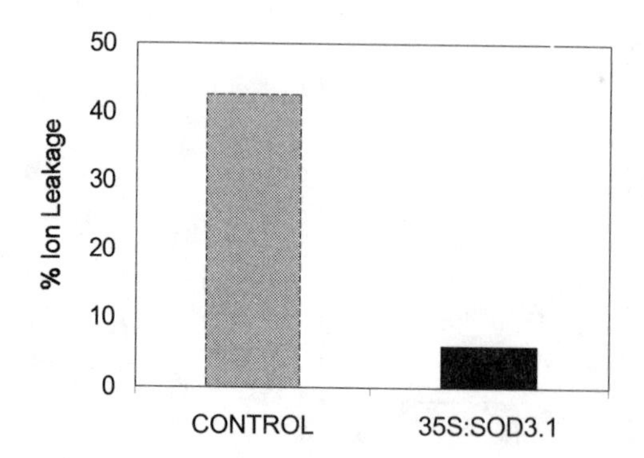

Figure 5. Heat tolerance of canola seedlings subjected to a heat stress of 42°C for sixteen hours.

In controlled heat tests, flax plants at the boll stage of development were subjected to 42 °C for sixteen hours. After testing, control plants had over 60% ion leakage versus 6% for 35S:SOD3.1 transformants (results not shown). The effect of the heat treatment on seed yield is shown in Figure 6. The control plant produced a few small seeds whereas the 35S:SOD3.1 transformant produced large, healthy, normal seeds. Positive results were also obtained for COR78:CBF1 and COR78:ROB5. In addition, fourteen lines of potato plants transformed with 35S:SOD3.1 were superior to the control plants following 42 °C for sixteen hours (results not shown).

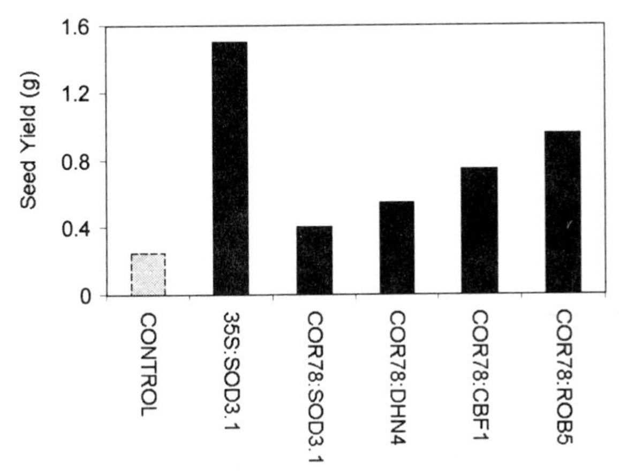

Figure 6. Seed yield of flax seedlings subjected to 42°C for sixteen hours.

3.3. Drought tolerance

Genetic engineering has resulted in significant increases in drought tolerance. Luchi et al. (2001), increased the drought tolerance of Arabidopsis by overexpressing 9-cis-epoxycartenoid dehydrogenase, a key enzyme in ABA biosynthesis. Overexpression of DREB1/CBF and DRE 2, a stress inducible transcriptional factor, also resulted in an increase in drought tolerance of Arabidopsis (Kasuga et al., 1999; Liu et al., 1998).

In our studies we determined the rate of water lost from canola seedlings growing in 4L pots. Canola overexpressing 35S:SOD3.1, COR78:DHN4, and COR78:ROB5 lost water at a slower rate compared to the control (see Fig. 7A). Plants transformed with COR78:SOD3.1 were able to recover following rewatering (see Fig. 7B). The reduced rate of water loss may be due to a reduction in stomatal aperture caused by elevated ABA levels. The gene encoding DHN4 (Close et al., 1989) and ROB5 (Robertson et al., 1994) are both ABA responsive, which may account for partial stomatal closure. Reactive oxygen species generally increase in response to an environmental stress such as drought and frost. Murata et al. (2001), suggest ABA induces reactive oxygen species activates calcium-permeable channels involved in stomatal closure. If this scenario exists during a drought stress, then SOD must protect plants via a different mechanism or does not completely scavenge all the reactive oxygen species involved in stomatal closure. Upon rewatering the SOD, DHN4, and ROB5 transformed plants regained turgor and resumed growth to produce plants that flowered and eventually set seed. In contrast, the drought stressed control plants did not regain similar growth levels and only produced a few pods containing seeds. Similar results were obtained for water stressed flax seedlings (results not shown).

In 2001, the driest season on record in the Saskatoon region, our plots received less than 30% of the normal effective rainfall. These conditions provided an excellent test to screen transformants for both heat and drought tolerance under natural conditions. The 1000 canola seed weight for the control and various transformants is shown in Figure 8. Plants transformed with 35S:SOD3.1 and COR78:ROB5 had the highest seed weights (4.7 g versus 3.75 g for the control). Similar results were obtained for flax (results not shown).

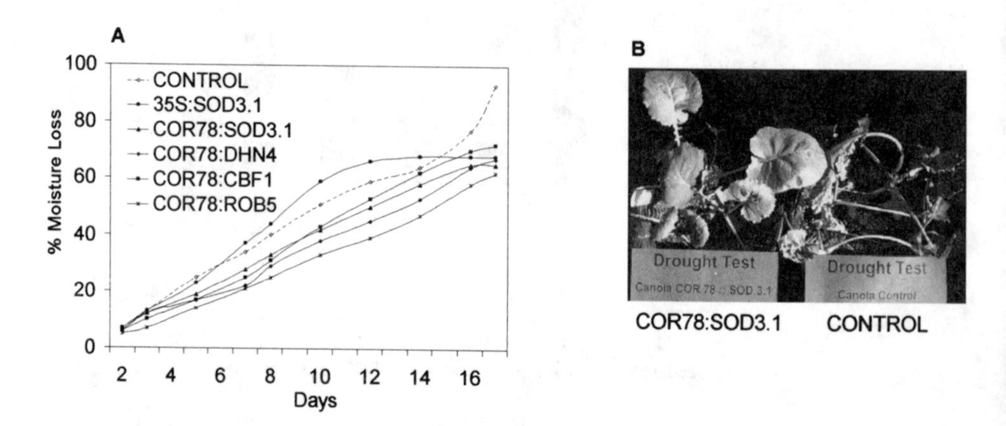

Figure 7. A. water loss from canola seedlings subjected to an artificial drought. **B.** Recovery of water stressed canola seedlings following rewatering.

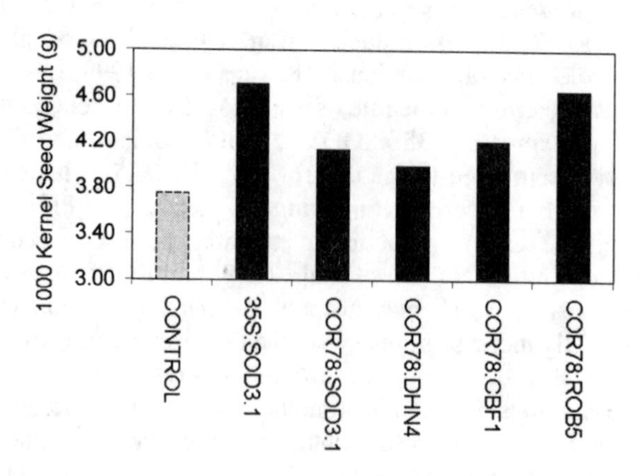

Figure 8. The 1000 seed weight (g) of canola plants harvested from the field.

4. PLANT GROWTH AND DEVELOPMENT

Plant growth and development comprises cell division and cell growth. A complete cell division cycle consists of four phases: DNA synthesis (S), two gap phases (G1 and G2), and mitosis. Progression through these phases is catalyzed by a family of cyclin-dependent kinases (CDK) that are both positively and negatively regulated, depending upon the environmental conditions (Stals et al., 2000). CDKs regulate a number of cell cycle functions including transcriptional regulators, cytoskeleton nuclear matrix, nuclear membrane proteins, and cycle proteins (Joubes et al., 2000). The cyclin dependent kinases play a critical role when abiotic stress (e.g. high and low temperatures, and drought) induce the formation of reactive oxygen species that impair G1/S transition, retard replication, and delay entry into mitosis (Reichheld et al., 1999). Inhibition of plant growth and development is one of the first events to occur as a result of an abiotic stress.

In field trials conducted in the spring of 2000, which was colder and drier than average, we were pleasantly surprised that some of our transformed canola lines emerged sooner and displayed increased seedling vigour compared to the controls. The number of seedlings that emerged was determined daily. Due to the extremely dry soil, emergence was delayed until early June. Over 80 seedlings for two COR78:SOD3.1 transformants emerged June 14, 2000 versus only approximately five seedlings for the control (results not shown). Similar results were obtained for field studies in 2001, which had a hot and dry spring. Plant seedling emergence and vigour, as determined by plants per row and plant development, were determined. COR78:SOD3.1 transformants emerged earlier than the controls in 2001 (see Fig. 9). Positive results were also obtained for canola transformed with COR78:ROB5, COR78:CBF1, and COR78:DHN4. There were over 200% more plants per row for these transformants compared to the control.

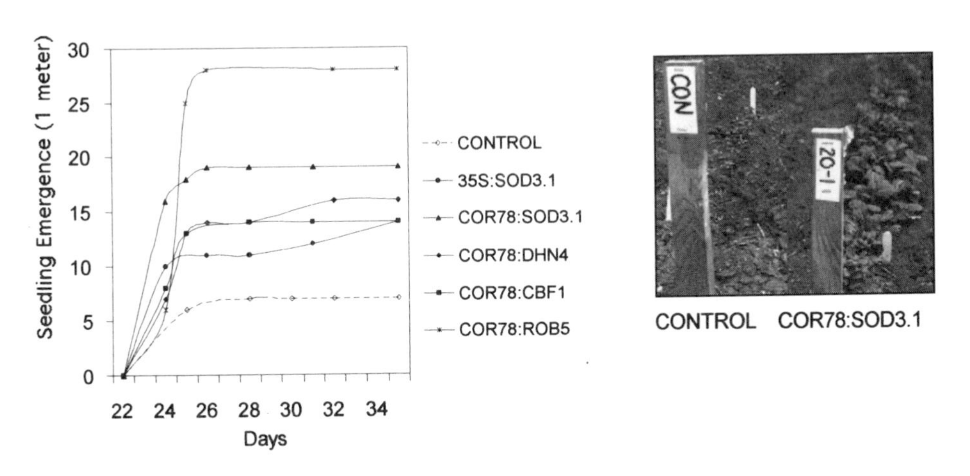

Figure 9. Seedling emergence and development of transformed canola lines grown in the field, 2000.

The most impressive results of genetic engineering were observed in the number of days for canola plants to flower (see Fig. 10). In the field, canola plants transformed with 35S:SOD3.1, COR78:SOD3.1, COR78:DHN4, and COR78:ROB5 flowered at least seven days earlier than the control. In controlled environment tests, conducted at 10 °C light and 8 °C dark with a 12% soil moisture content, canola plants transformed with either 35S:SOD3.1 or COR78:SOD3.1 emerged two days earlier than the control (see Fig. 11). In addition, ten days after planting, only 35% of the total control seedlings emerged versus 60 to 80% emergence for the transformants.

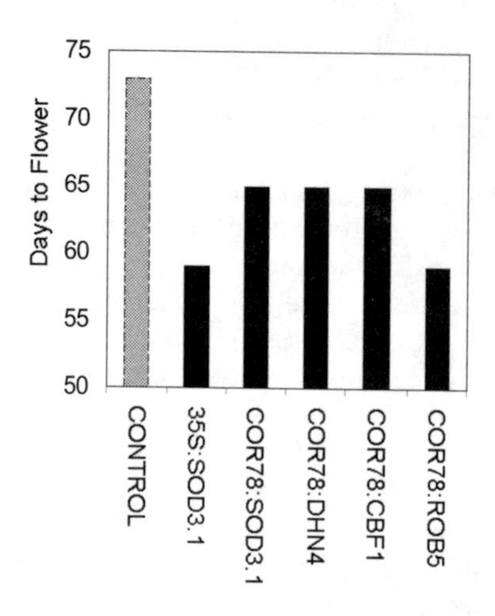

Figure 10. Enhanced flowering of transgenic canola lines grown in the field, 2000.

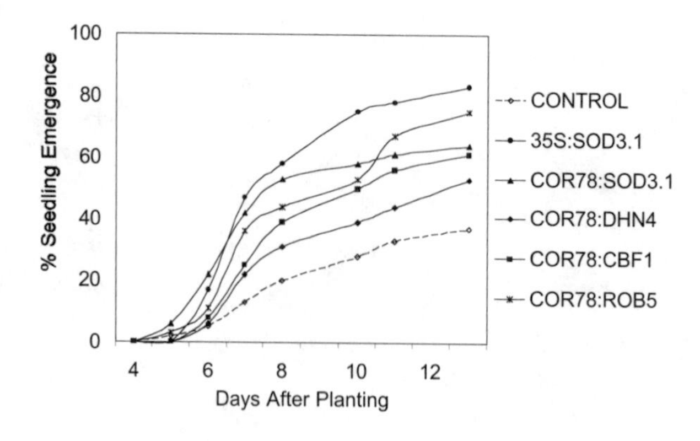

Figure 11. Rate of emergence of transformed canola seedlings at 10°C light and 8°C dark with a 12% soil moisture content.

Similar results for emergence, seedling vigour, number of plants per meter row, plant height, days to flower, and 1000 kernel weight were obtained for flax plants grown in the field (results not shown). As well, potato plants transformed with SOD3.1 flowered ten days earlier than the control under field conditions. In controlled environment studies conducted at 8 °C, potato 35S:SOD3.1 transformants grew at a faster rate compared to the control (see Fig. 12).

CONTROL 35S:SOD3.1

Figure 12. Growth and development of transformed potato seedlings subjected to 8°C.

5. SUMMARY

We have demonstrated via genetic engineering significant increases in frost, heat, and drought tolerance for canola, potato, and flax plants grown both in the field and in controlled environment chambers. In the future we plan to conduct large field tests at various locations with the most promising lines to determine if these characteristics are maintained in a commercial environment. Promising lines will be crossed with established cultivars to develop new transgenic cultivars.

Significant gains (2 to 4 °C) in freezing tolerance were observed for canola and flax while only an increase of 1 to 1.5 °C was obtained for potato. Large dramatic differences in heat tolerance were obtained for several canola, potato, and flax lines transformed with SOD3.1, DHN4, CBF1 and ROB5 constructs. We identified lines of superior drought tolerance both from controlled environment tests and from plants growing in the field. The year 2001, an extremely dry season, proved to be an excellent test year to select superior drought and heat tolerant lines. Compared to the control, several lines had larger seeds and higher 1000 kernel seed weight. We also observed enhanced emergence, larger plants and a reduction in days to flower in lines transformed with SOD3.1, DHN4, CBF1 and ROB5. Even though these tests are still in the preliminary stage, we have obtained optimistic results for genetic engineering of plants for enhanced stress tolerance.

6. ACKNOWLEDGEMENTS

This research was supported in part by an Agriculture and Food Innovation Fund (AFIF) grant awarded to Dr. L. V. Gusta.

REFERENCES

Angadi, S. V., Cutforth, H. W., Miller, P. R., McConkey, B. C., Entz, M. H., Brandt, S. A., and Volkmar, K. M., 2000, Responses of three *Brassica* species to high temperature stress during reproductive growth, *Can J Plant Sci*, **80**: 693-701.

Close, T. J., Kortt, A. A., and Chandler, P. M., 1989, A cDNA-based comparison of dehydration-induced proteins (dehydrins) in barley and corn, *Plant Mol Biol*, **13**: 95-108.

Gaxiola, R. A., Li, J., Undurrago, S., Dan, L. M., Allen, G., Alper, S., and Fink, G. R., 2001, Drought and salt-tolerant plants result from overexpression of the AVP H^+ pump, *Proc Natl Acad Sci USA*, **25**: 11444-11449.

Gilmour, S. J., Zarka, D. G., Stockinger, E. J., Salazar, M. P., Houghton, J. M., and Thomashow, M. F., 1998, Low temperature regulation of the Arabidopsis CBF family of APZ transcriptional activators as an early step in cold-induced COR gene expression, *Plant J*, **16**: 433-442.

Hall, A. E., 1992, Breeding for heat tolerance, *Plant Breed Rev*, **10**: 129-168.

Horvath, D. P., McLarney, B. K., and Thomashow, M. F., 1993, Regulation of Arabidopsis thaliana L. (Heyn) cor78 in response to low temperature, *Plant Physiol*, **103(4)**: 1047-1053.

Jaglo, K. R., Kleffs, S., Amundsen, K. L., Zhang, X., Haake, V., Zhang, J. Z., Deits, T., and Thomashow, M. F., 2001, Components of the Arabidopsis C-Repeat/dehydration-responsive element binding factor cold-responsive pathway are conserved in *Brassica napus* and other plant species, *Plant Physiol*, **127**: 910-917.

Joubes, J., Chevalier, D., Dudits, D., Heberle-Bors, E., Inze, D., Umeda, M., and Renaudi, J. P., 2000, CDK-related protein kinases in plants, *Plant Mol Biol*, **43**: 607-620.

Kasuga, M., Liu, Q., Miura, S., Yamaguchi-Shinozaki, K., and Shinozaki, K., 1999, Improving plant drought, salt and freezing tolerance by gene transfer of a single stress-inducible transcription factor, *Nature Biotech*, **17**: 287-291.

Liu, Q., Kasuga, M., Sakuma, Y., Abe, H., Miura, S., Yamaguchi-Shinozaki, and Shinozaki, K., 1998, Two transcriptional factors, DREB1 and DREB2 with an EREBP/AP2DNA binding domain separate two cellular signal transduction pathways in drought and low-temperature-responsive gene expression, respectively in Arabidopsis, *Plant Cell*, **10**: 1391-1406.

Luchi, S., Kobayashi, M., Taji, T., Naramoto, M., Sehi, M., Kato, T., Tabata, S., Kakubari, Y., Yamaguchi-Shinozaki, K., and Shinozabi, K., 2001, Regulation of drought tolerance by gene manipulation of 9-cis-epoxycarotenoid dehydrogenase, a key enzyme in abscisic acid biosynthesis in Arabidopsis, *Plant J*, **26**: 325-333.

McKersie, B.D., Chen, Y., deBeus, M., Bowley, S. R., Bowler, C., Inje, D., D'Halliun, K., and Botterman, J., 1993, Superoxide dismutase enhances tolerance of freezing stress in transgenic alfalfa (*Medicago sativa* L.), *Plant Physiol*, **103**: 1155-1163.

Murata, Y., Pei, Z-M., Mori, I., and Schroder, J., 2001, Abscisic acid activation of plasma membrane Ca^2 channels in general cells requires cytosolic NAD(P)H and is differentially disrupted upstream and downstream of reactive oxygen species production in abil-1 as abi 2-1 protein phosphatase 2C mutants, *Plant Cell*, **13**: 2513-2523.

Paulsen, G.M., 1994, High temperature response of crop plants, in: Physiology aand Determination of Crop Yield, Boote, K. J., Sinclair, T. R., and Paulsen, G. M. ed., ASA, CSSA, SSSA, Madison, WI, pp. 365-389.

Reichheld, J-P., Vernoux, T., Lardon, F., vanMoutagu, M., and Inze, D., 1999, Specific checkpoints regulate plant cell cycle progression in responses to oxidative stress, *Plant J*, **17**: 647-656.

Robertson, A. J., Ishikawa, M., Gusta, L. V., and MacKenzie, S. L., 1994, Abscisic acid-induced heat tolerance in bromegrass (*Bromus inermis* Leyss) cell cultures, *Plant Physiol*, **105**: 823-830.

Stals, H., Casteels, P., vanMoutagu, M., and Inze, D., 2000, Regulation of cyclin-dependent kinases in Arabidopsis thaliana, *Plant Mol Biol*, **43**: 583-593.

Vierling, E., 1991, The roles of heat shock proteins in plants, *Ann Rev Plant Physiol and Plant Mol Biol*, **42**: 579-620.

Yamaguchi-Shinozaki, K., and Shinozaki, K., 1993, Characterization of the expression of a desiccation-responsive rd29 gene of the Arabidopsis thaliana and analysis of its promotor in transgenic plants, *Mol Gen Genet*, **236(2-3)**: 331-340.

Zhang, H-X., and Blumwald, E., 2001, Transgenic salt-tolerant tomato plants accumulate salt in foliage but not in fruit, *Nature Biotech*, **19**: 765-768.

18

ENGINEERING TREHALOSE BIOSYNTHESIS IMPROVES STRESS TOLERANCE IN ARABIDOPSIS

Ilkka Tamminen[1], Tuula Puhakainen[1], Pirjo Mäkelä[1], Kjell-Ove Holmström[3], Joachim Müller[4], Pekka Heino[1,2] and E. Tapio Palva[1,2]

1. INTRODUCTION

Environmental stresses caused by drought and extremes of temperature are main factors limiting plant growth, productivity and distribution and up to 80 % of the total crop losses are caused by such climatic factors (Boyer, 1982). Consequently, increase in plant stress tolerance could have a major impact on agricultural productivity. Genetic engineering has recently been shown to provide new approaches for plant breeding and considerable efforts has been made to design strategies for genetic engineering of stress tolerance (Thomashow, 1999; Nuotio *et al.*, 2001). Unfortunately, developmental, structural and physiological adaptations to stresses are often based on complex mechanisms involving a number of different genes (McCue and Hanson, 1990) and therefore not amenable to genetic engineering. However, some of the responses to abiotic stress appear to be based on relatively simple metabolic traits governed by a limited number of genes.

One class of protective substances that accumulate in response to drought and low temperature stress include small organic molecules called compatible solutes (Yancey *et al.*, 1982; Smirnoff, 1998). Compatible solutes can accumulate to a high level in response to different stresses in most living organisms. These low-molecular-weight compounds are usually non-toxic at high concentrations, highly soluble and protect macromolecular structures in cell (reviewed in Hare *et al.*, 1998). Only a limited number of organic compounds are commonly employed as compatible solutes. These include quaternary ammonium and tertiary sulfonium compounds (e.g. glycine betaine, dimethylsulfoniopropionate, sugars (e.g. sucrose and trehalose), some amino acids (e.g.

[1]Department of Biosciences, Division of Genetics and [2]Institute of Biotechnology, Viikki Biocenter, P.O. Box 56, FIN-00014, University of Helsinki, Finland. [3]Department of Natural Sciences, University of Skövde, S-54128 Skövde, Sweden.[4]Friedrich-Meischer-Institute, P.O.B. 2543, CH-4002 Basel, Switzerland.

Plant Cold Hardiness, edited by Li and Palva
Kluwer Academic/Plenum Publishers, 2002

proline) and polyols (e.g. inositol and mannitol) (reviewed in McCue and Hanson, 1990; Bohnert *et al.*, 1995).

Recent reports have shown that engineering of osmolyte biosynthesis results in improved stress tolerance. Introduction of the bacterial *mtlD* gene into tobacco resulted in mannitol production and enhanced salt tolerance in transgenic plants (Tarczynski *et al.*, 1992, 1993). Fructan has been shown to improved drought tolerance in transgenic plants (Pilon-Smits *et al.*, 1995). Overexpression in tobacco of the mothbean gene for 1-pyrroline-5-carboxylate synthase, catalyzing a limiting step in proline biosynthesis pathway, resulted in enhanced tolerance of the transgenic plants to both salt and drought (Kishor *et al.*, 1995). Lilius *et al.* (1996) and Holmström *et al.* (2000) demonstrated that transgenic tobacco plants producing glycine betaine had improved salt resistance. Furthermore, introduction of the *A. globiformis codA* gene to Arabidopsis to produce glycine betaine resulted in transgenic plants that exhibited higher tolerance to salt and cold (Hayashi *et al.*, 1997; Alia *et al.*, 1998; Sakamoto *et al.*, 1998; 2000). In addition, genes for trehalose biosynthesis have been introduced to plants and this has led to increased drought tolerance (Holmström *et al.*, 1996; Romero *et al.*, 1997 Pilon-Smits *et al.*, 1998; Serrano *et al.*, 1999). Sofar there has not been any evidence for possible protective role of trehalose towards other abiotic stresses such as freezing. The rest of this discussion is focused on the osmolyte trehalose and its role as stress protectant in transgenic plants.

2. OVERVIEW OF TREHALOSE

Trehalose (α-D-glucopyranosyl-1,1-α-D-glucopyranoside) is a disaccharide composed of two molecules of glucose that are linked by their reducing carbons. This makes trehalose a very stable molecule and a non-reducing sugar. Because trehalose is so stable, it is chemically quite inert and biologically non-toxic (Roser and Colaco, 1993). For these properties trehalose is well documented as a potent drought- and cryo-protectant employed in many different biological, medical and industrial fields (Roser and Colaco, 1993).

The name trehalose derives from "*trehala manna*", the oval cocoon of a desert beetle (*Larinus* species), which contains a large amount of trehalose (20-30%). It is found on aromatic thorn bush in the Middle East and might have been the original ''manna from heaven'' mentioned in The Bible, Exodus 16: 15-30.

Even though trehalose biosynthesis is very similar to that of sucrose, its evolutionary origin is though to be more ancient, because it is present in all kingdoms. Trehalose is commonly found in a wide variety of organisms including bacteria, yeast, fungi, algae, insects and other invertebrates and non-vascular plants (Elbein, 1974; Hoekstra *et al.*, 1997; Yancey *et al.*, 1982). Trehalose is found in high concentrations in anhydrobiotic organisms (e.g. Baker's yeast) during drought stress, with trehalose levels up to 20% of their dry weight. Anhydrobiotic organisms, such as yeast can survive total dehydration and recover completely when rehydrated.

The presence of trehalose in angiosperms has been questioned (Gussin, 1972) and trehalose accumulation was originally only detected in the resurrection plant, *Myrothamnus flabellifolia* (Bianchi *et al.*, 1993; Drennan *et al.*, 1993). However, recent discoveries have demonstrated that angiosperms have genes for trehalose biosynthesis

(Vogel *et al.*, 1998; Blázquez *et al.*, 1998). The presence of such genes has been shown by expressing Arabidopsis-derived trehalose-6-phosphate synthase (*AtTPS1*) and trehalose-6-phosphate phosphatase (*AtTPPA* and *AtTPPB*) genes in yeast and complementing yeast mutants lacking these enzymes. In Arabidopsis *AtTPS1* is expressed in non-organ specific way, whereas *AtTPPA* and *AtTPPB* are expressed in flowers and young developing tissues (Vogel *et al.*, 1998; Blázquez *et al.*, 1998). These results strongly indicate that the capacity to synthesise trehalose is present in angiosperms. Müller *et al.* (2001) have recently reported that endogenous substance that had all the properties of trehalose also appears in Arabidopsis, and suggested that trehalose and trehalase (trehalose degrading enzyme) may play a role in regulating the carbohydrate allocation in plants.

2.1. Trehalose metabolism

Biosynthesis of trehalose is a two-step process starting with condensation of UDP-glucose and glucose-6-phosphate into trehalose-6-phosphate (Tre-6-P) catalyzed by enzyme trehalose-6-phosphate synthase (TPS1). Tre-6-P is then subsequently dephosphorylated into trehalose by the enzyme trehalose-6-phosphate phosphatase (TPS2) (Cabib and Leloir, 1958) (Figure 1). In plants, fungi, animals and bacteria trehalose can be hydrolyzed to glucose by the enzyme trehalase (reviewed in Goddijn and Smeekens, 1998).

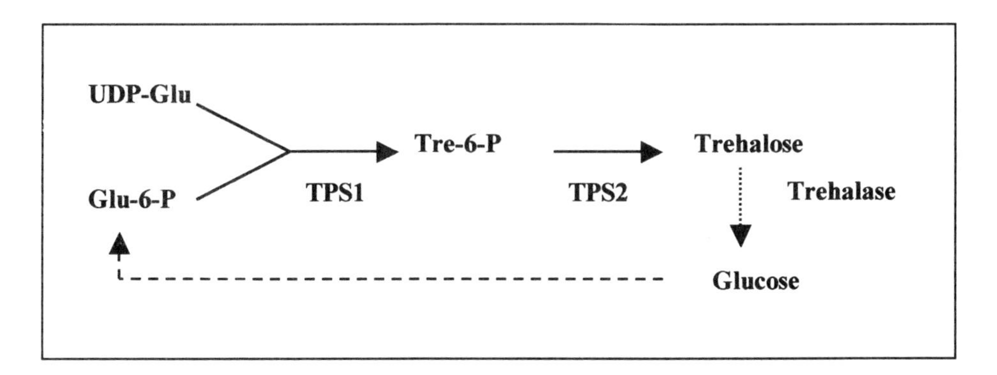

Figure 1. Simplified model of biosynthesis and degradation of trehalose.

Many organisms including higher plants and mammals, possess trehalase activity. In higher plants trehalase is present in all organs, with highest activities in flowers (Müller *et al.*, 1995a; Müller *et al.*, 2001). It is possible that the presence of trehalase activity in all plant tissues prevents the accumulation of trehalose and for that reason detection of even low amounts of trehalose would be extremely difficult. Only when a specific inhibitor of trehalase activity, Validamycin A, is applied during growth, small amounts of trehalose can be detected in wild-type tobacco and potato plants (Goddijn *et al.*, 1997).

In yeast the accumulation of trehalose has functions both as a protective metabolite under stress conditions and as a storage carbohydrate (reviewed in Singer and Lindquist, 1998). Also in *E. coli* trehalose is known to protect cells against osmotic stress (reviewed in Strøm and Kaasen, 1993). Trehalose has also been found to be more effective than most sugars at increasing lipid bilayer fluidity (Crowe *et al.*, 1984) and at preserving enzyme stability during drying (Colaço *et al.*, 1992). Because the potential function of trehalose in stress-protection, there has been a growing interest to introduce trehalose pathway to plants. Subsequently, trehalose biosynthesis has been engineered in different plants like tobacco (Holmström *et al.*, 1996; Goddijn *et al.*, 1997; Romero *et al.*, 1997; Pilon-Smits *et al.*, 1998) and potato (Goddijn *et al.*, 1997) by expression of trehalose pathway genes from yeast and *E. coli*.

3. TREHALOSE PRODUCTION IN TRANSGENIG ARABIDOPSIS

Our aim was to engineer trehalose biosynthesis in Arabidopsis for improved stress tolerance by expression of the *TPS1* and *TPS2* genes from yeast (*Saccharomyces cerevisiae*). In yeast, trehalose biosynthetic enzymes form a large trehalose synthase complex which appears to be composed of the Tre-6-P synthase (TPS1), Tre-6-P phosphathase (TPS2) and a regulatory subunit of 130 kD (TSL1) (Londesbrough and Vuorio, 1991). The regulatory subunit encoded by *TSL1* does not seem to be necessary for trehalose biosynthesis (Vuorio *et al.*, 1993) and thus engineering of trehalose biosynthesis in plants would only require introduction of the *TPS1* or *TPS1* and *TPS2* genes. We used two different constructs to achieve this (Figure 2). In both cases the gene/s were expressed under light responsive promoter (*RbcS1A*).

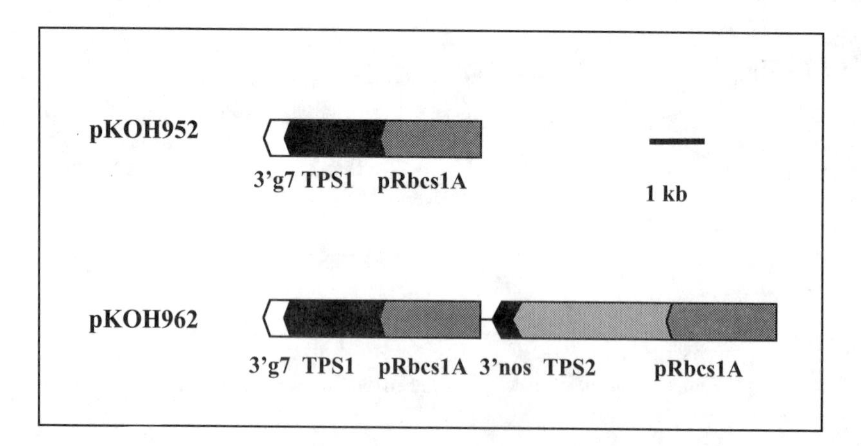

Figure 2. Chimeric gene constructs for expressing *TPS1* and *TPS2* in *Arabidopsis thaliana*. Genes encoding trehalose-6-phosphate synthase (TPS1) and trehalose-6-phosphate phosphatase (TPS2) from *S. cerevisiae* were transformed to Arabidopsis. Genes were expressed from Arabidopsis derived light stimulated ribulose-1,5-bisphosphate carboxylase (Rubisco) small subunit gene 1A promoter, p*RbsS1A*. Polyadenylation sequences were from two *Agrobacterium* genes, gene 7 (*g7*) and nopaline synthase (*nos*).

The gene/s were introduced to *Arabidopsis thaliana* (L.) ecotype Landsberg *erecta* by *Agrobacterium*-mediated transformation (Bechtold et al., 1993). The presence of TPS1 in transgenic lines was screened by Western blot analysis using antiserum raised against the purified yeast (56 kD) TPS1 polypeptide. Two transgenic lines for each construct were selected for further analysis. TPS1 polypeptide was strongly produced in both lines containing p*Rbcs1A*:*TPS1*, and also in lines containing both p*Rbcs1A*:*TPS1* and p*Rbcs1A*:*TPS2*.

3.1. Trehalose Accumulation in Transgenic Arabidopsis

Measurement of trehalose accumulation in the transgenic lines first showed that the concentration was below the detection limit. Several studies have previously shown that transgenic tobacco and potato plants producing enzymes for trehalose biosynthesis accumulated trehalose in concentrations, which were either very low or under the detection limit (Holmström *et al.*, 1996; Goddijn *et al.*, 1997; Romero *et al.*, 1997; Pilon-Smits *et al.*, 1998; Serrano *et al.*, 1999). Due to the occurrence of trehalase activity in Arabidopsis, it was possible that the produced trehalose could be degraded by this enzyme. To avoid this, we used the trehalase inhibitor validamycin A. After validamycin treatment we could clearly demonstrate detectable trehalose accumulation (about 0.4 mg g^{-1} DW) in TPS1 positive transgenic plants, in contrast to control plants. Our results indeed suggest that the presence of trehalase activity in plant tissues prevents the accumulation of trehalose and hampers detection of trehalose. This is in agreement with Müller *et al.* (1995b) who suggested that the lack of trehalose accumulation in transgenic plants may not be due to a malfunction of the transgene but to a degradation of the newly synthesized trehalose by plant-borne trehalases.

3.2. Transgenic Plants Have a Growth Phenotype

Trehalose biosynthesis has been previously associated with aberrant growth and leaf morphology in transgenic plants. It has been demonstrated that many of the transgenic tobacco plants expressing the *E. coli* trehalose biosynthesis genes (*otsA* and *otsB*) or yeast *TPS1* gene showed varying degrees of stunted growth and difference in leaf phenotype (Holmström *et al.*, 1996; Goddijn *et al.*, 1997; Romero *et al.*, 1997). Pilon-Smits *et al.* (1998) reported that trehalose producing tobaccos showed stunted growth only in early stages of growth. In contrast, there were no clear morphological alterations in transgenic Arabidopsis plants as compared to control plants. However, transgenic plants containing *TPS1* gene only or *TPS1* and *TPS2* genes were somewhat smaller than the wild- type plants, the effect being more clear with the single *TPS1* construct. It is possible that the high trehalase activity of Arabidopsis helps to ameliorate the negative effects of trehalose accumulation on e.g. leaf morphology.

4. IMPROVED STRESS TOLERANCE

4.1. Drought Tolerance in Arabidopsis

Previous studies have demonstrated that introduction of genes for trehalose biosynthesis resulted in enhanced drought tolerance (Holmström *et al.*, 1996; Romero *et al.*, 1997 Pilon-Smiths *et al.*, 1998; Serrano *et al.*, 1999). Similarly, TPS1-positive transgenic Arabidopsis lines showed clearly enhanced drought tolerance when compared with control plants. Transgenic plants recovered faster from drought treatment, continued growth, and subsequently, flowered. Control plants suffered from drought clearly more than transgenic plants and could not recover from drought treatment as well as transgenic plants.

4.2. Enhanced Freezing Tolerance in Transgenic Arabidopsis

During freezing the cell dehydrates in a manner similar to the situation when water is removed by evaporation during desiccation (Palta and Weiss, 1993). Thus, encouraged by the enhanced drought survival of transgenic Arabidopsis we also wanted to analyze their freezing tolerance. First non-acclimated plants were exposed to –5°C for 14 h and their survival was monitored. While control plants could not survive and died, the transgenic TPS1-positive plants showed some damage but recovered from the treatment, continued their growth and began to flower. This enhanced survival of the non-acclimated plants was confirmed by controlled freezing test using electrolyte leakage as the indicator for damage (Sukumaran and Weiser, 1972). Transgenic Arabidopsis could tolerate lower temperatures showing somewhat enhanced freezing tolerance when compared to control plants (Figure 3). Interestingly, this enhanced freezing tolerance was also evident in 4 d cold acclimated transgenic plants (Figure 3). This suggests that the increase in tolerance obtained by trehalose production is by a different mechanism than in low temperature acclimation, and that the two processes are additive.

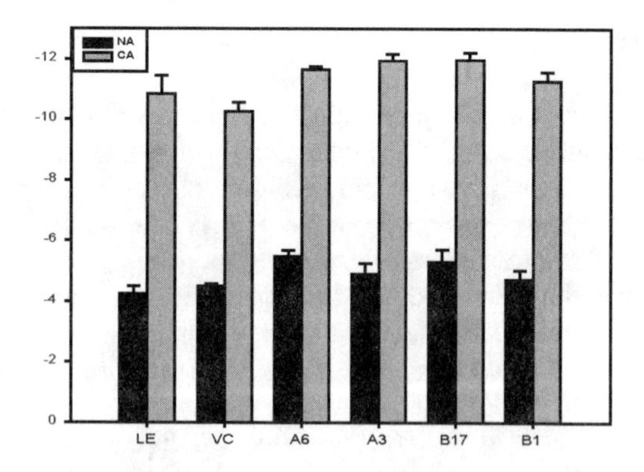

Figure 3. Freezing tolerance of non-acclimated and 4d cold acclimated (4°C) plants from wild type (**LE**), vector transformed (**VC**) and transgenic Arabidopsis plant lines expressing TPS1 (**A6**, **A3**) and TPS1 and TPS2 (**B17**, **B1**). Freezing tolerance (LT_{50}) was measured by an ion leakage assay.

4.3. Mechanism of Improved Tolerance

Expression of specific low temperature-responsive genes has been correlated with development of freezing tolerance (reviewed in Thomashow, 1999; Nuotio *et* al., 2001). To elucidate whether the improved freezing tolerance in TPS1-positive Arabidopsis was related to expression of these genes the expression of diagnostic *LTI/COR* genes was characterized both in non-acclimated and 4d cold acclimated plants. No difference in the accumulation of mRNAs corresponding to the four genes was found between the transgenic and control plants in either non-acclimating or acclimating conditions. These data suggest that *LTI/COR* gene expression does not play a role in trehalose mediated freezing tolerance. This is in agreement with the apparent additive effects of trehalose production and cold acclimation on freezing tolerance. Thus, trehalose must act by another mechanism.

Previous studies have indicated that osmolytes can be associated with protection of the photosynthetic apparatus from stress damage. To test this possibility chlorophyll fluorescence technique was used to detect changes in photosynthetic capacity due to various stresses. We examined the effect of low temperature on the activity of photosystem II of wild type and transgenic plants by determining the ratio of variable to maximum fluorescence of chlorophyll a. Plants were exposed to low temperature and recovery of photosystem II was monitored. Control plants had decreased photosystem II activity after –4 °C treatment whereas in transgenic plants the photosystem II activity was decreased only after exposure to –6°C. From these results we could conclude that the trehalose producing transgenic plants tolerate lower temperatures and recovered from low temperature stress better than non-trangenic plants.

5. CONCLUSIONS

Engineering trehalose biosynthesis in *Arabidopsis thaliana* clearly improves freezing and drought tolerance. The results suggest that trehalose accumulation has an additive effect on development of freezing tolerance during cold acclimation and that the effect of trehalose is not derived from alterations in the expression of endogenous stress related genes but functions through another, currently unknown mechanism. Protection of photosynthetic apparatus appears to be part of this mechanism. Further studies are required to elucidate the molecular basis of trehalose action in transgenics plants.

REFERENCES

Alia, Hayashi, H., Chen, T.H.H. and Murata, N., 1998, Transformation with a gene for choline oxidase enhances the cold tolerance of *Arabidopsis* during germination and early growth, *Plant Cell Environ.* **21**: 232-239.

Bechtold, N., Ellis, J. and Pelletier, G., 1993, In planta *Agrobacterium* mediated gene transfer by infiltration of adult *Arabidopsis thaliana* plants, *CR Acad. Sci. Paris, Life Sci.* **316**: 1194-1199.

Bianchi, G., Gamba, A., Limiroli, R., Pozzi, N., Elster, R., Salamini, F. and Bartels, D., 1993, The unusual sugar composition in leaves of the resurrection plant *Myrothamnus flabellifolia*, *Physiol Plant.* **87**: 223-226.

Blázquez, M. A., Santos, E., Flores, C. L., Martinezzapater, J. M., Salinas, J. and Gancedo, C., 1998, Isolation and molecular characterization of the *Arabidopsis* Tpsl gene, encoding trehalose-6-phosphate synthase, *Plant J.* **13**: 685-689.

Bohnert, H. J., Nelson, D. E. and Jensen, R. G., 1995, Adaptations to environmental stresses, *Plant Cell* **7**: 1099-1111.

Boyer, J. S., 1982, Plant productivity and environment, *Science* **218**: 443-448.

Cabib, E. and Leloir, L.F., 1958, The biosynthesis of trehalose phosphate, *J. Biol. Chem.* **231**: 259-275.

Colaco, C., Sen, S., Thangavelu, M., Pinder, S. and Roser, B., 1992, Extraordinary stability of enzymes dried in trehalose: simplified molecular biology, *Bio/Technology* **10**: 1007-1011.

Crowe, J.H., Crowe, L.M. and Chapman, D., 1984, Preservation of membranes in anhydrobiotic organisms: The role of trehalose, *Science* **223**: 701.

Crowe, J.H., Hoekstra, F.A. and Crowe, L.M.A., 1992, Anhydrobiosis, *Annu. Rev. Plant Physiol.* **54**: 579-599.

Drennan, P. M., Smith, M. T., Goldsworthy, D. and van Staden, J., 1993, The occurrence of trehalose in the leaves of the desiccation-tolerant angiosperm *Myrothamnus flabellifolius* Welw, *J. Plant Physiol.* **142**: 493-496.

Elbein, A., 1974, The metabolism of alpha-alpha-trehalose, *Adv. Carbohydr. Chem.* **30**: 227-256.

Goddijn, O. and Smeekens, S., 1998, Sensing trehalose biosynthesis in plants, *Plant J.* **14**: 143-146.

Goddijn, O. J. M., Verwoerd, T. C., Voogd, E., Krutwagen, R. W. H. H., de Graaf, P. T. H. M., Poels, J., Vandun, K., Ponstein, A. S., Damm, B. and Pen, J., 1997, Inhibition of trehalase activity enhances trehalose accumulation in transgenic plants, *Plant Physiol.* **113**:181-190.

Gussin, A. E. S., 1972, Does trehalose occur in Angiospermae? *Phytochemistry* **11**: 1827-1828.

Hanahan, D. (1983) Studies on transformation of *Escherichia coli* with plasmids. J. Mol. Biol. 166: 557-580.

Hare, P. D., Cress, W. A. and Van Staden, J., 1998, Dissecting the roles of osmolyte accumulation during stress, *Plant, Cell and Environ.* **21**: 535-553.

Hayashi, H., Mustardy, L., Deshium, P., Ida, M. and Murata, N., 1997, Transformation of *Arabidopsis thaliana* with the *codA* gene for choline oxidase – accumulation of glycine betaine and enhanced tolerance to salt and cold stress, *Plant J.* **12**: 133-142.

Hoekstra, F. A., Wolkers, W. F., Buitink, J., Golovina, E. A., Crowe, J. H. and Crowe, L. M., 1997, Membrane stabilization in the dry state, *Comp. Biochem. Physiol.* **117**: 335-341.

Holmström, K-O., Mäntylä, E., Welin, B., Mandal, A., Palva, E. T., Tunnela, O. E. and Londesborough, J., 1996, Drought tolerance in tobacco, *Nature* **379**: 683-684.

Kishor, P. B. K., Hong, Z., Miao, G. H., Hu, C. A. A. and Verma, D. P. S., 1995, Overexpression pf DELTA-1-pyrroline-5-carboxylate synthetase increases proline production and confers osmotolerance in transgenic plants, *Plant Physiol.* **108**:1387-1394.

Lilius, G., Holmberg, N. and Bülow, L. (1996) Enhanced NaCl stress tolerance in transgenic tobacco expressing bacterial choline dehydrogenase. Biotechnol. 14: 177-180.

Londesborough, J. and Vuorio, O., 1991, Trehalose-6-phpsphate synthase/phosphatase complex from baker´s yeast: purification of a proteolytically activated form, *J. Gen. Microbiol.* **137**: 323-330.

McCue, K. F. and Hanson, A. D., 1990, Drought and salt tolerance: towards understanding and application, *Trends Biotechnol.* **8**: 358-362.

Müller, J., Aeschbacher, R.A., Wingler, A., Boller, T. and Wiemken, A., 2001, Trehalose and trehalase in Arabidopsis, *Plant Physol.* **125**: 1086-1093.

Müller, J., Boller, T. and Wiemken, A., 1995a, Effects of validamycin A, a potent trehalase inhibitor, and phytohormones on trehalose metabolism in roots and root nodules of soybean and cowpea, *Planta* 197: 362-368.

Müller, J., Boller, T. and Wiemken, A., 1995b, Trehalose and trehalase in plants: recent developments, *Plant Science* 112: 1-9.

Nuotio, S., Heino, P. and Palva, E. T., 2001, Signal traansduction under low-temperature stress, in: *Crop Responses and Adaptations in Temperature Stress*, A. S. Basra, ed., Food Products Press, Binghamton, New York, pp. 151-176.

Palta, J. and Weiss, L.S., 1993, Ice formation and freezing injury: an overview on the survival mechanisms and molecular aspects of injury and cold acclimation in herbaceous plants, in: *Advances in plant cold hardiness*, P. H. Li and L. Christerrson, eds., CRC Press, Boca Raton, Florida, USA, pp. 143-176.

Pilon-Smits, E. A. H., Ebskamp, M. J. M, Paul, M. J., Jeuken, M. J. W., Weisbeek, P.J. and Smeekens, S. C. M., 1995, Improved performance of transgenic fructan-accumulation tobacco under drought stress, *Plant Physiol.* **107**: 125-130.

Pilon-Smits, E. A. H., Terry, N., Sears, T., Kim, H., Zayed, A., Hwang, S., Van Dun, K., Voogd, E., Verwoerd, T. C., Krutwagen, R. W. H. H. and Goddijn, O. J. M., 1998, Trehalose-producing transgenic tobacco plants show improved growth performance under drought stress, *J. Plant Physiol.* **152**: 525-532.

Romero, C., Bellés, J. M., Vayá, J. L., Serrano, R. and Culláñez-Macià, F. A., 1997, Expression of the yeast trehalose-6-phosphate synthase gene in transgenic tobacco plants: pleiotropic phenotypes include drought tolerance, *Planta* **201**: 293-297.

Roser, B. and Colaco, C., 1993, A sweeter way to fresher food, *New Scientist* **15**: 25-28.

Sakamoto, A., Alia and Murata, N., 1998, Metabolic engineering of rice leading to biosynthesis of glycinebetaine and tolerance to salt and cold, *Plant. Mol. Biol.* **38**: 1011-1019.

Sakamoto, A., Valverde, R., Alia, Chen, T.H.H. and Murata, N., 2000, Transformation of *Arabidopsis* with the *codA* gene for choline oxidase enhances freezing tolerance of plants, *Plant J.* **22**:449-453.

Serrano, R., Culiañz-Maciá, F.A. and Morena, V., 1999, Genetic engineering of salt and drought tolerance with yeast regulatory genes, *Sci. Hortic.* **78**: 261-269.

Singer, M. A. and Lindquist, S., 1998, Thermotolerance in *Saccharomyces cerevisiae*: the Yin and Yang of trehalose, *Trends Biotechnol.* **16**: 460-468.

Smirnoff, N., 1998, Plant resistance to environmental stress, *Curr. Opin. Biotechnol.* **9**: 214-219.

Strøm, A. R. and Kaasen, I., 1993, Trehalose metabolism in *Escherichia coli*: stress protection and stress regulation of gene expression, *Mol. Microbiol.* **8**: 205-210.

Sukumaran, N.P. and Weiser, C.J., 1972, An excised leaflet test for evaluating potato frost tolerance, *Hort. Sci.* 7, 467-468.

Tarczynski, M. C., Jensen, R. G. and Bohnert, H. J., 1992, Expression of a bacterial *mtlD* gene in transgenic tobacco leads to production and accumulation of mannitol, *Proc. Natl. Acad. Sci. USA* **89**: 2600-2604.

Tarczynski, M. C., Jensen, R. G. and Bohnert, H. J., 1993, Stress protection of transgenic tobacco by production of the osmolyte mannitol, *Science* **259**: 508-510.

Thomashow, M.F., 1999, Plant cold acclimation: Freezing tolerance genes and regulatory mechanisms, *Annu. Rev. Plant Physiol. Plant Mol. Biol.* **50**, 571-599.

Vogel, G., Aeschbacher, R. A., Müller, J., Boller, T. and Wiemken, A., 1998, Trehalose-6-phophate phosphatases from *Arabidopsis thaliana* – identification by functional complementation of yeast TPS2 mutant, *Plant J.* **13**: 673-683.

Vuorio, E., Kalkkinen, N. and Londesborought, J., 1993, Cloning of two related genes encoding the 56-kDa and 123-kDa subunits of trehalose synthase from yeast *Saccharomyces cerevisiae*, *Eur. J. Biochem.* **216**: 849-861.

Yancey, P. H., Clark, M. E., Hand, S. C., Bowlus, R. D. and Somero, G. N., 1982, Living with water stress: evolution of osmolyte systems, *Science* **217**:1214-1222.

19

ENHANCING COLD TOLERANCE IN PLANTS BY GENETIC ENGINEERING OF GLYCINEBETAINE SYNTHESIS

Raweewan Yuwansiri, Eung-Jun Park, Zoran Jeknić, and Tony H.H. Chen*

1. INTRODUCTION

Low-temperature stress, including freezing and chilling, is one of the primary factors that limits crop production around the world. A severe winter or unseasonable cold spell in a major agricultural area can reduce yield, delay harvest, lower the quality of the products, and/or cause crop failure. Low-temperature stress is not limited to particular geographic regions; producers throughout the world have suffered significant economic losses during the past decade. These losses can significantly influence the economic well-being of rural communities, and affect product availability and consumer prices. Stress can also have a great impact on the stability of world food supplies, and may subsequently influence commodity prices and the international trade balance.

Often, the distinction between a good yield and crop failure is determined by only a 2 to 3 °C variation in cold tolerance, or by a few days difference in the stage of crop development. Such subtle differences are highly amenable to research solutions. Furthermore, crop production is generally proportional to the length of the growing season, and can be considerably increased if planting is done earlier in the spring and harvesting later in the fall. Because of the low-temperature constraint, however, farmers are not always able to implement these unconventional practices because of the lack of hardier cultivars. Thus, there are ample economic incentives for introducing more cold-tolerant crop plants.

Developing germplasm with increased cold tolerance via traditional breeding techniques would provide a long-term solution to the problem of low-temperature stress.

* Raweewan Yuwansiri, Eung-Jun Park, Zoran Jeknić, and Tony H.H. Chen, Department of Horticulture, ALS 4017, Oregon State University, Corvallis, OR 97331.

Plant Cold Hardiness, edited by Li and Palva
Kluwer Academic/Plenum Publishers, 2002

However, because cold tolerance is controlled by many genes and is inherited as a quantitative trait, traditional breeding approaches have progressed very slowly. Considerable attention has been paid recently to the possible genetic engineering of plants tolerant to low-temperature stress. Various genes have been used to produce transgenic plants with increased cold tolerance (Table 1). In this paper, we will focus our discussion on enhancing cold tolerance through genetic engineering of glycinebetaine (GB) biosynthesis.

Table 1. Summary of engineered cold tolerance in plants

Transgenic species	Gene / Sources	Tolerance	Reference
	Regulatory genes		
Arabidopsis	*CBF1* (CRT/DRE binding factor) from *Arabidopsis*	Freezing	Jaglo-Ottosen et al., 1998
Arabidopsis	*CBF3* (CRT/DRE binding factor) from *Arabidopsis* (same as *DREB1A*)	Freezing	Gilmour et al., 2000
Arabidopsis	*DREB1A* (dehydration-responsive element binding protein) from *Arabidopsis*	Freezing	Liu et al., 1998
Arabidopsis	Antisense phosphatase 2C (AtPP2CA) from *Arabidopsis*	Freezing	Tahtiharju and Palva, 2001
Arabidopsis	*SCOF-1* (soybean cold-inducible factor-1) from soybean	Freezing	Kim et al., 2001
Arabidopsis	*ABI3* (seed-specific transcriptional activator) from *Arabidopsis*	Freezing	Tamminen et al., 2001
Rice	*OsCDPK7* (calcium-dependent protein kinase) from rice	Chilling	Saijo et al., 2000
Tobacco	*AtDBF2* (protein kinase) from *Arabidopsis*	Chilling	Lee et al., 1999
Tobacco	*ANP1* (mitogen-activated protein kinase kinase kinase) from *Arabidopsis*	Freezing	Kovtun et al., 2000
Tobacco	*SCOF-1* (soybean cold-inducible factor-1) from soybean	Chilling	Kim et al., 2001
	Cold-regulated (COR) genes		
Arabidopsis	*COR15a* from *Arabidopsis*	Freezing	Artus et al., 1996
Tobacco	*CAP160* (cold–acclimation protein) from spinach	Freezing	Kaye et al., 1998
Tobacco	*hiC6* (hardening-induced Chlorella) from *Chlorella vulgaris* C-27	Freezing	Honjoh et al., 2001
	Antifreeze protein genes		
Potato	Synthetic antifreeze protein (based on Type I AFP of winter flounder)	Freezing	Wallis et al., 1997
	Oxidative stress-related genes		
Alfalfa	*Mn-SOD* (Mn-Superoxide dismutase) from tobacco	Freezing	McKersie et al., 1999
Maize	*Mn-SOD* (Mn-Superoxide dismutase) from tobacco	Chilling	Breusegem et al., 1999
Tobacco	*Cu/Zn-SOD (Cu/Zn superoxide dismutase) from pea*	Chilling	Gupta et al., 1993

Table 1. (continued)

Transgenic species	Gene / Sources	Tolerance	Reference
	Lipid modifying genes		
Tobacco	Glycerol-3-phosphate acyltransferase from *Arabidopsis*	Chilling	Murata et al., 1992a
Tobacco	FAD7 (chloroplast ω-3 fatty acid desaturase) from *Arabidopsis*	Chilling	Kodama et al., 1994
Tobacco	Δ9 desaturase from cyanobacterium, *Anacystis nidulans*	Chilling	Ishizaki-Nishizawa et al., 1996
	Osmoprotectant synthesis genes		
Arabidopsis	Antisense of proline dehydrogenase from *Arabidopsis*	Freezing	Nanjo et al., 1999
Arabidopsis	COD (choline oxidase) from *Arthrobacter globiformis*	Chilling Chilling Freezing	Alia et al., 1998 Hayashi et al., 1997 Sakamoto et al., 2000
Arabidopsis	COX (choline oxidase) from *Arthrobacter pascens*	Freezing	Huang et al., 2000
Tobacco	CDH (choline dehydrogenase) from *E. coli*	Chilling	Holmström et al., 2000
Rice	COD (choline oxidase) from *Arthrobacter globiformis*	Chilling	Sakamoto et al., 1998

2. GLYCINEBETAINE BIOSYNTHESIS

Plants have evolved diverse acclimation and avoidance strategies to cope with adverse growing conditions. One common cellular mechanism for adapting to unfavorable environments is the accumulation of osmoprotectants, i.e., osmotically active, low-molecular-weight, nontoxic compounds, such as GB (Rhodes and Hanson, 1993). GB is widely distributed in higher plants, animals, and a large variety of microorganisms (Robinson and Jones, 1986; Rhodes and Hanson, 1993), and is sometimes accumulated in response to salt or cold stresses (Wyn Jones and Story, 1981). Whereas several taxonomically distant plants are accumulators of GB, others, such as *Arabidopsis*, rice (*Oryza sativa*), and tobacco (*Nicotiana tabacum*) are considered to be 'non-GB accumulators' (Wyn Jones and Story, 1981).

GB most likely protects cells from salt stress by maintaining an osmotic balance with the environment (Robinson and Jones, 1986), and by stabilizing the quaternary structure of complex proteins (Bernard et al., 1988; Incharoenskdi et al., 1986; Papageorgiou and Murata, 1995; Winzor et al., 1992). In photosynthetic systems, GB upholds the oxygen-evolving Photo system II (PSII) complex (Murata et al., 1992b; Papageorgiou and Murata, 1995; Papageorgiou et al., 1991) as well as Rubisco (Incharoenskdi et al., 1986) at high concentrations of NaCl. Because low-temperature stress is considered a form of dehydration (Levitt, 1980), compatible osmolytes may be regarded as water substitutes. These osmolytes maintain a hydrophilic environment for all types of cellular macromolecules, such as proteins and membranes, thereby reducing the sensitivity of molecular structures to a lack of water. Thus, compatible osmolytes are likely a "general" cold tolerance mechanism.

GB can be synthesized via two distinct pathways: from choline, by choline oxidation, or from glycine, through glycine *N*-methylation. For almost all biological systems, including most animals, plants, and microorganisms, GB synthesis is accomplished by converting choline to GB via the unstable intermediate, betaine aldehyde. Recently, however, a novel pathway for GB synthesis from glycine was identified in two extremely halophilic microorganisms, *Actinopolyspora halophila* and *Ectothiorhodospira halochloris* (Nyyssölä et al., 2000)

2.1. In *Escherichia coli*

The biosynthesis of GB in *E. coli* is regulated by a cluster of genes called '*bet* operon'. The *bet* operon consists of four genes: *betI,* a regulatory gene, and three structural genes, *betA*, *betB*, and *betT* (Andresen et al., 1988; Lamark et al., 1991, 1996). Genes *betA*, *betB*, and *betT* encode choline dehydrogenase [CDH; EC 1.1.99.1], betaine aldehyde dehydrogenase [BADH; EC 1.2.1.8], and a high-affinity choline transporter, respectively. The entire *bet* gene cluster has been sequenced (Lamark et al., 1991, 1996). DNA sequencing data has revealed that *betI*, *betB*, and *betA* are transcribed as a single transcript, whereas *betT* is transcribed as a separate transcript from the opposite orientation. The BetI protein (21.8 kDa) is a transcriptional repressor that negatively regulates *bet* gene expression in response to choline by binding at or near the promoter region (Rokenes et al., 1996). The *betT* encodes a 75.8-kDa protein, which is a proton-motive force-driven, high-affinity transport system for choline. At low external concentrations of choline, the main function of BetT is to uptake choline into *E. coli* cells (Lamark et al., 1996).

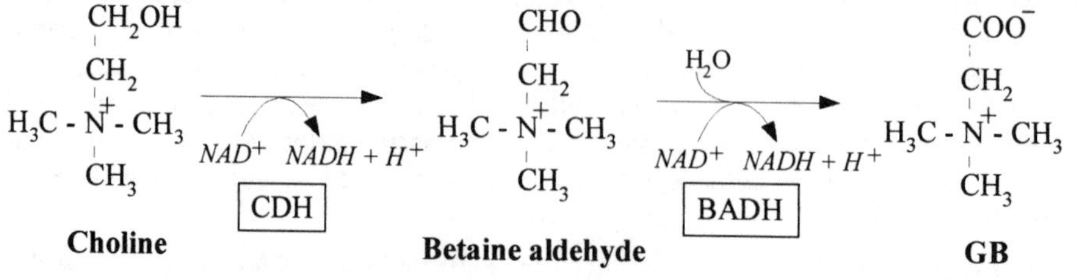

CH_2OH–CH_2–$N^+(CH_3)_3$ (Choline) $\xrightarrow[\text{CDH}]{NAD^+ \rightarrow NADH + H^+}$ CHO–CH_2–$N^+(CH_3)_3$ (Betaine aldehyde) $\xrightarrow[\text{BADH}]{H_2O;\ NAD^+ \rightarrow NADH + H^+}$ COO^-–CH_2–$N^+(CH_3)_3$ (GB)

Figure 1. Biosynthesis of GB in *E. coli* catalyzed by choline dehydrogenase (CDH) and betaine aldehyde dehydrogenase (BADH)

The *betA* gene encodes CDH, a 61.9-kDa flavoprotein containing an *N*-terminal *FAD*-binding region. It is a membrane-bound oxygen-dependent enzyme (Lamark et al., 1991). CDH also catalyzes the oxidation of betaine aldehyde to GB *in vitro*, but it is not certain whether this reaction occurs *in vivo*, since CDH has a much lower affinity for betaine aldehyde than BADH (Lamark et al., 1991). The *betB* gene encodes BADH, a 52.8-kDa soluble enzyme with a high affinity for betaine aldehyde and a strong preference for NAD^+ as an electron acceptor (Boyd et al., 1991). GB synthesis from choline is shown in Figure 1.

2.2. In Higher Plants

In higher plants, GB is synthesized from choline in a two-step oxidation reaction catalyzed by choline monooxygenase (CMO), a ferredoxin (Fd)-dependent Rieske-type [2Fe-2S] protein, and BADH, a soluble NAD^+-dependent enzyme (Rhodes and Hanson, 1993) (Figure 2). CMO, a soluble enzyme of 42.9 kDa, is localized in the chloroplast stroma, and catalyzes the first step of GB biosynthesis from choline to betaine aldehyde. The second step of the oxidation reaction from betaine aldehyde to GB is catalyzed by BADH. The majority of BADH activity is found in the chloroplast stroma, with minor activity detected as a cytosolic isozyme (Weretilnyk and Hanson, 1988). The stroma enzyme is a homodimer consisting of 60 to 63 kDa subunits encoded by a nuclear gene with an atypically short transit peptide (7 to 8 residues) that is able to target the BADH to chloroplasts in the leaves of transgenic plants (Rathinasabapathi et al., 1994). However, the BADH protein has also been found in peroxisomes of transgenic tobacco plants transformed with a barley BADH cDNA (Nakamura et al., 1997).

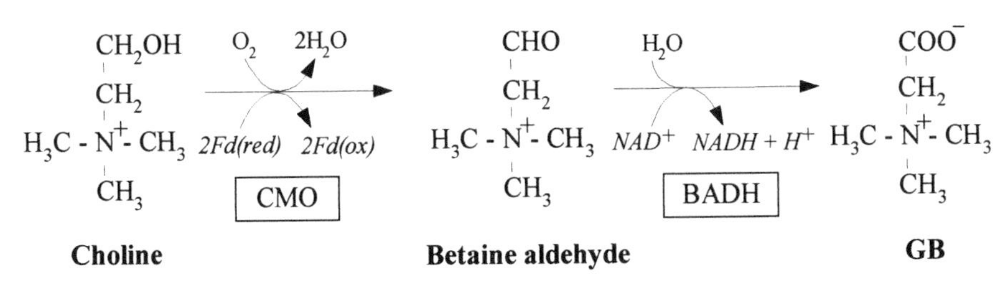

Figure 2. Biosynthesis of GB in plants catalyzed by choline monooxygenase (CMO) and betaine aldehyde dehydrogenase (BADH).

2.3. In *Arthrobacter spp.*

In the gram-positive soil bacteria *Arthrobacter spp.*, GB is synthesized from choline by a single enzyme, choline oxidase (E.C. 1.1.3.17) (Figure 3). Choline oxidase is a soluble enzyme that does not require cofactors, and it uses O_2 as the primary electron acceptor. This enzyme is highly specific for choline and betaine aldehyde (relative reaction velocities: choline, 100%; betaine aldehyde, 46%) (Ikuta et al., 1977). The oxidation of one mole of choline produces two moles of H_2O_2.

The cDNAs of the choline oxidase gene from two species of *Arthrobacter*, *A. pascens* and *A. globiformis*, have been independently isolated. The *cox* gene from *A. pascens* encodes a protein of approximately 66 kDa (Rozwadowski et al. (1991). This gene was first introduced into an *E. coli* mutant defective in GB biosynthesis. The transformed *E. coli* was then able to synthesize significant amounts of GB. The choline oxidase gene, *codA*, was cloned from *A. globiformis* by Deshnium et al. (1995).

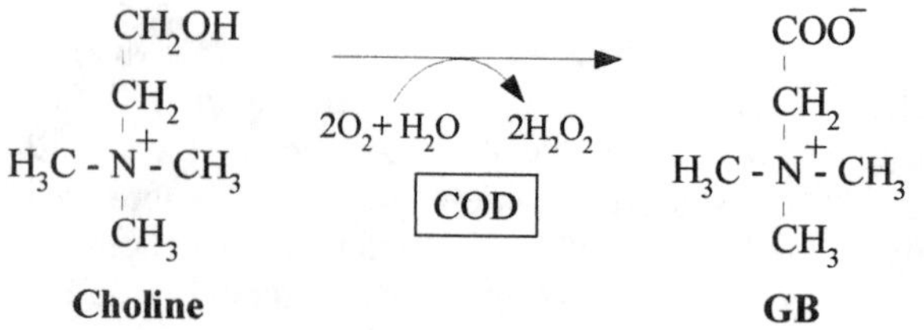

Figure 3. Biosynthesis of GB in *Arthrobacter globiformis* catalyzed by choline oxidase (COD).

2.4. In Halophilic Microorganisms

In the extreme halophytic phototrophic bacteria, *Actinopolyspora halophila* and *Ectothiorhodospira halochloris*, GB is synthesized from glycine (Nyyssölä et al., 2000). GB synthesis here is by a sequential three-step *N*-methylation reaction. First, glycine is methylated to sarcosine and then to dimethylglycine, and finally GB using S-adenosylmethionine (*SAM*) as the methyl group donor. These three *N*-methylations are catalyzed by two enzymes, glycine sarcosine methyltransferase (GSMT) and sarcosine dimethylglycine methyltransferase (SDMT), with partially overlapping substrate specificity (Nyyssölä et al., 2001). The glycine *N*-methylation pathway for GB biosynthesis is summarized in Figure 4.

Recently, Nyyssölä and Leisola (2001) found that *A. halophila* can also synthesize GB from choline. *A. halophila* is able to take up choline from medium and oxidize it to betaine aldehyde in a reaction in which H_2O_2 generation and oxygen consumption are coupled. Betaine aldehyde is then oxidized further to GB in a *NAD(P)*$^+$-dependant reaction. However, the genes involved in this pathway have not been identified.

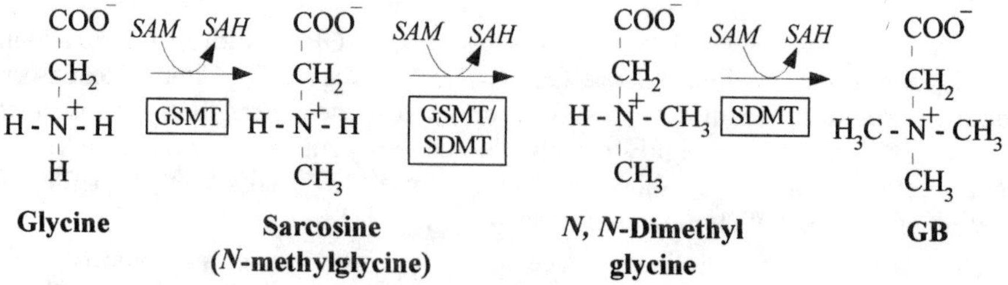

Figure 4. Biosynthesis of GB in *Actinopolyspora halophila* and *Ectothiorhodospira halochloris* catalyzed by glycine sarcosine methyltransferase (GSMT) and sarcosine dimethylglycine methyltransferase (SDMT).

2.5. In Other Species

Although other organisms can synthesize GB, their biosynthetic pathways have not yet been well characterized, nor the genes involved been cloned.

In *Bacillus subtilis*, GB is likely synthesized from choline by the two-step oxidation pathway. Two genes (*gbsA* and *gbsB*) encoding the *Bacillus subtilis* choline-betaine synthesis enzymes have been identified by functional complementation (Boch et al., 1996). The *gbsA* gene product shares strong sequence identity with aldehyde dehydrogenases, and the *gbsB* gene product exhibits significant similarity with the *NAD/NADP*-dependent alcohol dehydrogenases (Boch et al., 1996). However, the deduced gbsB protein shares no homology with the membrane-bound *FAD*-containing CDH (BetA) of *E. coli*, or with the COX of *Arthrobacter pascens*. This implies that *Bacillus subtilis* may possess a pair of novel enzymes for GB biosynthesis via choline oxidation.

Finally, in mammals, GB plays an important role as a compatible solute, particularly in the kidneys (Haubrich and Gerber, 1981). The pathway of GB synthesis resembles that in *E. coli*, and is catalyzed by a CDH localized in the mitochondrial inner membrane (Miller et al., 1996) and by a soluble BADH (Chern and Pietruszko, 1999). Chern and Pietruszko (1999) also found about 5% of the BADH activity in mitochondria and suggested BADH mitochondrial co-localization with CDH.

3. GLYCINEBETAINE AND COLD TOLERANCE

It is now generally accepted that GB is a critical determinant of cold tolerance. First, GB accumulation is induced during cold acclimation, with its level being proportional to the degree of cold tolerance. In barley (*Hordeum vulgare*), the amount of GB that accumulates in leaves increases in response to cold acclimation, and is closely correlated with the development of freezing tolerance (Kishitani et al., 1994). In winter wheat (*Triticum aestivum*) seedlings, endogenous GB contents increase during cold acclimation (Allard et al., 1998). By the end of that period, a hardier genotype may have accumulated significantly higher levels of GB (21.3 $\mu mol \cdot g^{-1}$ FW) than do less hardy genotypes (15.3 $\mu mol \cdot g^{-1}$ FW). In strawberry plants (F*ragaria X ananssa*), the level of GB increases nearly 2-fold after 4 weeks of cold acclimation (Rajashekar et al., 1999). In addition, exogenous applications of ABA (100 μM) to strawberries can trigger GB accumulation and increase cold tolerance (Rajashekar et al., 1999).

Second, exogenously applied GB increases cold tolerance. For example, the application of 250 mM GB can increase freezing tolerance in both spring wheat and winter wheat seedlings (Allard et al., 1998). Likewise, in alfalfa, Zhao et al. (1992) have reported that spraying alfalfa (*Medicago sativa*) seedlings with 0.2 M GB reduced the leakage of ions from shoot tissues when plants were subjected to freezing temperatures; this reduction was clearly differentiated at –6 °C. Furthermore, they found that 67% of the GB-treated plants could withstand exposure to –6 °C, whereas none of the control plants survived. In *Arabidopsis,* exogenous GB applications increased chilling tolerance during seed imbibition (Alia et al., 1998). Sakamoto et al. (2000) also found that exogenously applied GB brought about a significant increase in freezing tolerance of *Arabidopsis* plants. Furthermore, Chen et al. (2000) have reported that exogenous GB significantly increased chilling tolerance in suspension-cultured cells and seedlings of maize (*Zea mays* L. cv 'Black Mexican Sweet'), a genotype that never accumulates GB, even under salt stress.

Third, *in-vitro* studies have shown that, under high salt concentrations or extreme temperatures, GB is an effective compatible solute for stabilizing the quaternary structure of enzymes and complex proteins as well as the highly ordered state of membranes (Papageorgiou and Murata, 1995).

4. ENGINEERED GLYCINEBETAINE BIOSYNTHESIS

Because endogenous GB levels are generally correlated with cold tolerance and exogenous GB applications are effective in increasing that tolerance, researchers speculate that introducing the GB biosynthetic pathway into non-GB accumulators via genetic engineering may improve cold tolerance of transgenic plants.

Many genes involved in GB biosynthesis have been cloned, including CMO and BADH from higher plants (Rathinasabapathi et al., 1997); *betA* and *betB* from *E. coli* (Landfald and Strom, 1986); *codA* from *Arthrobacter globiformis* (Deshnium et al., 1995); *cox* from *A. pascens* (Rozwadowski et al., 1991); and GSMT and SDMT from *A. halophila* and *E. halochloris* (Nyyssölä et al., 2001). These genes, except GSMT and SDMT, have been used for genetic engineering of GB biosynthesis in a number of species. The resulting transgenic plants have accumulated various amounts of GB, and have shown enhanced tolerance to different abiotic stresses (Sakamoto and Murata, 2000, 2001a, b). Advances in metabolic engineering of GB for improved plant stress tolerance have been reviewed recently by Sakamoto and Murata (2000, 2001a, b).

4.1. Transgenic *Arabidopsis*

Hayashi et al. (1997) transformed *Arabidopsis* (ecotype Wassilewskija) with the *codA* gene under control of the CaMV35S promoter. This gene was preceded by a sequence that encoded the transit peptide of a small subunit of Rubisco from tobacco to direct COD into the chloroplasts. Maximum levels of accumulated GB were 1.2 $\mu mol \cdot g^{-1}$ fresh weight (FW) and 18.0 $\mu mol \cdot g^{-1}$ dry weight (DW) in the shoots and mature seeds, respectively. Transgenic plants expressing COD also showed enhanced resilience in cold temperatures. Tolerance to chilling stress was significantly increased during imbibition and germination of transgenic seeds, as indicated by higher frequencies and rates of germination compared with wild-type (WT) seeds (Alia et al., 1998). Furthermore, the transgenic seedlings were able to grow better at lower temperatures (10 to 15 °C), and mature plants were less susceptible to chilling-induced photoinhibition (Alia et al., 1998; Hayashi et al., 1997). GB accumulation also dramatically improved the survival of mature plants at freezing temperatures (Sakamoto et al., 2000). In these studies, however, several cold-regulated (*cor*) genes (*cor6.6*, *cor15a*, *cor47*, and *cor78*), which had been implicated in the development of freezing tolerance (Jaglo-Ottosen et al., 1998), apparently were not responsible for this enhanced tolerance, as indicated by a lack of any significant differences in expression levels of these genes between the WT and transgenic plants (Sakamoto et al., 2000). Transformation with *codA* also improved a plant's tolerance to high light intensity in combination with the chilling treatment (Alia et al., 1999). The presence of GB in the chloroplasts seems likely to have only enhanced recovery of the damaged PSII complex, rather than directly protecting it against photo-induced damage (Alia et al., 1999).

Huang et al. (2000) transformed *Arabidopsis* (ecotype RLD) with the *cox* gene driven by a double CaMV35S promoter with an AMV enhancer. Those plants accumulated GB up to a level of 18.6 $\mu mol \cdot g^{-1}$ DW. Transgenic plants were slightly more tolerant to freezing stress than were WT plants. The amount of GB increased significantly (over 600 $\mu mol \cdot g^{-1}$ DW) when choline (10 mM) was supplied exogenously. This suggests that the availability of substrate, i.e., choline, might be a limiting factor in GB synthesis. Transgenic plants appeared morphologically normal, which indicates that neither the elevated level of GB nor that of the by-product H_2O_2 had any adverse effect.

We have generated transgenic *Arabidopsis* plants in our lab that express COD in the cytosol, in chloroplasts, and in both the cytosol and the chloroplasts simultaneously (Sakamoto et al., 2001). However, the levels of GB accumulation in our transgenic lines are not much higher than those reported by Hayashi et al. (1997). The plants expressing COD in both cytosol and chloroplasts have accumulated the highest amount (1.54 $\mu mol \cdot g^{-1}$ FW). Transgenic *Arabidopsis* that expresses COD in only the cytosol or the chloroplasts accumulated GB at 0.80 or 0.54 $\mu mol \cdot g^{-1}$ FW, respectively. The transgenic lines with higher levels of GB accumulation generally are more tolerant of freezing than those accumulating lower amounts, regardless of the localization of COD.

4.2. Transgenic Rice

Sakamoto et al. (1998) transformed the japonica variety of rice with two chimeric constructs in which the *codA* gene was under control of the CaMV35S promoter. Both constructs contained an intron from a rice gene inserted between the CaMV35S promoter and the *N*-terminus of the *codA* gene to enhance gene expression. One of the constructs encoded COD with a signal for targeting into chloroplasts, while the other lacked such a signal sequence, thereby allowing COD to remain in the cytosol. This allowed for direct comparison of the effects of subcellular localization of COD on both GB accumulation and stress tolerance. The level of GB in the leaves of transgenic rice was about 1 $\mu mol \cdot g^{-1}$ FW with the chloroplast-targeting construct, but was 3 to 5 $\mu mol \cdot g^{-1}$ FW with the cytosol-targeting one. Although targeting COD into the chloroplasts resulted in lower levels of GB, the PSII complex was better protected under low-temperature conditions. When those plants were incubated at 5 °C under 1300 $\mu mol \cdot m^{-2} \cdot s^{-1}$ lights, activity of the PSII (expressed as the chlorophyll fluorescence ratio, F_v/F_m) was higher in the chloroplast-targeted plants than in either the WT or the cytosol-targeted plants.

Takabe et al. (1998) transformed japonica rice with a modified *E. coli betA* gene encoding choline dehydrogenase (CDH). A series of modifications was made to the *betA* gene to maximize CDH activity in transgenic plants. First, a signal sequence was placed for targeting the gene product to the mitochondria. Second, the coding region sequence was modified to eliminate possible polyadenylation signals, palindromic structures, and those codons rarely used in rice. Transgenic plants that produced CDH accumulated GB at levels as high as 5 and 1.2 μmol·g^{-1} FW in the leaves and roots, respectively. By contrast, neither the native gene nor the modified gene without the targeting signal led to detectable levels of accumulation in the transgenic plants. This indicates that either these genes were not efficiently expressed or, if they were expressed, appropriate cofactors for the catalytic reaction were not available in the cytosol. Transgenic plants that accumulated GB survived better than control plants during recovery from salt and drought stress. Furthermore, measurements of PSII activity revealed that the photosynthetic machinery was more tolerant in transgenic plants than in WT plants during salt stress.

4.3. Transgenic Tobacco

Holmström et al. (1994, 2000) transformed tobacco with either the *E. coli betA* gene or the *betB* gene. Transgenic plants expressing *betA* accumulated GB (35 nmol·g^{-1} FW), and exhibited improved tolerance to photoinhibition under low temperatures. In contrast, the *betB*-transgenic plants produced <1 nmol·g^{-1} FW of GB because BADH, the product of *betB*, could convert only betaine aldehyde but not choline into GB. When those two transgenic lines were crossed, the F1 hybrids expressed both *betA* and *betB* genes. Those plants were able to accumulate much higher levels of GB (about 66 nmol·g^{-1} FW) compared with either the WT or the *betA*-transgenic plants. They also showed improved tolerance to photoinhibition, as seen by a slower reduction in PSII efficiency (F_v/F_m) under a low temperature (4 °C).

4.4. Transgenic Tomato

Tomato (*Lycopersicon esculentum*) is a chilling-sensitive species that normally does not accumulate GB. We transformed the cultivar 'Moneymaker' with the *codA* gene via *Agrobacterium*-mediated transformation. The *codA* coding region was preceded by the transit peptide sequence of a small subunit of the tobacco Rubisco gene to direct COD into the chloroplasts. Primary transgenic plants (T_0) that expressed the *codA* gene were identified by western blot analysis and subsequently self-pollinated to produce homozygous T_2 and T_3 lines. We quantified the level of GB in their leaves using an HPLC method described by Naidu (1998). The transgenic leaves accumulated up to 0.65 μmol·g^{-1} FW, whereas the non-transformed controls contained no detectable levels of GB. However, GB concentrations in the leaves of transgenic tomato plants were lower than those reported in *Arabidopsis* (0.8 to 1.2 μmol·g^{-1} FW; Hayashi et al., 1997), *Brassica juncea* (0.64 to 0.82 μmol·g^{-1} FW; Prasad et al., 2000), or rice (1.1 to 5.3 μmol·g^{-1} FW; Sakamoto et al., 1998).

In further study, we placed the seeds of both the WT and two transgenic homozygous lines on a germination medium and incubated them for 2 weeks at 2 °C under 16 h light/ 8 h dark. This was followed by germination at 25 °C for 1 week. Under those conditions, only about 50% of the WT seeds were able to germinate compared with a germination rate of 95% for untreated seeds. In contrast to the performance of the WT seeds, 85 to

95% of those from the two transgenic lines germinated. In those two lines, the rates seemed to be related to GB levels: 95% in the 46-5 line with a GB content of 0.39 $\mu mol \cdot g^{-1}$ FW, and 85% in the 522-2 line that contained 0.33 $\mu mol \cdot g^{-1}$ FW GB.

Seedlings from both the WT and the transgenic homozygous lines were then incubated at 2 °C under either 16 h light/8 h dark or 24 h dark. After 2 weeks of chilling treatment, the plants were transferred to a growth chamber maintained at 25 °C with 16 h light/8 h dark for 1 week. Seedling growth for both WT and transgenic plants was completely inhibited at 2 °C and with a 16 h-photoperiod. However, when those plants were transferred to a 25 °C growth chamber, the transgenic seedlings immediately began to grow, whereas the WT seedlings became severely wilted and necrotic, with most of them soon senescing. During the chilling treatment in the dark, seedlings from the two transgenic lines grew slowly while no growth was observed in the WT seedlings. When the latter were transferred to conditions of 25 °C with a 16 h-photoperiod, development of the true leaves was greatly delayed. On the contrary, seedlings of the transgenic lines grew normally and developed new leaves after a similar transfer. These results indicate that the transgenic seedlings were more tolerant to chilling stress than were the WT.

Greenhouse-grown, 7-week-old WT and transgenic plants were exposed to a low temperature (2 °C) for 5 d, then moved back to the greenhouse. The WT plants were wilted and chlorotic, with 50% dying soon after. Stem collapse in those plants, just below the apical meristem, was probably responsible for their high mortality rates. In contrast, plants from the two homozygous lines showed no obvious symptoms of chilling injury under the same treatment. They recovered within a few days and resumed normal growth. This indicates that GB synthesis in transgenic plants enhanced chilling tolerance during very early development as well as in the later stages of the tomato life cycle.

Taken together, our results demonstrate that genetic engineering of GB biosynthesis is an effective means of enhancing cold tolerance in tomatoes.

5. PROBLEMS WITH GLYCINEBETAINE ENGINEERING

As presented above, metabolically engineering GB biosynthesis into non-GB accumulators appears to be an effective means for improving stress tolerance. However, the amounts of GB accumulated in transgenic plants are generally low (<5 $\mu mol \cdot g^{-1}$ FW) compared with levels observed in natural GB-accumulators (4 to 40 $\mu mol \cdot g^{-1}$ FW; Rhodes and Hanson., 1993) when under stress. This has been observed in plants transformed with the *E. coli betA* gene, with or without modified codon usage (Takabe et al. 1998); or with plant CMO and BADH genes (Rathinasabapathi et al. 1997; Nuccio et al., 1998); the *cox* gene (Huang et al., 2000), or the *cod*A genes (Hayashi et al., 1997), regardless of the plant host species used (Sakamoto and Murata, 2001a, b). Therefore, the availability of substrate (choline and/or phosphocholine), rather than the kind and origin of the choline-oxidation enzymes, may be what limits GB accumulation (McNeil et al., 1999; Huang et al., 2000; Rathinasabapathi et al., 1997; Nuccio et al., 1998).

Choline has been reported as a limiting factor for GB biosynthesis in transgenic plants. For example, Nuccio et al. (1998) showed that the GB in CMO-transgenic tobacco increased by at least 30-fold when 5-mM choline or phosphocholine was supplied in the culture medium. Likewise, Huang et al. (2000) found that choline-fed *cox*-transgenic plants synthesized substantially more GB, thereby enhancing GB levels up to 613 $\mu mol \cdot g^{-1}$ DW in *Arabidopsis* (10 mM choline), 250 $\mu mol \cdot g^{-1}$ DW in *Brassica napus* (20 mM choline), and 80 $\mu mol \cdot g^{-1}$ DW in tobacco (10 mM choline). Therefore, the endogenous choline supply appears to be a major limiting factor for high levels of GB accumulation in transgenic plants.

In many GB-accumulating transgenic plants, GB supposedly is accumulated in the chloroplasts. Because choline is synthesized mainly in the cytosol, it must be transported across chloroplast membranes to reach the site of active enzyme(s) for GB synthesis. Using transgenic tobacco plants, Nuccio et al. (2000) targeted the CMO gene into chloroplasts and the *cox* gene into the cytosol. When 5 mM choline was supplied, the *cox*-transgenic plants accumulated four times more GB (4.7 ± 1.8 $\mu mol \cdot g^{-1}$ FW) than did the CMO-transgenic plants. In addition, the levels of choline and phosphocholine in the *cox* plants were lower than the amount of GB, whereas they were 3.5-fold greater than the GB level in the CMO plants. These results indicate that inefficient import of choline into chloroplasts also was probably a limiting factor for GB synthesis in the chloroplasts due to the lack of available substrate. As mentioned before, the *betT* gene from *E. coli* encodes a high-affinity transport system for choline (Lamark et al., 1996), so use of this gene may facilitate choline transport into the chloroplast. We are currently exploring this possibility by transforming chloroplast-targeted *codA*-transgenic *Arabidopsis* plants with a *betT* construct.

McNeil et al. (2001) attempted to increase the availability of choline in transgenic plants. First, they cloned a gene encoding a spinach phosphoethanolamine *N*-methyltransferase (PEAMT), the key enzyme catalyzing all three methylation steps required for converting phosphoethanolamine to phosphocholine. The PEAMT gene, under control of the CaMV35S promoter and an enhancer sequence, was then overexpressed in CMO/BADH-transgenic tobacco plants. This overexpression resulted in a 5-fold increase in phosphocholine levels as well as a 50-fold increase in free choline. The increased amounts of choline and phosphocholine did not adversely affect either the phosphatidylcholine content or normal plant growth. In addition, the expanded choline pools led to a 40-fold increase in GB synthesis (from 0.04 $\mu mol \cdot g^{-1}$ FW to 1.81 $\mu mol \cdot g^{-1}$ FW).

The other important consideration is subcellular localization of GB. Sakamoto et al. (1998) reported that, although transgenic rice plants that expressed COD in the chloroplasts had lower GB contents (1 $\mu mol \cdot g^{-1}$ FW), their level of tolerance to salt and cold stresses was higher than for plants with cytosolic COD (5 $\mu mol \cdot g^{-1}$ FW). These results indicate that subcellular compartmentalization of GB biosynthesis may play a critical role in the efficient enhancement of stress tolerance in genetically engineered plants.

Surprisingly, a single gene construct for GB biosynthesis that is driven by a constitutive promoter can be differentially expressed in different plant hosts. For example, Huang et al. (2000) transformed *Arabidopsis*, *Brassica napus*, and tobacco with a *cox* construct. Transgenic tobacco expressed the lowest level of COX activity, approximately 60-fold lower than in the transformed *Arabidopsis*. The *cox* transgene was so poorly expressed in tobacco, in fact, that the *cox* gene product was not detectable by northern and western blot analyses. Currently, there is no good explanation for the differential expression level in different species. Even among the three solanaceous species transformed with the *codA* gene in our lab, the level of GB accumulation was: tomato>potato>tobacco. This may limit wider application of engineered GB biosynthesis to just some particular crop species.

6. CONCLUSIONS AND PERSPECTIVES

Genetic engineering of GB biosynthesis appears to be an effective approach for conferring tolerance to both chilling and freezing stresses at various stages of growth and development in many crop species. Although the amounts of accumulated GB in transgenic plants are much lower than those detected in native accumulators, they still significantly affect cold tolerance. Among the GB-producing transgenic lines, higher accumulations generally are correlated with a greater level of protection against cold stress. Therefore, one challenge to researchers is to increase the level of GB in transgenic plants. Fortunately, the major limitations have already been identified (McNeil et al. 2001). By overexpressing an additional PEAMT gene in GB-transgenic plants, we may be able to further increase the level of GB and the degree of cold tolerance.

As mentioned previously, subcellular localization of GB accumulation is important. By increasing the level of transgene expression and the availability of the substrate in a specific subcellular compartment, it may be possible to generate plants that produce elevated levels of GB and, consequently, promote maximum cold protection. In addition, we can identify a combination of genes that, individually, are effective in enhancing stress tolerance with different underlying modes of action. These target genes would include those involved in the biosynthesis of other types of compatible solutes; stress-induced genes, such as *cor* and *lea* genes; and regulatory proteins, such as stress-inducible transcription factors (see Table 1). By expressing various genes in a single plant, we may be able to simulate the multiple gene traits of cold tolerance found in nature. This could then have a significant impact on increasing the cold tolerance of agriculturally important crops.

REFERENCES

Alia, Hiyashi, H., Chen, T. H. H., and Murata, N., 1998, Transformation with a gene for choline oxidase enhances the cold tolerance of *Arabidopsis* during germination and early growth, *Plant Cell Environ.* **21**:232-239.

Alia, Kondo, Y., Sakamoto, A., Nonaka, H., Hayashi, H., Saradhi, P. P., Chen, T. H. H., and Murata, N., 1999, Enhanced tolerance to light stress of transgenic *Arabidopsis* plants that express the *codA* gene for a bacterial choline oxidase, *Plant Mol. Biol.* **40**:279-288.

Allard, F., Houde, M., Krol, M., Ivanov, A., Huner, N. P. A., and Sarhan, F., 1998, Betaine improves freezing tolerance in wheat, *Plant Cell Physiol.* **39**:1194-1202.

Andresen, P. A., Kaasen, I., Styrvold, O. B., Boulnois, G., and Strom, A. R., 1988, Molecular cloning, physical mapping and expression of the *bet* genes governing the osmoregulatory choline-glycine betaine pathway of *Escherichia coli*, *J. Gen. Microbiol.* **134**:1737-1746.

Artus, N. N., Uemura, M., Steponkus, P. L., Gilmour, S. J., Lin, C., and Thomashow, M. F., 1996, Constitutive expression of the cold-regulated *Arabidopsis thaliana* COR15a gene affects both chloroplast and protoplast freezing tolerance, *Proc. Natl. Acad. Sci. U.S.A.* **93**:13404-13409.

Bernard, T., Ayache, M., and Rudulier, D. L., 1988, Restoration of growth and enzymatic activities of *Escherichia coli Lac*⁻ mutants by glycinebetaine. C. R. Acad. Sci. III. 307:99-104.

Boch, J., Kempf, B., Schmid, R., and Bremer, E., 1996, Synthesis of the osmoprotectant glycine betaine in *Bacillus subtilis*: characterization of the *gbsAB* genes, *J. Bacteriol.* **178**:5121-5129.

Boyd, L.A., Adam, L., Pelcher, L.E., McHughen, A., Hirji, R. and Selvaraj, G., 1991, Characterization of an *Escherichia coli* gene encoding betaine aldehyde dehydrogenase (BADH): structural similarity to mammalian ALDHs and a plant BADH, *Gene.* **103**:45-52.

Breusegem, F. V., Slooten, L., Stassart, J., Botterman, J., Moens, T., van Montagu, M., and Inze, D., 1999, Effects of overproduction of tobacco *MnSOD* in maize chloroplasts on foliar tolerance to cold and oxidative stress, *J. Exp. Bot.* **50**:71-78.

Chen, W. P., Li, P. H., and Chen, T. H. H., 2000, Glycinebetaine increases chilling tolerance and reduces chilling-induced lipid peroxidation in *Zea mays* L., *Plant Cell Environ.* **23**:609-618.

Chern, M. K., and Pietruszko, R., 1999, Evidence for mitochondrial localization of betaine aldehyde dehydrogenase in rat liver: purification, characterization, and comparison with human cytoplasmic E3 isozyme, *Biochem. Cell Biol.* **77**:179-187.

Deshnium, P., Los, D. A., Hayashi, H., Mustardy, L., and Murata, N., 1995, Transformation of *Synechococcus* with a gene for choline oxidase enhances tolerance to salt stress, *Plant Mol. Biol.* **29**:897-907.

Gilmour, S. J., Sebolt, A. M., Salazar, M. P., Everard, J. D., and Thomashow, M. F., 2000, Overexpression of the *Arabidopsis* CBF3 transcriptional activator mimics multiple biochemical changes associated with cold acclimation, *Plant Physiol.* **124**:1854-1865.

Gupta, A. S., Heinen, J. L., Holaday, A. S., Burke, J. J., and Allen, R. D., 1993, Increased resistance to oxidative stress in transgenic plants that overexpress chloroplastic Cu/Zn superoxide dismutase, *Proc. Natl. Acad. Sci. U.S.A.* **90**:1629-1633.

Haubrich, D. R., and Gerber, N. H., 1981, Choline dehydrogenase. Assay, properties and inhibitors, *Biochem. Pharmacol.* **30**:2993-3000.

Hayashi, H., Alia, Mustardy, L., Deshnium, P., Ida, M., and Murata, N., 1997, Transformation of *Arabidopsis thaliana* with the codA gene for choline oxidase; accumulation of glycinebetaine and enhanced tolerance to salt and cold stress, *Plant J.* **12**:133-142.

Holmström, K. O., Somersalo, S., Mandal, A., Palva, T. E., and Welin, B., 2000, Improved tolerance to salinity and low temperature in transgenic tobacco producing glycine betaine, *J. Exp. Bot.* **51**:177-185.

Honjoh, K., Shimizu, H., Nagaishi, N., Matsumoto, H., Suga, K., Miyamoto, T., Iio, M., and Hatano, S., 2001, Improvement of freezing tolerance in transgenic tobacco leaves by expressing the hiC6 gene, *Biosci. Biotechnol. Biochem.* **65**:1796-1804.

Huang, J., Hirii, R., Adam, L., Rozwadowski, K. L., Harnmerlindl, J. K., Keller, W. A., and Selvaraj, G., 2000, Genetic engineering of glycinebetaine production toward enhancing stress tolerance in plants: metabolic limitations, *Plant Physiol.* **122**:747-756.

Ikuta, S., Imamura, S., Misaki, H., and Horiuti, Y., 1977, Purification and characterization of choline oxidase from *Arthrobacter globiformis*, *J. Biochem.* **82**:1741-1749.

Incharoenskdi, A., Takabe, T., and Akazawa, T., 1986, Effect of betaine on enzyme activity and subunit interaction of ribulose-1,5-bisphosphate carboxylase/oxygenase from *Aphanothece halophytica*, *Plant Physiol.* **81**:1044-1049.

Ishizaki-Nishizawa, O., Fujii, T., Azuma, M., Sekiguchi, K., Murata, N., Ohtani, T., and Toguri, T., 1996, Low-temperature resistance of higher plants is significantly enhanced by a nonspecific cyanobacterial desaturase, *Nature Biotech.* **14**:1003-1006.

Jaglo-Ottosen, K. R., Gilmour, S. J., Zarka, D. G., Schabenberger, O., and Thomashow, M. F., 1998, Arabidopsis CBF1 overexpression induces COR genes and enhances freezing tolerance, *Science.* **280**:104-106.

Kaye, C., Neven, L., Hofig, A., Li, Q. B., Haskell, D., and Guy, C., 1998, Characterization of a gene for spinach *CAP160* and expression of two spinach cold-acclimation proteins in tobacco, *Plant Physiol.* **116**:1367-1377.

Kim, J. C., Lee, S. H., Cheong, Y. H., Yoo, C. M., Lee, S. I., Chun, H. J., Yun, D. J., Hong, J. C., Lee, S. Y., Lim, C. O., and Cho, M. J., 2001, A novel cold-inducible zinc finger protein from soybean, *SCOF-1*, enhances cold tolerance in transgenic plants, *Plant J.* **25**:247-259.

Kishitani, S., Watanabe, K., Yasuda, S., Arakawa, K., and Takabe, T., 1994, Accumulation of glycinebetaine during cold acclimation and freezing tolerance in leaves of winter and spring barley plants, *Plant Cell Environ.* **17:** 89-95.

Kodama, H., Hamada, T., Horiguchi, G., Nishimura, M., and Iba, K., 1994, Genetic enhancement of cold tolerance by expression of a gene for chloroplast ω-3 fatty acid desaturase in transgenic tobacco, *Plant Physiol.* **105**:601-605.

Kovtun, Y., Chiu, W. L., Tena, G., and Sheen, J., 2000, Functional analysis of oxidative stress-activated mitogen-activated protein kinase cascade in plants, *Proc. Natl. Acad. Sci. U.S.A.* **97**:2940-2945.

Lamark, T., Rokenes, T. P., McDougall, J., and Strom, A. R., 1996, The complex *bet* promoters of *Escherichia coli:* regulation by oxygen (*ArcA*), choline (*BetI*), and Osmotic Stress, *J. Bacteriol.* **178**:1655-1662.

Lamark, T., Kaasen, I., Eshoo, M. W., Falkenberg, P., McDougall, J., and Strom, A. R., 1991, DNA sequence and analysis of the *bet* genes encoding the osmoregulatory choline-glycine betaine pathway of *Escherichia coli, Mol. Microbiol.* **5**:1049-1064.

Landfald, B., and Strom, A. R., 1986, Choline-glycine betaine pathway confers a high level of osmotic tolerance in *Escherichia coli*, *J. Bacteriol.* **165**:849-855.

Lee, J. H., van Montagu, M., and Verbruggen, N., 1999, A highly conserved kinase is an essential component for stress tolerance in yeast and plant cells, *Proc. Natl. Acad. Sci. U.S.A.* **96**:5873-5877.

Liu, Q., Kasuga, M., Sakuma, Y., Abe, H., Miura, S., Yamaguchi-Shinozaki, K., and Shinozaki, K., 1998, Two transcription factors, DREB1 and DREB2, with an EREBP/AP2 DNA binding domain separate two cellular signal transduction pathways in drought- and low-temperature-responsive gene expression, respectively, in Arabidopsis, *Plant Cell.* **10**:1391-1406.

Levitt, J., 1980, *Responses of Plants to Environmental Stresses*, Academic Press, New York.

McKersie, B. D., Bowley, S. R., and Jones, K. S., 1999, Winter survival of transgenic alfalfa overexpressing superoxide dismutase, *Plant Physiol.* **119**:839-849.

McNeil, S. D., Nuccio, M. L., Ziemak, M. J., and Hanson, A. D., 2001, Enhanced synthesis of choline and glycine betaine in transgenic tobacco plants that overexpress phosphoethanolamine N-methyltransferase, *Proc. Natl. Acad. Sci. U.S.A.* **98**:10001-10005.

McNeil, S. D., Nuccio, M. L., and Hanson, A. D., 1999, Betaines and related osmoprotectants. Targets for metabolic engineering of stress resistance. *Plant Physiol.* **120**:945-949.

Miller, B., Schmid, H., Chen, T. J., Schmolke, M., and Guder, W. G., 1996, Determination of choline dehydrogenase activity along the rat nephron, *Biol. Chem. Hoppe-Seyler.* **377**:129-137.

Murata, N., Ishizaki-Nishizawa, O., Higashi, S., Hayashi, H., Tasaka, Y., and Nishida, I., 1992a, Genetically engineered alteration in the chilling sensitivity of plants, *Nature.* **356**:710-712

Murata, N., Mohanty, P. S., Hayashi, H., and Papageorgiou G. C., 1992b, Glycinebetaine stabilizes the association of extrinsic proteins with the photosynthetic oxygen-evolving complex, *FEBS Lett.* **296**:187-189.

Naidu, B. P., 1998, Separation of sugars, polyols, proline analogues, and betaines in stressed plant extracts by high performance liquid chromatography and quantification by ultra violet detection, *Aust. J. Plant Physiol.* **25**:793-800.

Nanjo, T., Kobayashi, M., Yoshiba, Y., Kakubari, Y., Yamaguchi-Shinozaki, K., and Shinozakia, K., 1999, Antisense suppression of proline degradation improves tolerance to freezing and salinity in *Arabidopsis thaliana*, *FEBS Lett.* **461**:205-210.

Nakamura, T., Yokotaz, S., Muramoto, Y., Tsutsuil, K., Oguri, Y., Fukui, K., and Takabe, T., 1997, Expression of a betaine aldehyde dehydrogenase gene in rice, a glycinebetaine nonaccumulator, and possible localization of its protein in peroxisomes, *Plant J.* **11**:1115-1120.

Nuccio M.L., Russell, B.L., Nolte, K.D., Rathinasabapathi, B., Gage, D.A., Hanson, A.D., 1998, The endogenous choline supply limits glycine betaine synthesis in transgenic tobacco expressing choline monooxygenase, *Plant J.* **16**: 487-496.

Nuccio, M. L., McNeil, S. D., Ziemak, M. J., Hanson, A. D., Jain, R. K., and Selvaraj, G., 2000, Choline import into chloroplasts limits glycine betaine synthesis in tobacco: analysis of plants engineered with a chloroplastic or a cytosolic pathway, *Metab. Eng.* **2**:300-311.

Nyyssölä, A., Kerovuo, J., Kaukinen, P., Weymarn, N. V., and Reinikainen, T., 2000, Extreme halophiles synthesize betaine from glycine by methylation, *J. Biol. Chem.* **275**:22196-22201.

Nyyssölä, A., Reinikainen, T., and Leisola, M., 2001, Characterization of glycine sarcosine *N*-methyltransferase and sarcosine dimethylglycine *N*-methyltransferase, *Appl. Environ. Microbiol.* **67**:2044-2050.

Nyyssölä, A., and Leisola, M., 2001, *Actinopolyspora halophila* has two separate pathways for betaine synthesis, *Arch. Microbiol.* **176**:294-300.

Papageorgiou, G. C., and Murata N., 1995, The unusually strong stabilizing effects of glycinebetaine on the structure and function in the oxygen-evolving photosystem II complex, *Photosyn. Res.* **44**:243-252.

Papageorgiou, G. C., Fujimura, Y., and Murata, N., 1991, Protection of the oxygen-evolving photosystem II complex by glycinebetaine, *Biochim. Biophys. Acta.* **1057**:361-366.

Prasad, K. V. S. K., Sharmilal, P., Kumar, P. A., and Saradhi, P. P., 2000, Transformation of *Brassica juncea* (L.) Czern with bacterial *codA* gene enhances its tolerance to salt stress, *Molecular Breeding.* **6**:489-499.

Rajashekar, C. B., Zhou, H., Marcum, K. B., and Prakash, O., 1999, Glycine betaine accumulation and induction of cold tolerance in strawberry (*Fragaria X ananssa* Duch) plants, *Plant Sci.* **148**:175-183.

Rathinasabapathi, B., Burnet, M., Russell, B. L., Gage, D. A., Liao, P. C., Nye, G. J., Scott, P., Golbeck, J. H., and Hanson, A. D., 1997, Choline monooxygenase, an unusual iron-sulfur enzyme catalyzing the first step of glycine betaine synthesis in plants: prosthetic group characterization and cDNA cloning, *Proc. Natl. Acad. Sci. U.S.A.* **94**:3454-3458.

Rathinasabapathi, B., McCue, K. F., Gage, D. A., and Hanson, A. D., 1994, Metabolic engineering of glycine betaine synthesis: plant betaine aldehyde dehydrogenases lacking typical transit peptides are targeted to tobacco chloroplasts where they confer betaine aldehyde resistance, *Planta.* **193**:155-162.

Rhodes, D., and Hanson, A. D., 1993, Quaternary ammonium and tertiary sulfonium compounds in higher plants, *Annu. Rev. Plant Physiol. Plant Mol. Biol.* **44**:357-384.

Robinson, S. P., and Jones, G. P., 1986, Accumulation of glycinebetaine in chloroplasts provides osmotic adjustment during salt stress, *Aust. J. Plant Physiol.* **13**:659-668.

Rokenes, T. P., Lamark, T., and Strom, A.R., 1996, DNA-binding properties of the BetI repressor protein of *Escherichia coli*: the inducer choline stimulates BetI-DNA complex formation, *J.Bacteriol.* **178**:1663-1670.

Rozwadowski, K. L., Khachatourians, G. G., and Selvaraj, G., 1991, Choline oxidase, a catabolic enzyme in *Arthrobacter pascens,* facilitates adaptation to osmotic stress in *Escherichia coli*, *J. Bacteriol.* **173**:472-478.

Saijo, Y., Hata, S., Kyozuka, J., Shimamoto, K., and Izui, K., 2000, Over-expression of a single Ca2+-dependent protein kinase confers both cold and salt/drought tolerance on rice plants, *Plant J.* **23**:319-327.

Sakamoto, A., Alia, and Murata, N., 1998, Metabolic engineering of rice leading to biosynthesis of glycine betaine and tolerance to salt and cold, *Plant Mol. Biol.* **38**:1011-1019.

Sakamoto, A., and Murata, N, 2001a, The use of bacterial choline oxidase, a glycinebetaine-synthesizing enzyme, to create stress-resistant transgenic plants, *Plant Physiol.* **125**:180-188.

Sakamoto, A., and Murata, N., 2001b, The role of glycine betaine in the protection of plants from stress: clues from transgenic plants, *Plant Cell Environ.* **24** (in press).

Sakamoto, A., and Murata, N., 2000, Genetic engineering of glycinebetaine synthesis in plants: current status and implications for enhancement of stress tolerance, *J. Exp. Bot.* **51**:81-88.

Sakamoto, A., Valverde, R., Alia, Chen, T. H. H., and Murata, N., 2000, Transformation *of Arabidopsis* with the codA gene for choline oxidase enhances freezing tolerance of plants, *Plant J.* **22**:449-453.

Sakamoto, A., Jeknic Z., Yuwansiri, R., Chen, T. H. H., and Murata N., 2001, Role of glycinebetaine in freezing tolerance: a transgenic approach. *Plant Cell Physiol.* **42**- s142.

Takabe, T., Hayashi, Y., Tanaka, A., Takabe, T., and Kishitani, S., 1998, Evaluation of glycinebetaine accumulation for stress tolerance in transgenic rice plants, *Proceedings of international workshop on breeding and biotechnology for environmental stress in rice*, Hokkaido National Agricultural Experiment Station and Japan International Science and Technology Exchange Center, Sapporo, Japan, pp. 63-68.

Tamminen, I., Makela, P., Heino, P., and Palva, E. T., 2001, Ectopic expression of ABI3 gene enhances freezing tolerance in response to abscisic acid and low temperature in *Arabidopsis thaliana*, *Plant J.* **25**:1-8.

Tahtiharju, S., and Palva, T., 2001, Antisense inhibition of protein phosphatase 2C accelerates cold acclimation in Arabidopsis thaliana, *Plant J.* **26**:461-470.

Wallis, J. G., Wang, H., and Guerra, D. J., 1997, Expression of a synthetic antifreeze protein in potato reduces electrolyte release at freezing temperatures, *Plant Mol. Biol.* **35**:323-330.

Weretilnyk, E. A., and Hanson, A. D., 1988, Betaine aldehyde dehydrogenase polymorphism in spinach: Genetic and biochemical characterization, *Biochem. Genet.* **26**:143-151.

Winzor, C. L., Winzor, D. J., Paleg, L. G., Jones, G. P., and Naidu, B. P., 1992, Rationalization of the effects of compatible solutes on protein stability in terms of thermodynamic nonideality, *Arch. Biochem. Biophys.* **296**:102-107.

Wyn Jones, R. G., and Storey, R., 1981, Betaines. In Paleg, L. G., and Aspinall, D. (eds.), *The Physiology and Biochemistry of Drought Resistance in Plants*. Academic Press, Sydney, pp. 171-204.

Zhao, Y., Aspinall, D., and Paleg, L. G., 1992, Protection of membrane integrity in *Medicago sativa* L. by glycinebetaine against the effects of freezing, *J. Plant Physiol.* **140**:541-543.

WHEAT CATALASE EXPRESSED IN TRANSGENIC RICE PLANTS CAN IMPROVE TOLERANCE AGAINST LOW TEMPERATURE INJURY

Takeshi Matsumuraa, Noriko Tabayashib, Yasuyo Kamagata, Chihiro Souma and Haruo Saruyama*

1. ABSTRACTS

We isolated wheat catalase (CAT: EC 1.11.1.6) cDNA and introduced it into rice (*Oryza sativa* L.). Some of 56 transgenic regenerated rice plants were confirmed to express the wheat catalase by the native PAGE with catalase activity staining. The wheat catalase detected in leaf, root, anther, seed germ and seed endosperm suggested that wheat catalase under 35S CaMV promoter is expressed in these tissues. In the transgenic rice plants, the catalase activities in leaf at 25 °C and at 5 °C were increased from 2 to 5 fold and 4 to 15 fold respectively, compared to that of non-transgenic rice. The transgenic rice CT 2-6-4, which showed the highest catalase activity among the transgenic rice plants indicated an increased resistance to low temperature stress. These results were obtained by comparing the damage in leaves of withering (curling) due to chilling at 5 °C. This chilling treatment resulted in the decrease of catalase activities in both types of plant, but the transgenic plants indicated higher catalase activities retained in the leaves than those of non-transgenic ones. The transgenic CT 2-6-4 also contained lower concentration of hydrogen peroxide in leaves than the control rice plant. During the chilling, the H_2O_2 concentration of non-transgenic rice increased, but the increase was depressed in the transgenic rice. Therefore, enhanced activities of catalase in genetically engineered rice plants lead to the effective detoxification of H_2O_2, which improves tolerance against low temperature injury.

---

* Haruo Saruyama, Hokkaido Green-Bio Institute, Higashi 5 Kita 15, Yubari-gun, Naganuma, Hokkaido 069-1301, Japan.
Noriko Tabayashib, Hokkai Sankyo Co., Ltd. 27-4, Kitanosato, Kitahiroshima, Hokkaido 061-1111, Japan.
Yasuyo Kamagata, Chihiro Souma, Hokkaido Green-Bio Institute, Higashi 5 Kita 15, Naganuma, Hokkaido 069-1301, Japan.Takeshi Matsumura, Research Institute of Biological Resources, National Institute of Advanced Industrial and Science Technology, 2-17-1, 2, Tsukisamuhigashi, Toyohira-ku, Sapporo 062-8517, Japan

2. INTRODUCTION

The electron transport system in mitochondria and the photosystems in chloroplasts are well known as process where active oxygen species, such as hydrogen peroxide (H_2O_2), superoxide (O_2^-) and hydroxyl radicals ($OH^·$) are produced (Asada, 1992; Bowler et al., 1992; Foyer et al., 1994; Puntarulo et al., 1991). These active oxygens can react very rapidly with proteins, membrane lipids and nucleic acids, which causes severe cellular damage (Rice-Evans et al., 1991). The damage to plants is attributed to active oxygens arising under stress conditions such as air pollutants like O_3 (Reich and Amundson, 1985; Landry and Pell, 1993; Torsethaugen et al., 1997), low temperatures (Omran, 1980; Clare et al., 1984; Hodgson and Raison, 1991; Prasad et al., 1994; Wise and Naylor, 1987), freezing (Kendall and McKersie, 1989), high light intensity (Krause, 1988), UV-irradiation (Boldt and Scandalios, 1997) and water stress (Baisak et al., 1994; Leprince et al., 1990; Senaratna et al., 1985). Active oxygen scavenging system composed of superoxide dismutase (SOD: EC 1.15.1.1), catalase (CAT) and ascorbate peroxidase (APx; EC 1.11.1.11) (Asada, 1992; Scandalios, 1990) have been developed in order to protect plants from oxidative injury. The functional stability of scavenger enzymes and/or the effective response of the enzyme-synthesizing system to such stress conditions, are suggested having a critical effect on plant growth.

Gene transfer technology has made it possible to improve stress tolerance in plants, so as to increase the functional capacity of active oxygen scavenging enzymes (Allen, 1995). Cu/Zn SOD transgenic potato (*Solanum tuberosum* cv Desiree) (Perl et al., 1993) and tobacco (*N. tabacum* cv Xanthi) (Sen Gupta et al., 1993) showed the increased tolerance to photodamage by methyl viologen (MV). In contrast, overexpression of mitochondrial Mn SOD from tobacco (*Nicotiana plumbaginifolia*) was found to offer protection against MV-mediated damage in tobacco (Bowler et al., 1991; Slooten et al., 1995), freezing damage in alfalfa (*Medicago sativa*) (McKersie et al., 1993) and chilling damage in cotton (*Gossypium hirsuitum*) (Trolinder and Allen, 1994; Allen, 1995). Over expressed Arabidopsis peroxisomal APx in transgenic tobacco (*N. tobacum* cv Xanthi) indicated increased protection against oxidative stress (Wang et al., 1999). Transgenic tobacco (*N. tabacum* SR1) that expressed glutathione reductase (GR) from *Escherichia coli* displayed increased GR activity and enhanced tolerance to paraquat stress (Aono et al., 1991; 1993). GR from pea expressed in transgenic tobacco indicated more effective protection against paraquat than that from *E. coli* (Broadbent et al., 1995).

As far as catalase is concerned, its important protective qualities under stress conditions was reported in tobacco plants. Resistance against photooxidation by drought stress at the high-light intensity was improved by the expression of E. coli catalase in the transgenic tobacco (*N. tabacum* cv. Xanthi) (Shikanai et al., 1998). In crop plants, however, there has been no study examining whether or not the overexpression of catalase can offer plants tolerance to low temperature stress. We have already confirmed that catalase in rice is responsible for germination ability under low temperatures (Tanida and Saruyama, 1995; Tanida, 1996) and recovery from cold injury (Saruyama and Tanida, 1995). Therefore, in this study, we isolated a cDNA encoded catalase from wheat, and tried to improve tolerance in rice against low temperature stress by overproduction of wheat catalase.

3. MATERIALS AND METHODS

3.1. Plant Transformation

First, we isolated a cDNA clone for wheat (*Triticum aestivum* L.) catalase (Saruyama and Matsumura, 1999). The cDNA was then inserted into the expression vector of pBI 221 (CLONTECH Laboratories, Inc. CA, USA) in place of the β-glucuronidase gene that had been excised. The resulting plasmid was transferred into rice (Oryza sativa L. cv Yuukara or Matsumae) protoplasts (4 x 106 cells/ml electroporation buffer), and isolated from the suspension culture by electroporation (Tada et al., 1990). The electroporated protoplasts were incubated according to Kyozuka et al. (1987) for 14 days, and then hygromycin was added to the medium (20 μg/ml). Colonies displaying resistance to hygromycin were further incubated in the regenerated medium according to Fujimura et al (1985), and transgenic rice plants were developed. The regenerated plants expressing wheat catalase cDNA were confirmed by native PAGE with catalase activity staining as described below.

3.2. Protein Extraction

Crude extracts of leaves, roots and anthers of rice plants, and germ and endosperm of rice seeds were prepared as reported previously (Saruyama and Tanida, 1995). Protein concentration was determined according to the method of Bradford (1976) using the Bio-Rad protein assay kit (Bio-Rad Laboratories, CA., USA).

3.3. Enzyme Assays

Catalase (EX 1.11.1.6) activity was measured according to Aebi (1983) in a reaction mixture containing 50 mM potassium phosphate, pH 7.0, 45 mM H_2O_2 and the crude extract using a Shimadzu UV-2100 spectrophotometer. The activity was determined by the decrease of absorbance at 240 nm due to H_2O_2 consumption.

Superoxide dismutase (EX 1.15.1.1) activity was measured by the nitro blue tetrazolium (NBT) method according to Beyer and Fridovich (1987) using a Wako SOD kit (Wako Chemical Co., Japan). The reaction was started by the addition of xanthine oxidase as the superoxide generation, and NBT reduction was measured at 560 nm. One unit of SOD activity causes 50% inhibition of the rate of reduction of NBT. The specific activity of SOD was expressed as units mg^{-1} protein.

Ascorbate peroxidase (EX 1.11.1.11) activity was determined according to Chen and Asada (1989) with minor modification. The reaction mixture was composed of 50 mM potassium phosphate (pH 7.0) containing 1 mM NaN_3, 0.5 mM ascorbate, 1.54 mM H_2O_2, and the enzyme fraction. The oxidation of ascorbate was started by H_2O_2 and the decrease in the absorbance at 290 nm was monitored.

3.4. H_2O_2 Concentration in Leaves

Leaves (0.5 g) were ground with a mortar and pestle in liquid nitrogen. A sample was then suspended in 1 ml of 0.2 N perchloric acid and centrifuged at 10,000 g for 10 min. To remove the perchloric acid, the supernatant was neutralized to pH 7.0 with 4 N potassium hydroxide and the solution was centrifuged at 10,000 g for 10 min. Supernatant was applied to a 1 ml column of anion-exchange resin (AG 1-X2, Bio-Rad) and the column was washed with 3 ml of distilled water. The eluate was used for the measurement of H_2O_2 concentration. The reaction mixture contained 750 μl of the eluate, 300 μl of 12.5 mM 3-(dimetylamino) benzoic acid (DMAB) in 0.375 M phosphate buffer (pH 6.5), 60 μl of 1.3 mM 3-metyl-2-benzothiazoline hydrazone (MBTH), and 15μl of horseradish peroxidase (20 units) (Okuda et al. 1991). The reaction was started by the addition of peroxidase at 25°C. The absorbance at 590 nm after 3 min was measured by spectrophotometer (Shimadzu UV-2500PC).

3.5. Native-PAGE and Activity Staining of Catalase Isozymes

Native polyacrylamide gels were prepared using a separation gel of 10% acrylamide and a stacking gel of 4% acrylamide according to the procedure of Laemmli (1970) except that SDS was omitted. Separation was done at 15 mA for 5 to 6 hrs, then the gel was soaked in 0.03% H_2O_2 solution for 2 min, washed with water and then incubated in a solution composed of 12.2 mM potassium ferricyanide and 14.8 mM ferric chloride. The catalase activity was detected as yellow bands on a blue-green background.

3.6. Chilling Treatment

The effects of chilling on the transgenic rice plants (CT 2-6-4 and CT 2-6-12), non-transgenic regenerated rice plants (TN control) and the control non-transgenic plant (Yuukara) were examined. These transgenic plants were confirmed to express wheat catalase by native-PAGE, combined with activity staining as mentioned above. These plants grown to 5th leaf stage in a greenhouse were transferred into a cold room at 5 °C under light (40 $\mu mol\ m^{-2}\ s^{-1}$) for 8 days (chilling). Then, the chilled plants were incubated in a greenhouse again (re-warming). During these treatments of chilling and re-warming, the degrees of the phenotypic change, withering and curling of leaves were compared.

4. RESULTS AND DISCUSSION

4.1. Plant Transformation and Expression of Wheat Catalase in Transgenic Plants

The cDNA clone of wheat catalase consists of 1573 nucleotides and encodes a polypeptide of 492 amino acids residues. The composition of overall deduced amino acid sequence showed the higher homologies of 81,81,81, and 84% for barley CAT-1, rice CAT B, maize CAT-1 and maize CAT-2, respectively. All amino acid residues at active sites and heme-binding sites are conserved among plants, animals and prokaryotes (Ossowskie et al. 1993). In the case of wheat catalase, we also found the same amino acid residues at the active site of His-65, Ser-104 and Asn-138, proximal heme-binding sites of Pro-326, Arg-344 and Tyr-348 and distal heme-binding sites of Val-63, Thr-105, Phe-143 and Phe-346. These results suggested that the amino acid residues for the functionally important sites were also conserved in wheat catalase.

The expression vector containing wheat cDNA was transferred into rice protoplasts, and we got 56 regenerated transgenic plants (40 and 16 plants of cultivar, from Yuukara and Matsumae, respectively) after screening for hygromycin resistance. Among them, the expression of wheat catalase in leaf samples was examined by the native-PAGE combined with the catalase activity staining method. Rice and wheat catalases can be distinguished by the difference in their mobilities on native-PAGE. A catalase isozyme band in the transgenic running faster than that from control rice plant, was observed at the same position as the wheat catalase one (Figure 1). Thus, we considered that these transgenic plants expressed the wheat catalase. Ultimately, 12 transgenic plants were identified as expressing wheat catalase. In addition, wheat catalase cDNA was confirmed to be transferred into these 12 transgenic rice plants by genomic PCR.

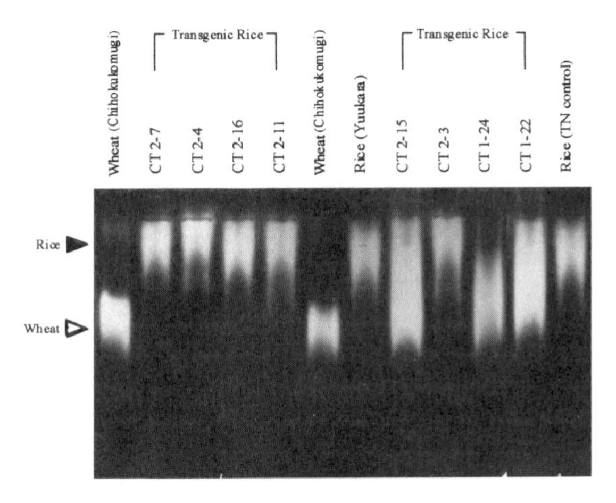

Figure 1. Expression of wheat catalase in leaves of the transgenic rice plants. Total protein (20 μg) from leaves was analyzed by native-PAGE combined with activity staining method. Arrows indicate the position of the catalase isozyme band of rice and wheat leaves.

4.2. Distribution of Wheat Catalase Expressed in Transgenic Rice Plants

We examined whether wheat leaf catalase was expressed in various other tissues of the transgenic rice plants, since wheat catalase cDNA was expressed under the control of 35S CaMV promoter. Crude extracts prepared from not only leaf, but also root, anther, seed germ, and seed endosperm were analyzed by native-PAGE with the catalase activity staining method as mentioned above. In the root, anther, seed germ and seed endosperm of transgenic rice plant, the catalase band was also observed clearly at the same position of wheat catalase, just as in the case of the leaf sample. Therefore, wheat catalase under the control of the 35S CaMV promoter was expressed in these tissues.

4.3. Effect of Low Temperature on Catalase in Transgenic Plants

We analyzed catalase activities in the 56 transgenic rice plants. Increased activities were observed in the transgenic plants compared to the non-transgenic control rice plants (Yuukara, Matsumae and TN control) and wheat (Chihokukomugi). Among them, transgenic CT 2-6, CT 2-10, CT 2-13, CT 2-14, CT 2-15 and CT 2-17 showed higher activities; 1.8 - 4.2 fold that of the control, Yuukara. In the case of the transgenic plants in which wheat catalase was not expressed, the catalase activities were around 78-140 % of the control value.

The effect of temperature on the catalase activities in leaves was examined. Activities at 25, 20, 15, 10 and 5 °C were measured. All plants examined showed a decrease of catalase activities, in response to the decrease of temperatures from 25 to 5 °C. Activities of wheat catalase at 5 °C was about 60% of that at 25 °C, and in these transgenic plants, from 40 to 70%, of the activities at 25 °C were retained at 5 °C. In contrast, rice (Yuukara) and CT 1-19 (wheat catalase was not expressed) displayed only 20 and 25 % respectively of the residual activity at 5 °C. Furthermore, catalase activities at 5 °C in transgenic CT 1-24, CT 2-6, CT 2-10, CT 2-13, and CT 2-17 were as high as 470, 1,600, 1,240, 830 and 630%, respectively of the values of the non-transgenic plant (Yuukara).

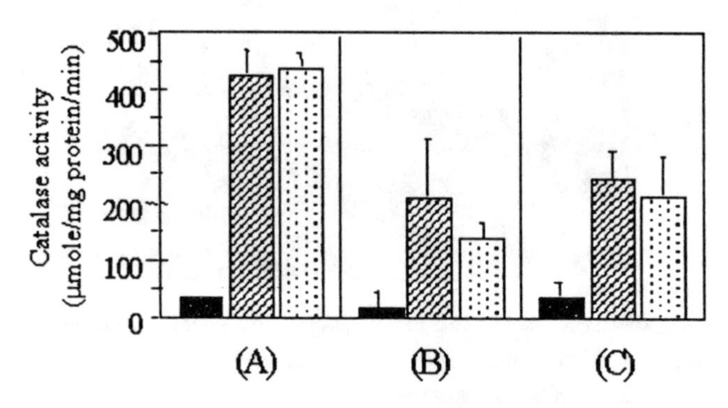

Figure 2. Effects of chilling on the activity of catalase in leaves. Rice plants grown at 25°C for 10 days (A) were chilled at 5°C for 10 days (B), then re-warmed at 25°C and incubated for a further 2 days (C). The plants tested were: Non-transgenic rice cultivar Yuukara ( ), transgenic plant CT 2-6-12 ( ) and transgenic plant CT 2-6-4 ( ).

Next, we examined the effect of the chilling on CAT activities during the chilling at 5 °C and re-warming at 25 °C (Fig. 2). The catalase activity of the control plants grown in a greenhouse was around 40 (µmol mg^{-1} protein min^{-1}), but the transgenic rice plants indicated this value as 420 and 430 in CT 2-6-4 and CT 2-6-12 respectively, (T3 plants of CT 2-6 that were confirmed to express wheat catalase,respectively). Then after 10 days chilling, the activity was decreased to 20 in non-transgenic rice and 210 and 150 in CT 2-6 12 and CT 2-6-4, respectively. After the re-warming at 25 °C, the activity increased to 35 in non-transgenic, and 250 and 230 in transgenic rice plants, CT 2-6-12 and CT 2-6-4, respectively. Therefore, higher catalase activities remained at 5 °C in these transgenic plants strongly suggested that H_2O_2 might be detoxified effectively under the low temperature environments.

4.4. Effect of Chilling on Transgenic and Non-transgenic Plants

Next we examined whether the transgenic plants that showed the accelerated catalase activities at both 25 °C and 5 °C would show improved tolerance to low temperature stress. Transgenic CT 2-6-4 and CT 2-6-12 were used to compare the effect of the chilling on the plants since they showed the highest catalase activity among the 12 transgenic rice plants at both 25 °C and 5 °C. The plants were incubated at 5 °C for 8 days and visible damage during the chilling was compared, by counting the number of plants, which showed the damage as withering (curling) of 5th or 4th leaf. Non-transgenic plants showed the increase of % value of rolling in the 5th leaf just after the start of the chilling. 50 and 70% of the control plants, Yuukara and TN control, respectively indicated the damage at the 5th leaf after chilling for 1 day. Finally 90 and 100% of Yuukara and TN control plants respectively were damaged after 5-days chilling. In contrast, the transgenic plants showed the gradual increase of damage due to chilling, and 90% of both transgenic plants indicated the 5th leaf curling after 8 days chilling. In the case of the 4th leaf, the number of plants showing the curled leaves also increased in accordance to the chilling time. However, transgenic plants were more resistant to the chilling, since more than 90% of non-transgenic plants showed damage due to chilling for 8 days, but both CT 2-6-4 and CT 2-6-12 indicated the values as 55 and 60%, respectively. When the plants chilled at 5 °C for 8 days were then returned to 25 °C, the damage to both Yuukara and TN control became even more severe after 1 day of incubation. In transgenic CT 2-6-12 and CT 2-6-16, however, there was substantially lower chilling injury than in those of Yuukara and TN control plants. From these results, we judged that the transgenic plants CT 2-6-4 and CT 2-6-12 are more cold resistant than the non-transgenic rice plants.

4.5. Effect of the Chilling on H_2O_2 Concentration

Since the transgenic plants indicated the higher catalase activities during the chilling, H_2O_2 was suggested to be detoxified effectively. Therefore, we measured the concentration of H_2O_2 during the chilling (Fig. 3). Lower concentration of H_2O_2 in the transgenic rice plants compared with that of control plants was observed. The concentration of H_2O_2 just before the chilling (0 time) were 2.7 and 1.5 in the control plant (Yuukara) and CT 2-6-4, respectively. Both plant indicated the increase of H_2O_2 concentration due to chilling. Non-transgenic Yuukara showed the increase of H_2O_2 concentration from 2.7 to 4.0 due to chilling, however transgenic plants showed the suppression of the increase of the concentrations, especially in CT 2-6-4, the concentration remained at levels between 1.5 to 2.5 μmol g^{-1} fresh weight. The higher activities of catalase found in transgenic plants strongly suggest that the increased catalase is useful for scavenging the H_2O_2 produced during low temperature circumstances. Thus, gene transfer of active oxygen scavenging enzymes offers the possibility for improving cold resistance in the economically important crop plant of rice.

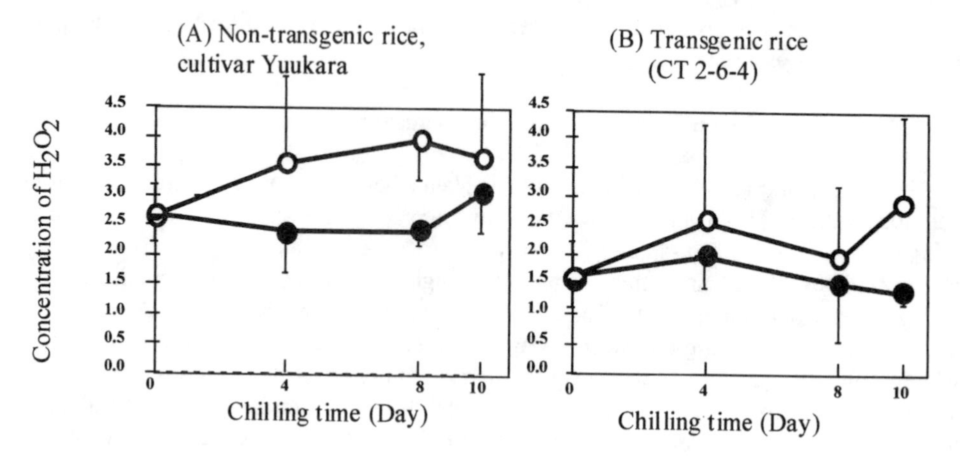

Figure 3. Effect of the chilling on the concentration of H_2O_2 in leaves. Non-transgenic rice, cultivar Yuukara (A) and the transgenic rice, CT 2-6-4 (B) were grown at 25 °C for 10 days, then chilled at 5 °C for 10 days. After the chilling for 4, 8 and 10 days, the concentration of H_2O_2 in leaves were measured. Each value is the mean± SD at least 3 separated experiments. ○ : under standard condition (grown at 25 °C), ● : under chilling (grown at 25 °C, then chilled at 5 °C for 10 days).

REFERENCES

Aebi, H. E., 1983, Catalase., in: H. U. Bergmeyer, J. Bergmeyer and M. Graßl, eds., *Methods of Enzymatic Analysis*, Ed 3 Vol 3. Verlag Chemie GmbH, Weinheim, pp.273-286.

Allen, R. D., 1995, Dissection of oxidative stress tolerance using transgenic plants, *Plant Physiol.* **107**: 1049-1054.

Anderson, M. D., Prasad, T. K. and Stewart, C. R., 1995, Changes in isozyme profiles of catalase, peroxidase, and glutathione reductase during acclimation to chilling in mesocotyls of maize seedlings, *Plant Physiol.* **109**: 1247-1257.

Aono, M., Kubo, A., Saji, H., Natori, T., Tanaka, K. and Kondo, N., 1991, Resistance to active oxygen toxicity of transgenic *Nicotiana tabacum* that expresses the gene for glutathione reductase from *Escherichia coli, Plant Cell Physiol.* **32**: 691-697.

Aono, M., Kubo, A., Saji, H., Tanaka, K. and Kondo, N., 1993, Enhanced tolerance to photooxidative stress of transgenic *Nicotiana tabacum* with high chloroplastic glutathione reductase activity. *Plant Cell Physiol.* **34**: 129-135.

Asada. K., 1992, Ascorbate peroxidase - a hydrogen peroxide-scavenging enzyme in plants, *Physiol. Plant.* **85**: 235-241.

Baisak, R., Rana, D., Acharya, P. B. B. and Kar, M., 1994, Alterations in the activities of active oxygen scavenging enzymes of wheat leaves subjected to water stress, *Plant Cell Physiol.* **35**: 489-495.

Beyer, W. F. and Fridovich, I., 1987, Assaying for superoxide dismutase activity: some large consequences of minor changes in conditions, *Anal. Biochem.* **161**: 559-566.

Boldt, R. and Scandalios, J. G., 1997, Influence of UV-light on the expression of the Cat2 and Cat 3 calatase genes in maize, *Free Rad.Biol.Med.* **23**: 505-514.

Bowler, C., Slooten, L., Vandenbranden, S., De Rycke, R., Botterman, J., Sybesma, C., Van Montagu, M., Inzé, D., 1991, Manganese superoxide dismutase can reduce cellular damage mediated by oxygen radicals in transgenic plants, EMBO J. **10**: 1723-1732.

Bowler, C., Van Montagu, M. and Inzé, D., 1992, Superoxide dismutase and stress tolerance. *Annu. Rev. Plant Physiol. Plant Mol. Biol.* **43**: 83-116.

Bradford, M. M., 1976, A rapid and sensitive method for the quantitation of microgram quantities of protein utilizing the principle of protein-dye binding. *Anal. Biochem.* **72**: 248-254.

Broadbent, P., Creissen, G. P., Kular, B., Wellburn, A. R. and Mullineaux, P. M., 1995, Oxidative stress responses in transgenic tobacco containing altered levels of glutathione reductase activity, *Plant J.* **8**: 247-255.

Chen, G. –X. and Asada, K., 1989, Ascorbate peroxidase in tea leaves: occurrence of two isozymes and the differences in their enzymatic and molecular properties, *Plant Cell Physiol.* **30**: 987-998.

Clare, D. A., Rabinowitch, H. D. and Fridovich, I., 1984, Superoxide dismutase and chilling injury in *Chlorella ellipsoidea*, *Arch. Biochem. Biophys.* **231**:158-163.

Creissen, G., Firmin, J., Fryer, M., Kular, B., Leyland, N., Reynolds, H., Pastori, G., Wellburn, F., Baker, N., Wellburn, A. and Mullineaux, P., 1999, Elevated glutathione biosynthetic capacity in the chloroplasts of transgenic tobacco plants paradoxically causes increased oxidative stress, *Plant Cell* **11**: 1277-1291.

Foyer, C. H., Lelandais, M. and Kunert, K. J., 1994, Photooxidative stress in plants. *Physiol. Plant.* **92**: 696-717.

Fujimura, T., Sakurai, M., Akagi, H., Negishi, T. and Hirose, A., 1985, Regeneration of rice plants from protoplasts, *Plant Tissue Culture Lett.* **2**: 74-75.

Hodgson, R. A. J. and Raison, J. K., 1991, Superoxide production by thylakoids during chilling and its implication in the susceptibility of plants to chilling-induced photoinhibition, *Planta* **183**: 222-228.

Hossain, M. A., Nakano, Y. and Asada, K., 1984, Monodehydroascorbate reductase in spinach chloroplasts and its participation in regeneration of ascorbate for scavenging hydrogen peroxide. *Plant Cell Physiol.* **5**: 385-395.

Kendall, E. J. and McKersie, B. D., 1989, Free radical and freezing injury to cell membranes of winter wheat. *Physiol. Plant.* **76**: 86-94.

Krause, G. H., 1988, Photoinhibition of photosynthesis. An evaluation of damaging and protective mechanisms. *Physiol. Plant.* **74**: 566-574.

Kyozuka, J., Hayashi, Y. and Shimamoto, K., 1987, High frequency plant regeneration from rice protoplasts by novel nurse culture methods, *Mol. Gen. Genet.* **206**: 408-413.

Laemmli, U. K., 1970, Cleavage of structural proteins during the assembly of the head of bacteriophage T4, *Nature* **227**: 680-685.

Landry, L. G., Pell, E. J., 1993, Modification of Rubisco and altered proteolytic activity in O_3-stressed hybrid poplar (*Populus maximowizii* x *trichocarpa*), *Plant Physiol.* **101**: 1355-1362.

Leprince, O., Deltour, R., Thorpe, P. C., Atherton, N. M., Hendry, G. A. F., 1990, The role of free radicals and radical processing systems in loss of desiccation tolerance in germinating maize (*Zea mays* L.), *New Phytol.* **116**: 573-580.

McKersie, B. D., Chen, Y., De Beus, M., Bowley, S. R., Bowler, C., Inzé, D., D'Halluin, K. and Botterman, J., 1993, Superoxide dismutase enhances tolerance of freezing stress in transgenic alfalfa (*Medicago sativa* L.). *Plant Physiol.* **103**: 1155-1163.

Mishra, N. P., Mishra, R. K. and Singhal, G. S., 1993, Changes in the activities of anti-oxidant enzymes during exposure of intact wheat leaves to strong visible light at different temperatures in the presence of protein synthesis inhibitors, *Plant Physiol.* **102**: 903-910.

Ni, W. and Trelease, R. N., 1991, Post-transcriptional regulation of catalase isozyme expression in cotton seeds, *Plant Cell* **3**: 737-744.

Okuda, T., Matsuda, Y., Yamanaka, A. and Sagisaka, S., 1991, Abrupt increase in the level of hydrogen peroxide in leaves of winter wheat is caused by cold treatment, *Plant Physiol.* **97**: 1265-1267.

Omran, R. G., 1980, Peroxide levels and the activities of catalase, peroxidase, and indoleacetic acid oxidase during and after chilling cucumber seedlings, *Plant Physiol.* **65**: 407-408.

Perl, A, Perl-Treves, R., Galili, S., Aviv, D., Shalgi, E., Malkin, S. and Galun, E., 1993, Enhanced oxidative-stress defense in transgenic potato expressing tomato Cu, Zn superoxide dismutases, *Theor. Appl. Genet.* **85**: 568-576.

Pitcher, L. H., Brennan, E., Hurley, A., Dunsmuir, P., Tepperman, J. M. and Zilinskas, B. A., 1991, Overproduction of petunia chloroplastic copper/zinc superoxide dismutase does not confer ozone tolerance in transgenic tobacco, *Plant Physiol.* **7**: 452-455.

Polle, A., Chakrabarti, K., Schürmann, W. and Rennenberg, H., 1990, Composition and properties of hydrogen peroxide decomposing systems in extracelluar and total extracts from needles of Norway spruce (*Picea abies* L., Karst.), *Plant Physiol.* **94**: 312-319.

Prasad, T. K., 1997, Role of catalase in inducing chilling tolerance in pre-emergent maize seedlings, *Plant Physiol.* **114**: 1369-1376.

Prasad, T. K., Anderson, M. D., Martin, B. A. and Stewart, C. R., 1994, Evidence for chilling-induced oxidative stress in maize seedlings and a regulatory role for hydrogen peroxide, *Plant Cell* **6**: 65-74.

Puntarulo, S., Galleano, M., Sanchez, R. A. and Boveris, A., 1991, Superoxide anion and hydrogen peroxide metabolism in soybean embryonic axes during germination, *Biochim. Biophys. Acta* **1074**: 277-283.

Reich, P. B. and Amundson, R. G., 1985, Ambient levels of ozone reduce net photosynthesis in tree and crop species, *Science* **230**: 566-570.

Rice-Evans, C. A., Diplock, A. T., Symons, M. C. R., 1991, Mechanisms of radical production, in: R. H. Burdon and P. H. van Knippenberg, eds, *Laboratory Techniques in Biochemistry and Molecular Biology*, Vol. 22. Elsevier, Amsterdam, pp19-50.

Saruyama, H. and Matsumura, T., 1999, Cloning and characterization of a cDNA encoding catalase in wheat, *DNA Sequence* **10**: 31-35.

Saruyama, H. and Tanida, M., 1995, Effect of chilling on activated oxygen-scavenging enzymes in low temperature-sensitive and -tolerant cultivars of rice (*Oryza sativa* L.), *Plant Sci.* 109: 105-113.

Scandalios, J. G., 1990, Response of plant antioxidant defense genes to environmental stress, in: J. G. Scandalios and T. R. F. Wright, eds, *Advances in Genetics.* Vol. 28. Academic Press, San Diego, pp 1-41.

Senaratna, T., McKersie, B. D. and Stinson, R. H., 1985, Simulation of dehydration injury to membranes from soybean axes by free radicals, *Plant Physiol.* 77: 472-474.

Sen Gupta, A., Heinen, J. L., Holaday, A. S., Burke, J. J. and Allen, R. D., 1993, Increased resistance to oxidative stress in transgenic plants that overexpress chloroplastic Cu/Zn superoxide dismutase. *Proc. Natl. Acad. Sci. USA* **90**: 1629-1633.

Sen Gupta A., Webb, R. P., Holaday, A. S. and Allen, R. D., 1993b, Overexpression of superoxide dismutase protects plants from oxidative stress, *Plant Physiol.* **103**: 1067-1073.

Shikanai, T., Takeda, T., Yamauchi, H., Sano, S., Tomizawa, K. –I., Yokota, A. and Shigeoka, S., 1998, Inhibition of ascorbate peroxidase under oxidative stress in tobacco having bacterial catalase in chloroplasts, *FEES Lett.* **428**: 47-51.

Slooten, L., Capiau, K., Van Camp, W., Van Montagu, M., Sybesma, C. and Inzé, D., 1995, Factors affecting the enhancement of oxidative stress tolerance in transgenic tobacco overexpressing manganese superoxide dismutase in the chloroplasts, *Plant Physiol.* **107**: 737-750.

Tada, Y., Sakamoto, M., and Fujimura, T., 1990, Efficient gene introduction into rice by electroplating and analysis of transgenic plants: use of electroplating buffer lacking chloride ions, *Theor. Appl. Genet.* **80**: 475-480.

Tanida, M., 1996, Catalase activity of rice seed embryo and its relation to germination rate at a low temperature, *Breeding Sci.* **46**: 23-27.

Tanida, M. and Saruyama, H., 1995, Catalase activity in embryos and its relation to germination ability of rice and wheat seeds, in: K. Noda and D. J. Mares, eds., *Seventh International Symposium on Pre-Harvest Sprouting in Cereals 1995*, Center for Academic Societies Japan, Osaka, pp. 357-361.

Tepperman, J. M., Dunsmuir, P., 1990, Transformed plants with elevated levels of chloroplastic SOD are not more resistant to superoxide toxicity, *Plant Mol. Biol.* **14**: 501-511.

Torsethaugen, G., Pitcher, L. H., Zilinskas, B. A. and Pell, E. J., 1997, Overproduction of ascorbate peroxidase in the tobacco chloroplast does not provide protection against ozone. *Plant Physiol.***114**: 529-537.

Trolinder, N. L. and Allen, R. D., 1994, Expression of chloroplast localized Mn SOD in transgenic cotton. *J. Cell Biochem.* **18A**: 97.

Wang, J., Zhang, H., and Allen, R. D., 1999, Overexpression of an Arabidopsis peroxisomal ascrobat peroxidase gene in tobacco increases protection against oxidative stress, *Plant Cell Physiol.* **40**: 725-732.

Willekens, H., Chamnongpol, S., Davey, M., Schraudner, M., Langebartels, C., Van Montagu, M., Inzé, D., and Van Camp, W., 1997, Catalase is a sink for H_2O_2 and is indispensable for stress defence in C3 plants, *EMBO J.* **16**: 4806-4816.

Wise, R. R. and Naylor, A. W., 1987a, Chilling-enhanced photooxidation. The peroxidative destruction of lipids during chilling injury to photosynthesis and ultrastructure, *Plant Physiol.* 83: 272-277.

INDEX